In 1944 the war in the east had become critical for the Germans. Determined defence, reinforced with local counterattacks, typified their fighting all along the front. Panzer division 'fire brigades' were rushed from one hot spot to another. German Generals such as von Manstein, Hube, and von Saucken showed expert skill holding back the Red Army against overwhelming odds, extracting their forces from encirclement, holding Soviet breakthroughs, and delaying the enemy advance at important bridgeheads. They were joined by their allies the Hungarians and Finns, who defended their homelands from the Soviet onslaught fighting alongside the German forces.

GREY WOLF

AXIS FORCES ON THE EASTERN FRONT, JANUARY 1944 - FEBRUARY 1945

Grey Wolf was compiled from the following books:
Stalin's Onslaught, Hammer And Sickle,
River Of Heroes, Stalin's Europe and *Fortress Europe.*
These books were originally produced by
the Battlefront team and friends.

Compiled and edited by: Wayne Turner

New material written by:
Heath Alexander, Steve Bernich Scott Elaurant, Van Norton,
Jyrki Saari, Ken Camel, Michael Haught, and Phil Yates.

*Many thanks to all of the playtesters and proofreaders
who have made this compilation possible.*

Editors: Peter Simunovich, John-Paul Brisigotti
Graphic Design: Victor Pesch
Terrain, Modelling and Photography: Battlefront Studio
Painting: James Brown, Blake Coster, Casey Davies,
Mike Haught, Jeremy Painter, Matt Parkes,
Victor Pesch, Chris Townley, and Wayne Turner
Cover Art: Vincent Wai
Internal Art: Randy Elliott, Warren Mahy,
Vincent Wai, and Ben Wooten
Miniatures Design: Evan Allen, Karl Cederman, and Seth Nash

CONTENTS

This is a supplement for *Flames Of War*, the World War II miniatures game.
A copy of the rulebook for *Flames Of War* is necessary to fully use the contents of this book.

© Copyright Battlefront Miniatures Ltd., 2011. ISBN: 978-0-9864661-4-4

INTRODUCTION

FLAMES OF WAR

In *Flames Of War* you take on the role of a company commander manoeuvring your troops across the battlefields of World War II. This classic period of warfare is brought to life in your own game room. *Grey Wolf* provides the core armies in the form of Intelligence Briefings. These Intelligence Briefings allow you to field the German, Hungarian and Finnish forces that clashed with the Soviet Red Army and its allies on the Eastern Front in 1944 and early 1945.

To play *Flames Of War* you'll also need the *Flames Of War* rulebook. The rulebook contains all the rules that you need to fight miniature World War II battles.

WHY COLLECT A FORCE FROM GREY WOLF?

Grey Wolf provides *Flames Of War* you with the German, Hungarian and Finnish forces needed to recreate the battles on the Eastern Front from January 1944 to February 1945.

With your German force, start by holding back the Soviet assaults to pry Leningrad from your grasp, before staunchly holding off the red tide on the Narva line in Estonia. Fight the brutal winter battles of the Korsun and Kamenets-Podolsky Pockets, escaping each in turn through determined defence and powerful counterattacks.

Fight to hold the *Festerplatz* fortified cities in Byelorussia as the Red Army throws all its power at your German forces during Operation Bagration. If they breach your line, delay them at every turn with your blocking forces before turning back their mobile forces with your panzers and Tiger tanks.

Finally bring an end to the Soviet assault on the banks of the Vistula River with elite SS armoured forces. Hold them on the river and bring their massive offensive to a stop. Throw the red scourge back across the river with counterattacks from the armoured troops of *3. SS-Panzerdivision 'Totenkopf'* and *5. SS-Panzerdivision 'Wiking'*.

Refight the battles to hold Romania and its precious oil from the Red Army. Fight with tough infantry forces or armoured forces as the red claw makes its grab for Romania.

Defend Transylvania with Hungarian and German forces as the traitorous Romanians switch sides and make to seize Hungarian Transylvania. Fight the tank battles of the Hungarian plains in October and November 1944. Hold the crossings of the Tisza River, while German and Hungarian forces re-group to fight the gruelling battle for Budapest.

In 100-days of bloody battle, fight through the streets of the Hungarian capital. Hold every street with barricades and other fortifications with mixed forces of Hungarians, dismounted SS-Cavalry and elite *Feldherrnhalle* Panzergrenadiers.

HOW THIS BOOK WORKS

The Germans in this book are divided into sections based on battles or operations. Each section has a number of forces appropriate for the battles featured. These include intelligence briefings for tank, armoured infantry, or infantry companies and other unique variations on these themes. In addition to the German section there is also a Hungarian section and a Finnish section including intelligence briefings for tank, motorised infantry, or infantry companies.

Each company has easy-to-use charts, highlighting available platoons which represent the fighting units of your company. Each platoon diagram visually displays what troops are included in the platoon.

A detailed arsenal at the end of each national section contains the specific ratings for your units' vehicles and the weapons they use. From tanks to infantry, this book contains everything you need to get started on your army!

In addition to the technical information you'll need to build your force, this book contains plenty of inspiring pictures and history to help you capture the flavour of your new *Flames Of War* force!

All of the forces in this book are based on historical examples that fought on the Eastern Front from January 1944 to February 1945.

To find out more, visit your local game store, or visit our informative website at ***www.FlamesOfWar.com***.

Flames Of War uses a point system when setting up and playing games. Typical games are around 1500 or 1750 points, but are certainly not limited to any value! Play any point value you and your opponent decide. You can play small 600 point games in an hour or you can play mammoth games using armies that are 3000 or 5000 points or more!

In most *Flames Of War* games you will command a company with several platoons.

---

3. SS-PANZERDIVISION 'TOTENKOPF'
SS-PANZERKAMPFGRUPPE

HQ

(A) **SS-Panzerkampfgruppe HQ**
2 Panzer IV H   200 points

COMBAT PLATOONS

(B) **Biermeyer's 3. SS-Panzer Platoon**
5 Panzer IV H   325 points

(C) **SS-Panzer Platoon**
3 Panther A   640 points

(D) **Gepanzerte SS-Panzergrenadier Platoon**
3 Panzergrenadier Squads   260 points

WEAPONS PLATOONS

(E) **Armoured SS-Anti-aircraft Gun Platoon**
2 Armoured Sd Kfz 7/1 (Quad 2cm)   125 points

(F) **Heavy Anti-tank Gun Platoon**
2 8.8cm PaK43   210 points

TOTAL: 1760 POINTS

COMPANY HQ

The company headquarters platoon is required—you have no command without it! When you choose your force the first thing you must purchase with your points is your company headquarters platoon.

COMBAT PLATOONS

Usually, at least two combat platoons are required. Whatever your battle plan requires, the Combat Platoons are the ones you rely upon to get the job done!

WEAPONS PLATOONS

Weapons platoons come from your own battalion. They are not required, but can offer your company excellent support, such as machine-guns, mortars, recon and anti-tank weapons.

SUPPORT PLATOONS

Support platoons are loaned to your company by the regiment, brigade, division, or corps. These platoons give you extra support in many forms ranging from tanks to artillery.

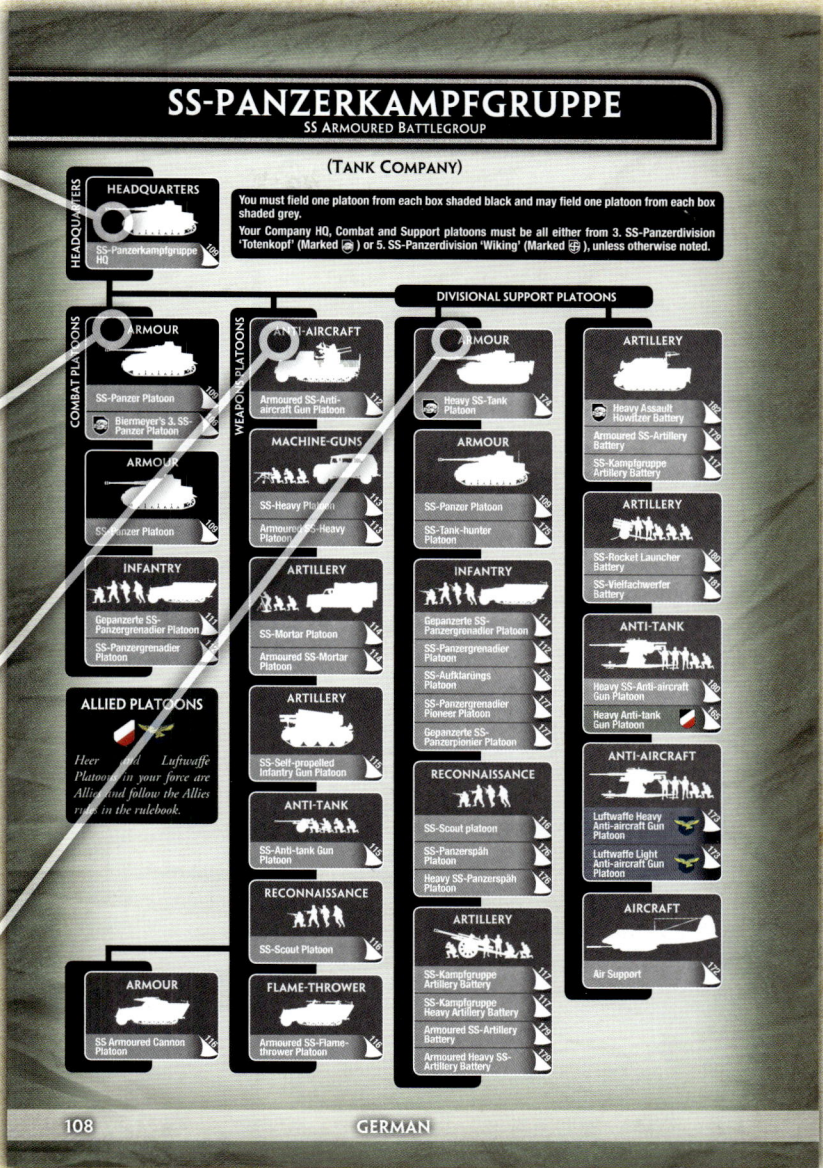

SS-PANZERKAMPFGRUPPE
SS Armoured Battlegroup

(Tank Company)

You must field one platoon from each box shaded black and may field one platoon from each box shaded grey.

Your Company HQ, Combat and Support platoons must be all either from 3. SS-Panzerdivision 'Totenkopf' (Marked ⊕) or 5. SS-Panzerdivision 'Wiking' (Marked ⊕), unless otherwise noted.

HEADQUARTERS
- SS-Panzerkampfgruppe HQ

COMBAT PLATOONS
- ARMOUR — SS-Panzer Platoon / Biermeyer's 3. SS-Panzer Platoon
- ARMOUR — SS-Panzer Platoon
- INFANTRY — Gepanzerte SS-Panzergrenadier Platoon / SS-Panzergrenadier Platoon
- ALLIED PLATOONS — Heer and Luftwaffe Platoons in your force are Allied and follow the Allies rules in the rulebook.

WEAPONS PLATOONS
- ANTI-AIRCRAFT — Armoured SS-Anti-aircraft Gun Platoon
- MACHINE-GUNS — SS-Heavy Platoon / Armoured SS-Heavy Platoon
- ARTILLERY — SS-Mortar Platoon / Armoured SS-Mortar Platoon
- ARTILLERY — SS-Self-propelled Infantry Gun Platoon
- ANTI-TANK — SS-Anti-tank Gun Platoon
- RECONNAISSANCE — SS-Scout Platoon
- FLAME-THROWER — Armoured SS-Flame-thrower Platoon

DIVISIONAL SUPPORT PLATOONS
- ARMOUR — Heavy SS-Tank Platoon
- ARMOUR — SS-Panzer Platoon / SS-Tank-hunter Platoon
- INFANTRY — Gepanzerte SS-Panzergrenadier Platoon / SS-Panzergrenadier Platoon / SS-Aufklärungs Platoon / SS-Panzergrenadier Pioneer Platoon / Gepanzerte SS-Panzerpionier Platoon
- RECONNAISSANCE — SS-Scout platoon / SS-Panzerspäh Platoon / Heavy SS-Panzerspäh Platoon
- ARTILLERY — SS-Kampfgruppe Artillery Battery / SS-Kampfgruppe Heavy Artillery Battery / Armoured SS-Artillery Battery / Armoured Heavy SS-Artillery Battery

DIVISIONAL SUPPORT (continued)
- ARTILLERY — Heavy Assault Howitzer Battery / Armoured SS-Artillery Battery / SS-Kampfgruppe Artillery Battery
- ARTILLERY — SS-Rocket Launcher Battery / SS-Vielfachwerfer Battery
- ANTI-TANK — Heavy SS-Anti-tank Gun Platoon / Heavy Anti-tank Gun Platoon
- ANTI-AIRCRAFT — Luftwaffe Heavy Anti-aircraft Gun Platoon / Luftwaffe Light Anti-aircraft Gun Platoon
- AIRCRAFT — Air Support

ARMOUR — SS Armoured Cannon Platoon

3. SS-PANZERDIVISION 'TOTENKOPF'
SS-PANZERKAMPFGRUPPE
SS Armoured Battlegroup

USING A COMPANY DIAGRAM

Each force begins with the Company Diagram, which demonstrates the company organisation graphically in an easy-to-read format. This simple diagram will help you create your *Flames Of War* company and get you playing in no time at all!

CHOOSING YOUR COMPANY

This first step is entirely up to you. There are a wide variety of companies, including infantry, reconnaissance, mechanised, and tank forces. Have a read through the backgrounds. Find a force that you find interesting and you're off!

INSTRUCTIONS

Once you have decided which company to build, have a look at the small box that contains the instructions you will need to build your force.

It will also detail any divisional variants that the company might have. The example to the right demonstrates a list with two variants: *3. SS-Panzerdivision* and *5. SS-Panzerdivision*. When building a force with variants, choose one variant and follow the instructions for variant forces below.

VARIANT FORCES

Some companies are based on a single division, like the Grenadierkompanie. However, other companies have several variants based on specific divisions, such as the SS-Panzerkampfgruppe representing *3. SS-Panzerdivision* and *5. SS-Panzerdivision*.

When building a force that has several division variants, you must choose one of the variants and stick to it. Your Company HQ and all of your Combat, Weapons, and a few Support platoons must match the division you have chosen. To help, we have included divisional symbols to distinguish the variants. Simply use the same divisional symbol when calculating points or choosing platoon options.

BUILDING YOUR COMPANY

You will notice that the Company Diagram consists of several black and grey boxes, each containing a silhouette of soldiers, guns, or vehicles. Each of these boxes are platoons available to your force.

The black boxes are the core of your force. When building your company, you must field one platoon from each box shaded black. The grey boxes are optional platoons. You may field one platoon from each box shaded grey.

PAGE REFERENCE

Each platoon box will have a page number. Use this to find the platoon and its points value.

ALLIED PLATOONS

Some platoons available to your company are from a different nation or branch of armed services. These can be easily identified by special national or branch symbols found to the right of the platoon title in the company diagram. Allied platoons have to follow Allied rules found in the *Flames Of War* rulebook.

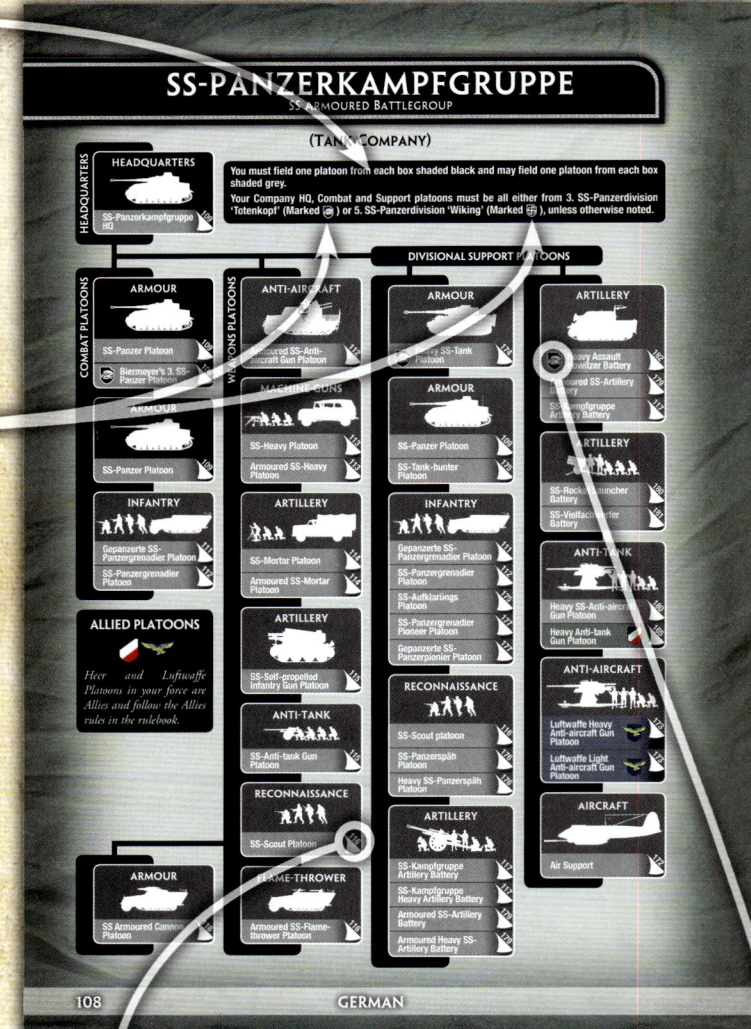

SPECIAL PLATOON OPTIONS

Sometimes a platoon box in the company diagram will have special platoon options available to a specific variant. These are always marked with the variant's symbol to the left of the platoon name. Only forces based on the same variant as the special option may take that platoon.

The example below (from the SS-Panzerkampfgruppe on page 108) demonstrates that while either variant may field an Armoured SS-Artillery Battery or SS-Kampfgruppe Artillery Battery, the *3. SS-Panzerdivision* ( ) can also take a Heavy Assault Howitzer Battery.

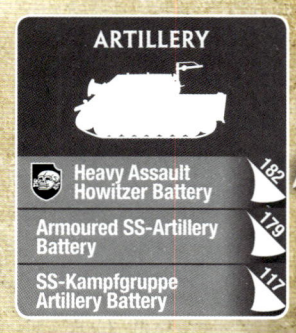

BUILDING A 3. SS-PANZERDIVISION SS-PANZERKAMPFGRUPPE

I have chosen to field a SS-Panzerkampfgruppe (found on page 108) using the *3. SS-Panzerdivision 'Totenkopf'* () variant. I chose a company from *3. SS-Panzerdivision 'Totenkopf'* because of their elite status.

I chose *3. SS-Panzerdivision 'Totenkopf'* () as my division variant, so all of my combat and weapons platoons should match that symbol or have no symbol.

Reading the instructions, I need to field at least a Company HQ and two Armour Combat Platoons in the black boxes. From there I'll add some of the optional platoons in grey boxes. One of my compulsory Combat Platoon boxes contains two options. I can either take a SS-Panzer Platoon or Biermeyer's 3. SS-Panzer Platoon, which comes with its own special rules. I decided to take Biermeyer and his platoon as one combat Platoon while my second Combat Platoon will be a standard SS-Panzer Platoon.

For my optional Combat Platoon I would like to take a Gepanzerte SS-Panzergrenadier Platoon (page 111). This means I need to use the points listed in the column marked () to match my division variant. There are two available variants: *3. SS-Panzerdivision,* and *5. SS-Panzerdivision.* Since I have chosen *3. SS-Panzerdivision* as my variant, I must use the column marked with my division symbol (see the following example).

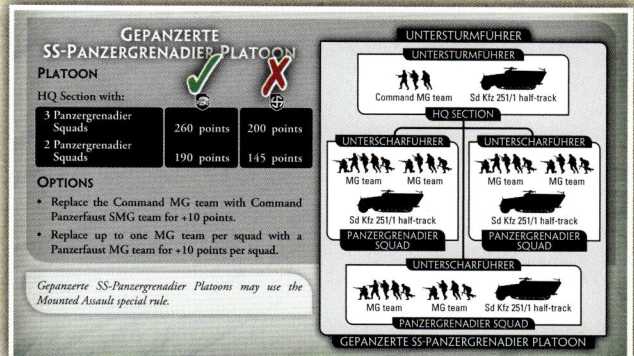

When building my 3. SS-Panzerdivision force I use the points in the left column.

I'll need some anti-aircraft guns to protect my tanks from the Red Air Force, so I've selected an Armoured SS-Anti-aircraft Gun Platoon (page 112). Again, there are two available variants, but I have to choose the *3. SS-Panzerdivision* variant.

Next, I would also like to take a Heavy Anti-tank Gun Platoon (page 165) to deal with enemy heavy tanks. This platoon is from the German *Heer* (army) and is an Allied Platoon. It has no divisional symbols, but since the company diagram allows it as an option, I can take it in my company (see the following example).

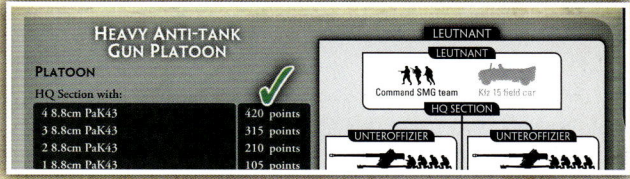

Because the Heavy Anti-tank Gun Platoon has no divisional symbol it can be taken as an option for any SS-Panzerkampfgruppe.

ADDING WARRIORS TO YOUR FORCE

There are many Warriors throughout this book. These are heroic soldiers who can join your force and help it to victory.

Warriors are available from the *Flames Of War* website *www.FlamesOfWar.com* and independent retailers as special order items (GSO### or FSO###,), blister packs (GE### or HU###), or in boxes (GBX##). Use the table below to find the relevant product code for each Warrior.

Pictured below is the Warrior Sturmbannführer Fritz Biermeyer in his Panzer IV H tank.

Warrior name and page number	Product code
Oberstleutnant Dr. Franz Bäke (Page 21)	GE894
Feldwebel Dietrich Uthoff (Page 51)	GSO30
Feldwebel Ludwig Windgruber (Page 68)	GBX28
Gereral der Panzertruppen Dietrich von Saucken (Page 69)	GE888
Sturmbannführer Fritz Biermeyer (Page 106)	GE889
Panzer Kanonen (Page 107)	GBX32
Fohadnagy Ervin Tarczay (Page 204)	HU880
Korpraali Toivo Ilomäki (Page 236)	FI520
Kapteeni Lauri Törni (Page 237)	FI703

BASING YOUR ARMY

Battlefront Miniatures packages *Flames Of War* products to give you everything you need to assemble your force as quickly and accurately as possible. Our blisters and box sets are packaged to give you all of the options available to build your army.

HOW TO BASE DIFFERENT TYPES OF UNITS

Every army organises its platoons differently, and the organisation diagrams reflect this. For example, a Panzergrenadier squad in a Gepanzerte Panzergrenadier Platoon is comprised of nine men split into two teams, one of four soldiers and one of five, yet a Panzergrenadier squad in a motorised Panzergrenadier Platoon has eight men split into two teams of four soldiers. Of course, units in combat rarely maintain their theoretical strength. We reflect this by allowing you to take fewer squads.

INFANTRY TEAMS

The fundamental building blocks of an infantry platoon are the various types of infantry teams. The most common ones are shown below with a brief description of their function and organisation.

COMMAND TEAMS

A Command team is made up of an officer, an NCO and a rifleman on a small base. There are often options to upgrade your Command team with a different weapon. To do so, simply replace the rifleman with the chosen upgrade. You can see an example of this on the following page.

RIFLE TEAMS

Rifle teams are the basic form of infantry. All the miniatures in a rifle team will normally be armed with rifles. Some squads may have a single machine-gun, but its effect is diluted by the number of rifles in the squad. Base your rifle teams on a medium base.

RIFLE/MG TEAMS

Rifle/MG teams are organised like rifle teams, except that every squad of two teams have a machine-gun. Base Rifle/MG teams on a medium base with the second base normally modelled with a crew-fed machine-gun.

MG TEAMS

MG teams are better armed than Rifle/MG teams. Every MG team has a machine-gun. Base MG teams with a crew-fed machine-gun and two to three riflemen on a medium base.

SMG TEAMS

Some nations equipped entire platoons with submachine-guns. SMG teams are made up of miniatures armed exclusively with submachine-guns. Base SMG teams on a medium base.

PIONEER TEAMS

A Pioneer team retains the normal characteristics and basing of its type, e.g. a Rifle team on a medium base, and gains combat engineering characteristics and abilities such as an increased anti-tank rating in assault and the ability to clear mines and demolish fortifications.

LIGHT MORTAR TEAMS

Light Mortar teams are made up of a miniature armed with a light mortar and a loader on a small base.

ANTI-TANK INFANTRY TEAMS

Anti-tank Infantry teams are infantry teams made up of two miniatures armed with a weapon like a Panzerschreck anti-tank rocket launcher and two loaders on a medium base.

GUN TEAMS

Artillery batteries and machine-gun, anti-tank gun, infantry gun platoons combine command infantry teams with gun teams. Information on basing gun teams can be found in Basing Your Miniatures in the rulebook. Essentially, Man-packed gun teams are mounted like infantry teams on a medium base, anti-tank and infantry guns are mounted on a medium base facing the narrow end, and artillery is mounted on a large base facing the narrow end.

PLATOON DIAGRAMS

Each platoon diagram indicates the required squads and teams you must have to make that unit combat-worthy. Troops and vehicles in black are the core of the unit. Troops and vehicles in grey are options that you can add to give them more punch or mobility. Many platoons also include options allowing you to improve the equipment or capabilities of some of the teams. The platoon entry will also list the special rules that the platoon follows.

Each platoon entry in a *Flames Of War* book reflects the historical make-up of the platoon, and tells you how to base the blister or box set that represents that particular unit. The following example shows a platoon of Panzergrenadiers and how to assemble it using the contents of the *GBX09 Panzergrenadier Platoon (Late)* box.

THE PLATOON ENTRY FROM THE BOOK (PAGE 77)

GEPANZERTE PANZERGRENADIER PLATOON

PLATOON

HQ Section with:

3 Panzergrenadier Squads	220 points
2 Panzergrenadier Squads	155 points

OPTIONS

- Replace the Command MG team with a Command Panzerknacker SMG team for +5 points or a Command Panzerfaust SMG team for +10 points.
- Replace Sd Kfz 251/1 half-track in HQ Section with a Sd Kfz 251/10 (3.7cm) half-track at no cost.

Gepanzerte Panzergrenadier Platoons may use the Mounted Assault special rule.

BUILDING THE PLATOON

Use the diagram below to base your platoon. By replacing the rifleman on the Command MG team with an Panzerknacker or Panzerfaust armed miniature, the team is upgraded to a Command Panzerknacker or Panzerfaust SMG team.

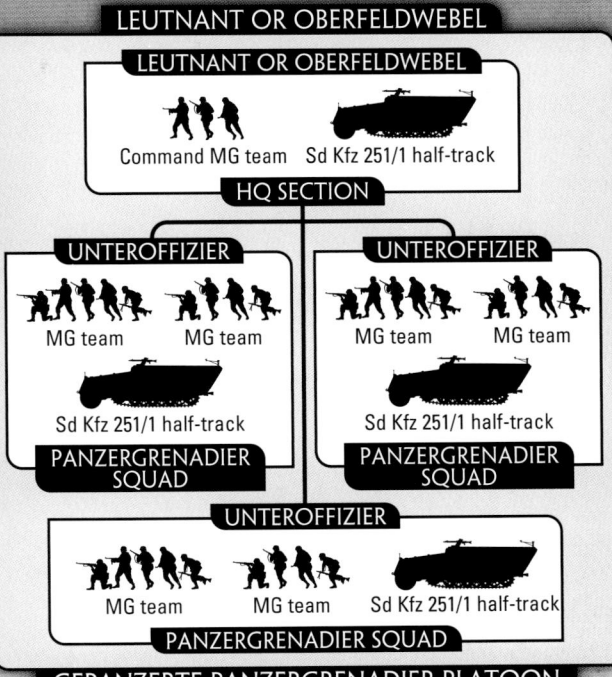

LEUTNANT OR OBERFELDWEBEL

LEUTNANT OR OBERFELDWEBEL

Command MG team — Sd Kfz 251/1 half-track

HQ SECTION

UNTEROFFIZIER — MG team / MG team — Sd Kfz 251/1 half-track — PANZERGRENADIER SQUAD

UNTEROFFIZIER — MG team / MG team — Sd Kfz 251/1 half-track — PANZERGRENADIER SQUAD

UNTEROFFIZIER — MG team / MG team — Sd Kfz 251/1 half-track — PANZERGRENADIER SQUAD

GEPANZERTE PANZERGRENADIER PLATOON

BUILDING THE PLATOON

LEUTNANT OR OBERFELDWEBEL

LEUTNANT OR OBERFELDWEBEL

Command MG team OR Command Panzerknacker SMG team OR Command Panzerfaust SMG team   Sd Kfz 251/10 (3.7cm) half-track

HQ SECTION

UNTEROFFIZIER

MG team

MG team

Sd Kfz 251/1 half-track

PANZERGRENADIER SQUAD

UNTEROFFIZIER

MG team

MG team

Sd Kfz 251/1 half-track

PANZERGRENADIER SQUAD

UNTEROFFIZIER

MG team

MG team

Sd Kfz 251/1 half-track

PANZERGRENADIER SQUAD

GEPANZERTE PANZERGRENADIER PLATOON

THE EASTERN FRONT

The German defeat at the Battle of Kursk in July 1943 marked the beginning of the end for the *Wehrmacht* on the Eastern Front. Both sides suffered huge losses in men and machines. However, they were losses the Soviets could make good, but from which the Germans would never fully recover. In August 1943 the Red Army began a series of massive counteroffensives that gathered in momentum, driving the invaders before them.

THE SOVIET COUNTEROFFENSIVES

Even while fighting continued at Kursk, the Red Army launched a series of counterattacks throughout the second half of 1943 that eventually led them to the Dnepr River.

Generalfeldmarschall Manstein scrambled to plug the holes, pleading for reinforcements for his battered Army Group South. Reluctantly, Hitler released *48. Panzerkorps* to reinforce the crumbling German lines. This allowed the beleaguered Germans to hold on until the autumn rains created a quagmire, halting Soviet attacks.

COUNTERATTACK AT ZHITOMIR

In December the bitter cold of the harsh Russian winter had frozen the mud solid. In a bold move that caught the Soviets by surprise, *4. Panzerarmee* attacked north of Zhitomir. The Soviet forces reeled before the ferocity of the assault and Red Army reserves were rushed west to bolster the lines. Army Group South recaptured Korosten and bought itself a little time to regroup.

THE KORSUN-CHERKASSY POCKET

On Christmas Eve Vatutin's First Ukrainian Front attacked. Most of Army Group South fell back to the *Panther-Wotan* line, but by early January large numbers of German troops still remained in exposed positions. Manstein sought to pull the exposed troops back to safety but Hitler, reluctant to take another backward step, ordered them to hold fast. The Soviet First and Second Ukrainian Fronts quickly enveloped the Germans, trapping 56,000 men from six divisions in the Korsun-Cherkassy pocket.

A relief force was hastily assembled to rescue the trapped divisions. Belatedly, Hitler authorised a withdrawal from the pocket. Soviet artillery shells and rockets pounded the retreating German columns mercilessly. The retreat became a rout and the troops abandoned almost all of their equipment.

RELIEVING LENINGRAD

Even as Army Group South suffered defeat in the Ukraine, far to the north troops from the Soviet Leningrad and Volkhov Fronts attacked the fortified lines south of Lake Ladoga. The offensive surprised the German Eighteenth Army besieging Leningrad and they were steadily driven back from the approaches to the city by wave after wave of assaulting infantry. On 18 January Red Army troops finally broke the siege that had lasted nearly 900 days.

DISASTER IN THE CRIMEA

Back in the south the Red Army maintained its momentum as troops of the Third Ukrainian Front swept across the Southern Ukraine, reaching the borders of Romania by 12 May. The Soviet advance had cut off the German and Romanian troops under General Janecke in the Crimean peninsula.

On 8 April Soviet troops attacked and quickly overwhelmed the Axis defences. On 9 May the German garrison in Sevastopol surrendered.

OPERATION BAGRATION

By May the Red Army had recaptured large areas of the motherland from the invaders. The Crimea and the Ukraine were back in Soviet hands and the Germans had been driven back from the approaches to Moscow and Leningrad, yet still large swathes of Byelorussia remained under German control.

On 22 June the Soviets launched Operation Bagration with the First Baltic, Second and Third Byelorussian Fronts smashing into the German defences between the cities of Vitebsk and Bobruisk, centred on Orsha. The Germans took a heavy toll on the attacking Soviet forces, but the Red Army still had more to give. By 25 June German forces were retreating back to the fortified cities, before Soviet assaults drove the Germans first out of Vitebsk, then Orsha, while encircling Mogilov and Bobruisk.

As German forces retreated towards Minsk, Berlin sent *Generalfeldmarschall* Model to stop the Soviet advance. Every available soldier was thrown into the battle in improvised blocking forces. Two panzer divisions and a heavy tank battalion were thrown against the Red Army's spearheads, delaying the onslaught bravely. More infantry, cavalry and armour were sent against them, but eventually the weight of the red tide told. Minsk fell on 3 July and the Germans withdrew to prepare for the next assault.

ON TO THE VISTULA

The initial Soviet spearheads of Operation Bagration had reached their objectives by mid July. They had retaken Byelorussia and now found themselves engulfing the *Festerplatz* (fortified) cities in eastern Poland. Soviet forces began to establish bridgeheads over the Vistula River. Clashing upon the banks of the Vistula the Soviets locked horns with German forces holding the river and the doorway to Germany and Berlin. The battles raged from Sandomierz in the south to the villages north of Warsaw. However, vigorous counterattacks by *SS*, *Luftwaffe* and *Heer* armoured forces finally halted the Red Army and Operation Bagration finally ended.

ASSAULT ON ROMANIA

When the Crimea fell to the Red Army in May 1944, only a handful of Romanians were left to be evacuated. In the meantime the Romanians were rebuilding, only returning to the fight when the advancing Soviets reached Romanian territory. In June 1944, the Soviets pushed the Romanians and Germans back across the Bug River and entered Bessarabia. The Red Army was on the very doorstep of Romania.

The Germans and Romanians established a defensive line along the Bessarabian frontier. This held until the Soviets launched a massive offensive on 20 August 1944. Two Soviet thrusts were directed at the cities of Iasi and Chisinau. Powerful Soviet armoured forces smashed through the Romanian lines and the small Romanian and German armoured reserves couldn't stem the tide.

Shortly afterwards, to save Romania from total destruction, an agreement was reached with the Soviet Union for Romania to swap sides and join the Allies. Romanian troops were soon in action defending the Transylvanian border from Hungarians trying to capture the vital passes through the Carpathian Mountains.

THE INVASION OF HUNGARY

After the fall of Romania three massive Soviet armies turned on Hungary which was desperately defended by six German and Hungarian armies. Hitler rushed fresh armoured forces

to Hungary to deal with the threat and a massive clash of armoured forces delayed the Soviet offensive at Debrecen. However, the respite was temporary as the great red sledgehammer was soon ready to swing again. The Red Army drove on towards Budapest, the Hungarian capital.

German and Hungarian forces attempted to stop the encirclement of the city, and a mixed corps of German and Hungarian troops held the city from repeated assaults by Soviet and Romanian forces. This savage and bitter struggle lasted for 100 gruelling days, but the city was finally taken on 13 February 1945.

Meanwhile the fight for Hungary continued as three German counter-offensives failed to break the siege. Eventually the assaults were called off on 27 January 1945 and the priority became the delay of the Soviet push on Austria.

GERMAN FORCES

ARMOURED FORCES

KAMPFGRUPPE BÄKE
(BATTLE GROUP BÄKE) PG. 22

In late 1943 a collection of units were flung together to form a *kampfgruppe* (battle group). The experienced *Oberstleutnant* Dr Franz Bäke was placed in command and they were hurled into action. Kampfgruppe Bäke is a fully self-contained force and although primarily a panzer unit of Panther and Tiger tanks, it had its own supporting infantry and artillery.

STUG BATTERIE
(ASSAULT GUN BATTERY) PG. 58

Every well-equipped German infantry division had a battalion of assault guns, this is especially true of *78. Sturmdivision*. The *189. Sturmgeschutzabteilung* (189th Assault Gun Battalion) follows the *sturm* grenadiers' assaults, taking out subborn Soviet resistance. Riding into battle with them are the *Begliet* (tank escorts), who fight from the StuG G assault guns with STG44 Assault Rifles, protecting the StuGs from enemy infantry.

SCHWERE PANZERKOMPANIE
(HEAVY TANK COMPANY) PG. 70

The heavy Tiger I E tanks of *502. Schwere Panzerabteilung* and *505. Schwere Panzerabteilung* provide the power to engage and destroy even the most tank heavy Soviet battalions. Hitler has also released the first battalion of forty-five new Tiger II heavy tanks, which were delivered to *501. Schwere Panzerabteilung* to defeat the Soviets at Sandomierz, Poland in August 1944.

PANZERKOMPANIE
(TANK COMPANY) PG. 72

5. Panzerdivision and *12. Panzerdivision* were rushed into the line to halt the Soviet breakthrough at the start of Operation Bagration. Other panzer divisions fought up and down the Eastern Front throughout 1944. The panzers of a Panzerkompanie have undergone a constant evolution throughout the war. They are heavier, better armed and more mobile than those of your enemies.

SS-PANZERKAMPFGRUPPE
(SS ARMOURED BATTLE GROUP) PG. 108

SS tank forces provide ample striking power to decimate enemy units. Newly refurbished with fresh recruits, the *Kanonen* (Tank Aces) of the Panzer companies provide the leadership needed to strike a deadly blow to advancing Soviet tanks.

FELDHERRNNHALLE PANZERKAMPFGRUPPE
(FELDHERRNNHALLE ARMOURED BATTLE GROUP) PG. 136

The *Panzerkampfgruppe* is the powerful strike arm of the *Panzerdivision*. By assembling the panzers and armoured panzergrenadiers together the division had a potent weapon with which to blunt the enemy's advance or overrun them.

PANZERJÄGERKOMPANIE
(TANK-HUNTER COMPANY) PG. 144

In the latter half of 1944 the Hetzer tank-hunter became available in increasing numbers and these were used to equip both assault gun and tank-hunter units. Many of these newly formed or refitted units went into battle supporting infantry forces where the Hetzer tank-hunter performed well in its dual role as an assault gun and tank-hunter.

MOTORISED FORCES

SS-FREIWILLIGEN-PANZERGRENADIERKOMPANIE
(SS VOLUNTEER MOTORISED INFANTRY COMPANY) PG. 38

While the Panzer divisions seize enemy held ground in sweeping counterattacks it is the role of Panzergrenadier divisions to quickly follow up and hold the new positions. However, on the Narva River the foreign volunteer SS-Panzergrenadiers hold the line against vigorous Red Army assaults, before organising counterattacks to retake ground.

GEPANZERTE PANZERGRENADIERKOMPANIE
(ARMOURED INFANTRY COMPANY) PG. 76

The Gepanzerte Panzergrenadierkompanie are the elite of the mechanised infantry. Instead of trucks they are equipped with armoured Sd Kfz 251 half-tracks giving them the mobility and armour to accompany the tanks into the fighting.

PANZERGRENADIERKOMPANIE
(MOTORISED INFANTRY COMPANY) PG. 80

The Panzergrenadierkompanie uses trucks to bring its veteran soldiers to the battlefield where they dismount and fight on foot. Well-supported by Panzers, assault guns and mobile artillery their usual role is to attack defences to create a breakthrough for the Panzers, or to hold the ground they take with the Panzers as a counterattack reserve.

GEPANZERTE PANZERPIONIERKOMPANIE
(ARMOURED ENGINEER COMPANY) PG. 86

A Panzer division's armoured pioneers are heavily-armed assault specialists. Equipped with armoured half-tracks the Gepanzerte Panzerpionier Platoon's combat engineers are well protected during the opening phase of close assaults.

PANZERPIONIERKOMPANIE
(MOTORISED ENGINEER COMPANY) PG. 88

Most of the combat engineers of the *Panzerdivision* are motorised in trucks. The Panzerpionier troops are called forward when enemy fortified positions need to be cleared. In defence, they are equally good at creating barriers and obstacles around strategic positions. Well armed with rifles, machine-guns, mines and demolition carriers, these pioneers are able to take on anything that the Soviets can throw against them.

SS-PANZERGRENADIERKAMPFGRUPPE
(SS MOTORISED INFANTRY BATTLE GROUP)   PG. 110

The unique make up of the *'Wiking'* division allows you to field either Danish or Flemish troops. The *Totenkopf* (Death's Head) division has its own Tiger company, as well as Sturmtiger assault howitzers, in support to assist them in erasing Soviet tanks or assault groups from the battlefield.

FELDHERNNHALLE PANZERGRENADIERKOMPANIE
(FELDHERNNHALLE MOTORISED INFANTRY COMPANY)   PG. 140

The *Feldhernnhalle* Panzer division's panzergrenadiers are excellent infantry provided with transport for fast deployment. In Budapest they mostly fight on foot from street fortifications. They are supported by weapons platoons from their own division as well as allied Hungarian platoons.

INFANTRY FORCES

GRENADIERKOMPANIE
(INFANTRY COMPANY)   PG. 26

A Grenadierkompanie is the finest infantry force you can command. The troops are all experienced veterans and they are backed by the best weapons that the German army can provide, from StuG assault guns to the feared Tiger, your Grenadierkompanie will never be out-gunned or out-classed.

FALLSCHIRMJÄGERKOMPANIE
(PARACHUTE RIFLE COMPANY)   PG. 32

The Fallschirmjäger are Germany's paratroopers. Like all paratroopers, they are tough, tenacious and independent fighters. Though no longer making air assaults, they are Germany's finest light infantry, quite capable of holding their position in the most dire situations.

STURMKOMPANIE
(ASSAULT COMPANY)   PG. 52

The *78. Sturmdivision* was organised in early 1943 as a specialised assault division. The concept of the *Sturmkompanie* went beyond the normal grenadier organisation by giving it a lot more heavy equipment, such as its own integral 7.5cm PaK40 anti-tank guns, a heavy platoon and the support from the divisions own StuG Batterie, which lends immediate armoured support whenever needed.

PIONIERKOMPANIE
(ENGINEER COMPANY)   PG. 60

The Pionierkompanie is composed of tough well-trained combat engineers ready to do two things: undertake dangerous engineering assignments while under fire, and storm enemy positions by close assault.

SPERRVERBAND
(BLOCKING FORCE)   PG. 94

In the desperate days of the Bagration breakthrough Sperrverband forces organised a core of veterans and supported them with the best troops available, allowing them time to dig in, prepare solid defences, and halt the Soviet tank hordes in their tracks.

ERSATZ PIONIERKOMPANIE
(RESERVE ENGINEER COMPANY)   PG. 98

If you see the need for more fortifications and engineering in your defence then try the Ersatz Pioniere (Reserve Pioneers). They provide the manpower and expertise you need to solidly hold the line until you can counterattack with your Panzers.

SICHERUNGSKOMPANIE
(SECURITY COMPANY)   PG. 118

The Sicherungskompanie was primarily used to hunt down partisans. However, as the front lines in Byelorussia crumbled German security forces were thrust into the front lines. While fighting partisans, or guarding the rail and roadways behind the lines, swiftly advancing Soviet spearheads sometimes forced security forces to hold until support could arrive.

SS-KAVALLERIESCHWADRON
(SS CAVALRY SQUADRON)   PG. 146

During the siege of Budapest the *8. SS-Kavalleriedivision 'Florian Geyer'* and *22. SS-Freiwilligen-Kavalleriedivision 'Maria Theresia'* cavalry divisions were trapped inside the city. They fought furiously against the red onslaught on foot alongside the Hungarians to save their glittering capital. Like all good modern cavalry they easily adapted to fighting on foot, making use of the good defensive positions.

RECONNAISSANCE FORCES

PANZERSPÄHKOMPANIE
(ARMOURED CAR COMPANY)   PG. 90

Armoured cars do not generally engage the enemy directly except when confronted by a weak enemy or at the point of attack. Armoured car patrols are the true reconnaissance troops of an *Aufklärungsabteilung* (reconnaissance battalion). While the rest of the battalion is beating a path through the enemy the armoured cars goal is to exploit the gap in the lines to get behind the enemy and scout out his positions.

HUNGARIAN FORCES

ARMOURED FORCES

HARCKOCSIZÓ SZÁZAD
(TANK COMPANY)   PG. 206

The armoured troops are highly motivated and well trained. They fight with a mix of Hungarian and German tanks and call on support from their infantry, anti-aircraft, and artillery. In addition they fight side-by-side with the Germans and are supported by German armour, artillery and aircraft.

ROHAMÁGYÚS ÜTEG
(ASSAULT GUN BATTERY)   PG. 214

The assault gun battalions are the elite mobile arm of the Hungarian artillery. They use their own native Zrínyi assault howitzer, or German StuG G or Hetzer assault guns. They are supported by the divisions they fight with, which includes Hungarian tanks, anti-tank, infantry, reconnaissance, anti-aircraft, artillery and aircraft, as well as German panzers, panzergrenadiers, and aircraft.

MOTORISED FORCES

GÉPKOCSIZÓ LÖVESZ SZÁZAD
(MOTORISED INFANTRY COMPANY)   PG. 210

The motorised infantry of 1st and 2nd Armoured Division are commanded by veterans of the battles on the Don in 1942 to 1943. They are some of the best troops in the Hungarian armies. These skilled riflemen keep pace with the rapid movements of the tanks to provide support when it is needed, ready to take captured positions and hold them against counterattack.

INFANTRY FORCES

PUSKÁS SZÁZAD
(RIFLE COMPANY)   PG. 216

The bulk of the Hungarian Army, or *Honvéd* (pronounced hon-veed), was made up of the infantry of the *Puskás Század* (Rifle Company). The Hungarians have built up their infantry forces rapidly as the threat of Soviet invasion loomed. Some divisions have been raised from reservists and new recruits, while other divisions call on the experience of previous battles. The Hungarians had also been reforming their organisation and arming their troops with more modern heavy weapons such as German 75mm anti-tank guns, 120mm mortars and German Panzerfaust and Panzershreck anti-tank weapons. They can even call on the support of their own assault guns in armoured support.

ÖNKÉNTES PUSKÁS SZÁZAD
(VOLUNTEER RIFLE COMPANY)   PG. 220

The riflemen of the Budapest defence infantry come from a variety of sources, from highly motivated volunteers to professional police and regular soldiers. All the different forces trapped inside Budapest fought together and supported each other. In addition to the machine-guns of their weapons platoons, they are backed by mortars, assault guns, panzers, anti-tank, infantry, reconnaissance, heavy mortars, anti-aircraft, artillery and aircraft.

FINNISH FORCES

ARMOURED FORCES

PANSSARIKOMPPANIA
(TANK COMPANY)   PG. 238

Finnish *Panssari* (Tank or Armour) units are seasoned veterans, who have had up to two years training. They have learnt to survive operating obsolete Soviet light and medium tanks and are now full of confidence in heavier T-34 tanks. Some *Panssari* companies are a mixture of the captured Soviet medium tanks. However, most brave crews still soldier on in T-26 and T-28 tanks The T-26 tank remains the most numerous tank in the Finnish arsenal.

STURMIKOMPPANIA
(ASSAULT GUN COMPANY)   PG. 240

With new German-supplied Sturmi (StuG III G) assault guns, the *Panssari* division now packs quite a punch. The Sturmi crews have been training hard, are led by selected officers picked from *Jääkäri* units, and are ready for battle against the toughest foes. Their 75mm guns, combined with the skill of their Finnish crews, have enabled them to achieve impressive kill ratios, even against Soviet heavy armour.

INFANTRY FORCES

JÄÄKÄRIKOMPPANIA
(LIGHT INFANTRY COMPANY)   PG. 242

The *Jääkäri* (light infantry) remain the best of the Finnish infantry, and rigorous training and patrolling has kept them so during the quiet of 1943. Their tactics emphasize speed and mobility. A high proportion of their leaders are volunteers or former regular army officers and often the men are from rural areas with excellent fieldcraft skills. For them the best form of reconnaissance is attack.

JALKAVÄKIKOMPPANIA
(INFANTRY COMPANY)   PG. 246

The *Jalkaväki* (infantry) are superbly trained and experienced, although some of the troops grow weary from five years of war. *Jalkaväki* regiments are balanced with infantry, light infantry (*Jääkäri* scouts), machine guns and mortars. Each regiment has an anti-tank unit. Divisional support is better than ever, with anti-aircraft, anti-tank and artillery support readily available and even modern assault guns – the Sturmi.

PIONEERIKOMPPANIA
(ENGINEER COMPANY)   PG. 250

The Pioneerikomppania (Engineer Company) provide the Finnish divisions with experts in field engineering and mine clearing. They also serve as excellent assault troops for taking our enemy's fortifications. In attack their satchel charges, flame-throwers and pioneer skills are invaluable. In defence they can lay the minefields that have so effectively slowed the Soviet advance.

Tiger IE tanks of Kampfgruppe Bäke grind through the snow to relieve the trapped forces.

Grenadiers hold fast as Soviet Strelkovy put the squeeze on their positions.

Kampfgruppe Bäke Panther tanks had vital support from attached Gebirgsjäger infantry.

The armoured anti-aircraft half-tracks are the last line of defence against attack from the air.

BATTLE FOR THE NARVA LINE

Germany's Army Group North had been besieging Leningrad for over two years. In January 1944, the Red Army launched an offensive that threatened to encircle the besieging army. Through a series of skilful rear guard actions, the *III SS-Panzerkorps* was able to withdraw to the natural defensive positions on the Narva River near the city of Narva.

As they arrived, the German forces found that the defensive fortifications of the Panther Line existed only on paper. There were a few emplacements from the invasion in 1941, but generally these were in poor locations for the current battle. The ground was frozen making the preparation of new trench lines difficult. However, the terrain offered a choke point with Lake Peipus in the south and the Gulf of Finland to the north connected from south to north by the Narva River.

The river represents the border between Russia and the Baltic nation of Estonia. If the Red Army could drive the Germans from Narva, they would remove them from the whole region. This would score both a strategic and political victory for Stalin's armies. The river was icy and the bridges were the key to both the defence of Narva and the success of the Soviet attacks.

The Red Army did not wait for the Germans to settle. On 2 February 1944 the Second Shock Army immediately attacked the German bridgehead on the east side of the Narva River. This bridgehead was a seven mile long line stretching from the village of Lilienback in the north to the village of Dolpaya Niva in the south.

Manning this position were troops from *4. SS-Freiwilligen Panzergrenadier-Brigade Nederland* and *11. SS-Freiwilligen Panzergrenadierdivision Nordland*. These units were unusual in that they consisted of volunteers from European countries other than Germany (*Freiwilligen* means volunteers).

On 3 February the Soviet attack began with a thrust through the defences on the eastern side of the river. Soviet tanks broke through and threatened to establish their own bridgehead on the western side of the river. The *Nordland* SS-Divison's Panzer Battalion, named Hermann von Salza, and the Tiger tanks of *502. Schwere Panzerabteilung* (502nd Heavy Tank Battalion), which included Tiger ace Leutnant Otto Carius, joined the battle. They defeated the Soviet armour, and with their support, the *Nordland* infantry was able to recapture the position on the eastern side of the river.

On 11 February the Soviet 43rd Rifle Corps attacked north of the city, but was met strongly by the *227. Grenadierdivision* and the *Nederland* SS-Brigade. In the south, attacks by the Red Army's 109th and 122nd Rifle Corps were somewhat more successful advancing 7½ miles before being stopped by *Nordland* SS-Division, *170. Grenadierdivision* and the *Panzergrenadierdivision Feldherrnhalle*.

Soviet assaults continued up and down the Narva line. In the north the Soviets established a bridgehead on the western side of the river near the village of Siivertsi. After nine days of heavy fighting the Soviets were again forced back to the east side of the river.

On 14 February the Soviet hight command (Stavka) issued new orders to General Govorov commanding the Red Army on the Narva front. A fresh Soviet assault south of Narva by the 30th Guard Rifle Corps broke through the defensive line and established a bridgehead. By 24 February the assault forces had swung in a wide arc around Narva, reaching the rail line supporting the city and threatening to encircle the defenders. It was stopped when *Panzergrenadierdivision Feldherrnhalle* and *502. Schwere Panzerabteilung* counter attacked.

After a week of relative quiet, March saw a renewed offensive by the Soviets. At dawn on 1 March, the Red Army began a 20-minute artillery barrage against the German positions. This was followed by the advance of units from three rifle corps. The artillery barrage proved inadequate against the well dug in defenders and the advance halted after heavy casualties. Accurate and timely German artillery played a large part in breaking up the assault, as did the *Luftwaffe* even though the Soviets enjoyed a roughly 3 to 1 advantage in air power. The Soviet assault had gained little.

General Govorov attempted to revive the assault on 2 March by bringing fresh forces to bear against the Narva defences. These attacks also floundered. Rather than allow the Soviets to consolidate the few gains they had achieved, the German forces launched a series of vicious counterattacks between 4 and 6 March. These attacks recaptured the lost ground and returned the front to its February positions.

In response, on the night of 6/7 March, the Red Air Force made a massive bombing raid on the city of Narva, reducing it to ruins. The surviving civilians fled west, but the defenders positions remained intact.

The Soviets continued attacking up and down the Narva Line probing for weaknesses. One attack against Nederland's *49. SS-Freiwilligen Panzergrenadier Regiment* broke through towards the river. The Soviets committed their tank reserves to seize the bridges across the river into the city. The defenders counterattacked with Nordland's Herman von Salza *Panzer Abteilung*. The panzers stopped the advancing tanks, but they were prevented from exploiting their advantage by heavy anti-tank fire from the eastern side of the river.

On 23 March Hilter ordered the creation of *Festung Narwa* (Fortress Narva) to be held at all costs. The Germans launched a series of counterattacks beginning on 26 March. These attacks were designed to eliminate the Soviet bridgehead gained by the 30th Guard Rifle Corps in February on the western side of the river.

They drove the Red forces back to the river, but not across it. The attacks could not be sustained with the equipment and manpower available, but the threat was sufficient that General Govorov ordered the construction of extensive defensive works on the eastern side of the river to prevent a possible breakout.

By the end of March, the Red Army was not able to defeat the Germans at Narva. The Germans had succeeded in rapidly reinforcing their position, while the Soviets were hampered in their attacks both by the terrain and a poor cooperation between the various units in the sector. The thaw in April and the *rasputitsa* (muddy season) forced a halt all major operations until late May.

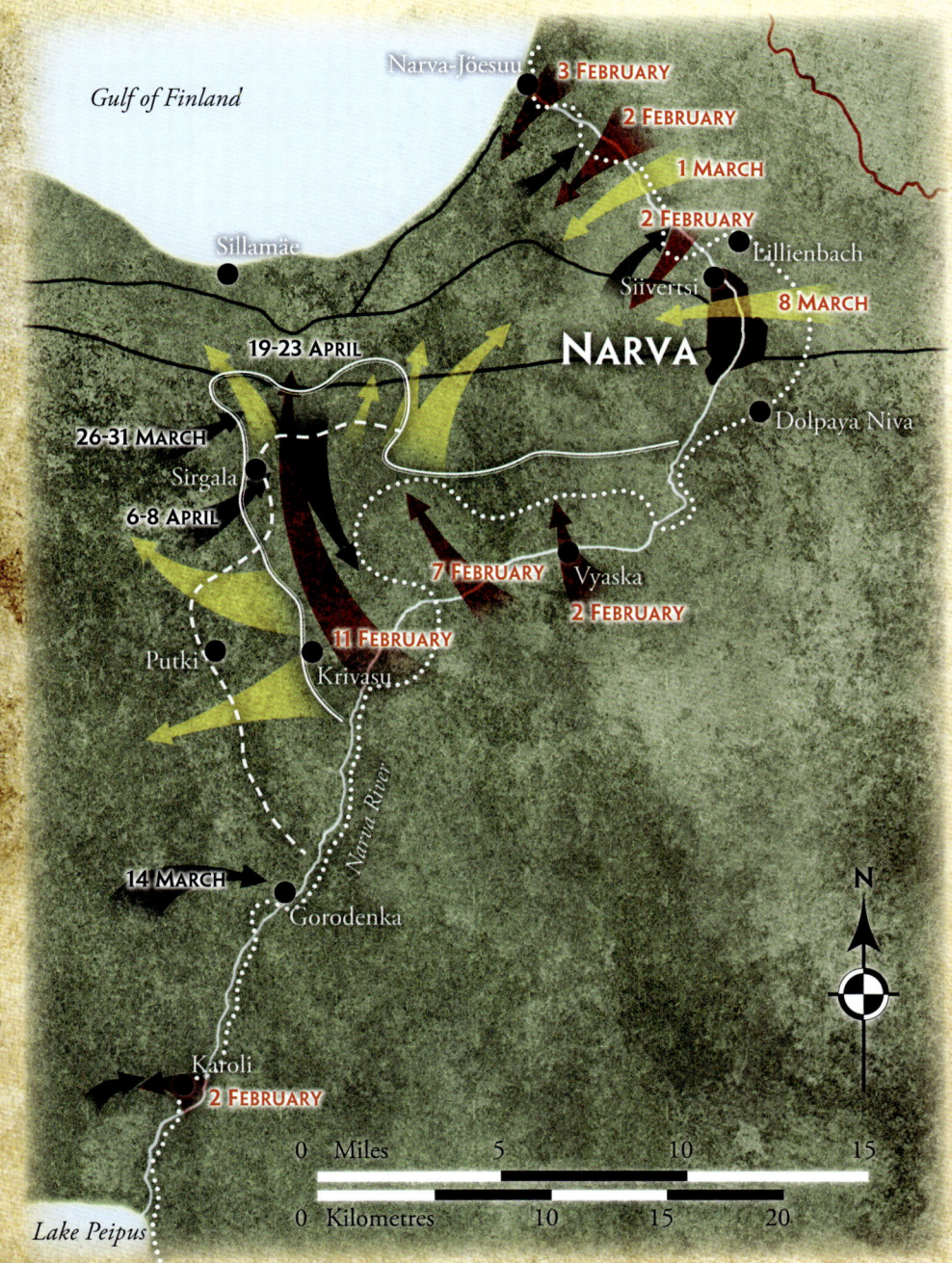

KORSUN POCKET

The battle for the Korsun Pocket was a huge complex battle involving 400,000 men fighting over an area 100 square kilometres for three weeks.

On 24 January 1944 the Soviet First and Second Ukrainian Fronts, having regrouped after their successful campaigns at the end of 1943, attacked south of Kiev hitting the positions of the German First Panzer Army near the town of Korsun-Shevchenkovskiy (100 miles southeast of Kiev). The aim was to trap the Germans in a pincer formed by the two fronts, taking advantage of the desire of the Germans to hold some part of the Dnepr River, rather than withdraw.

The Soviet First Ukrainian Front attacks met serious resistance the following day and suffered heavy casualties from concentrated German artillery. The battle around Korsun raged through the following days and by 28 January 1944 the Soviets had encircled six German divisions from the XI and XLII Korps (Eighth Army). Hitler insisted that the surrounded positions on the Dnepr River be held and forbad any breakout attempt. In response Field Marshal von Manstein, commanding Army Group South, began to assemble an armoured force to break into the encirclement.

By 4 February Von Manstein had massed four panzer divisions and an ad hoc heavy tank regiment under Dr. Franz Bäke (one battalion each of Tigers and Panthers) and began his counterattack to relieve forces trapped around Korsun. The force was denuded of the full strength *24. Panzerdivision* at the last minute when Hitler personally intervened to send it south to help at Sixth Army at Nikopol.

The German forces inside the pocket were designated *Kampfgruppe* Stemmermann (after the senior commander) on 5 February. The Soviets stepped-up attacks to reduce the pocket the same day. To compound difficulties for both sides, the temperature plummeted to well below zero throughout the battle area. By 7 February *Kampfgruppe* Stemmermann, while under continuous pressure from Soviet attacks, contracted its perimeter, abandoning Gorodische and Yanovka, and began to prepare for a breakout attack. Meanwhile, von Manstein's break-in operation continued against very heavy resistance.

On 8 February the Soviets offered *Kampfgruppe* Stemmermann the opportunity to surrender, but it refused. The Germans redoubled their efforts to fly supplies into the pocket and exit flights were able to evacuate some of the seriously wounded.

Having finally assembled an effective force, *III Panzerkorps* began its relief attacks on 11 February. They get within 10 miles of the pocket, but Soviet resistance stiffened in the following days.

On 14 February Soviet forces captured Korsun-Shevchenkovskiy against the determined resistance of the SS Walloon Brigade, of *5. SS-Panzerdivision Wiking*. As the perimeter of the pocket continued to shrink Hitler grudgingly gave permission to attempt a breakout on 15 February. The following day the *III Panzerkorps* was stopped 12 miles from the perimeter of the pocket. The relief forces were exhausted and it was concluded that any further action would be fruitless.

Led by the remnants of *5. SS-Panzerdivision Wiking*, *Kampfgruppe* Stemmermann launched their attempt to break out on 17 February. In a blinding snowstorm they managed to find a seam in the Soviet defences. At dawn, the weather cleared and the Soviet cavalry and aircraft pounced on the columns of fleeing Germans. Having abandoned their heavy equipment, the breakout turned into a rout as desperate men fled from the carnage. General Stemmermann was killed along with many of his men. However, around 35,000 men escaped. They had little more than their personal weapons, if that. However, if Stalin had expected a battle of annihilation like at Stalingrad, he was mistaken. The Germans had effected their escape and the pocket was empty.

The Korsun Pocket was a Soviet victory, though it came at a high price and the Germans could claim that most of their men escaped the trap. The intensity of the battles was born out in the casualty figures. In two months the Soviets lost 25,000 killed, 56,000 wounded (from 240,000) and half their 500 tanks. For the Germans claims vary, but as a minimum 20,000 Germans in the pocket were killed or captured, and another 11,000 escapees were wounded (from 55,000). In addition a similar number of killed and wounded were lost from the relieving forces, which also lost over 100 tanks (from 80,000 men and 300 tanks).

KAMENETS-PODOLSKY (HUBE'S) POCKET.

On 4 March 1944 the Soviets began another offensive aimed at pushing Army Group South out of the Ukraine. Taking advantage of the weakening of the southern German positions by moving panzer troops northwards for the Korsun Pocket battles, the Soviets drove the First and Second Ukrainian Fronts to the north and south of the *1. Panzerarmee*, eventually encircling them by 25 March.

Leading the *1. Panzerarmee* was *Generaloberst* Hans Valentin Hube, a standout general, known as '*Der Mensch*' ('The Man') among the soldiers. Having experienced the encirclement of *6. Armee* at Stalingrad he took steps to limit a potential disaster by ordering all administrative and non-combat personnel to the rear and reducing the number of non-combat vehicles in the area to save fuel.

The Soviet encirclement was complete by 25 March with six Soviet armies applying pressure to Hube's force. At the same time the external encirclement of another three Soviet armies, was designed to force the German *4. Panzerarmee* and *8. Armee* back and stop any rescue operation.

With the *1. Panzerarmee* still holding the bridge over the Dniester at Khotin, it seemed obvious to the Soviets that at some point Hube would attempt an offensive south. The two Soviet fronts began to shift forces southward to prepare for this break out.

Manstein and Hube, seeing the strengthening of the Soviet southern forces, began to devise a counter plan. In spite of the strength of the forces to the north and west of the *1. Panzerarmee* and the distance of 125 miles to the *4. Panzerarmee* lines, Manstein wanted Hube to breakout west. This would force the *1. Panzerarmee* to fight its way out of encirclement against some of the strongest forces surrounding it.

On 25 March Hitler gave Hube authority to break out as well as giving Manstein command of *II SS-Panzerkorps*, consisting of *9.* and *10. SS-Panzer* divisions (*Hohenstaufen* and

Frundsberg), *100. Jäger* and *367. Infanterie* divisions, which would fight their way from the *4. Panzerarmee* lines to try and reach the encircled troops.

Hube repositioned his forces into three distinct *kampfgruppen*, the first and second being northern and southern attack groups. The *kampfgruppen* were to carry out the initial break out west while a rearguard group would fight aggressive withdrawing engagements. The North and South spearheads were led by *Panzerkampfgruppe* Bäke and his Panthers as well as the other panzer divisions. The Tiger companies of the *509. Schwere Panzerabteilung* dominated the rearguard.

Early afternoon radio reports of 28 March and reconnaissance indicated to the Soviets that the *1. Panzerarmee* was making preparations to cross the bridgehead and break out south. Believing the break out was heading south the Soviets relaxed their other armies and focused all attention on the southern units.

This played perfectly into Hube's plan. On the night of 28/29 March, *Panzerkampfgruppe* Bäke broke west with *1. Panzer* army's entire force behind it. A blizzard began the night of 28 March blanketing everything in thick white snow, a mixed blessing that slowed movement, but also provided concealment. By 29 March, the first Panther battalions of *1. Panzerarmee* had seized three separate bridgeheads across the Zbrucz River and were advancing against Soviet defenders in the valley between the Zbrucz and the Seret.

It was at this point that the Soviet commanders realised they'd been tricked. The southern break out was simply deception. Immediately, intense attacks were launched against the *1. Panzer* army's rearguard. However, they held off the Soviet attacks, halting the assault by 31 March.

The Soviets were masters of keeping their advance moving in the winter conditions, but the Germans were equal to the task. The *Luftwaffe* kept *1. Panzerarmee* fuelled and supplied enough for the constant moving and fighting, both day and night, over the two week running battle.

II SS-Panzerkorps slammed into the weakly held lines around Podgaitsy, smashing through on 4 April. The next day, advance elements of *1. Panzer* army's breakout force met elements of the *II SS-Panzerkorps*. Over the next few days, elements of *1. Panzerarmee* pushed forward into *4. Panzer* army's lines at and around Buczacz on the Strypa River. The breakout had been secured, but fighting continued for another two weeks into April as all of *1. Panzer* army's divisions and men were successfully withdrawn.

The *1. Panzerarmee* had escaped with most of its manpower and remained a viable fighting force. They re-established a new front line from the Dniester River up to the town of Brody. They suffered minimal casualties but lost most of their heavy equipment throughout the almost month long battle. Having saved the *1. Panzerarmee* Generaloberst Hube, after receiving Diamonds for his Knight's Cross with Oak Leaves and Swords, tragically died when his plane crashed in the Austrian Alps after leaving his medal ceremony.

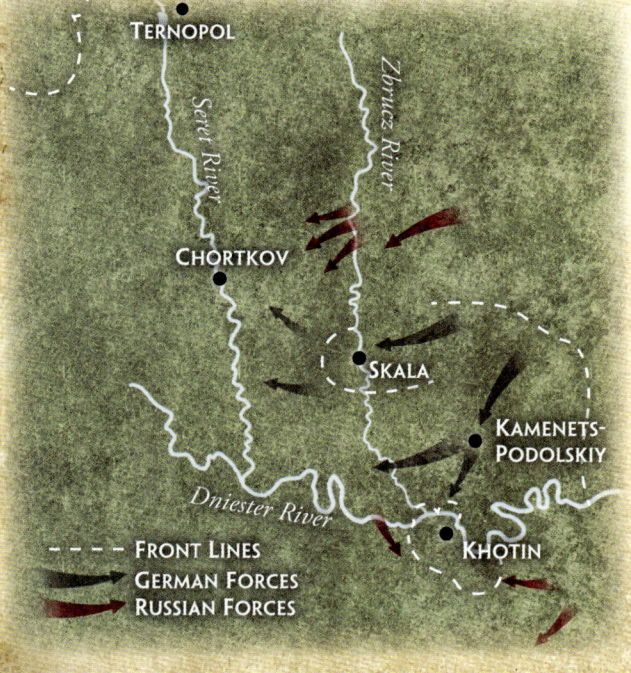

FRONT LINES
GERMAN FORCES
RUSSIAN FORCES

SCHWERE PANZERREGIMENT BÄKE

The year 1943 drew to a close in the cold of the Ukrainian winter, with German Army Group South in crisis. Retreating everywhere since the defeat at Kursk, a succession of Soviet offensives saw the liberation of Kharkov, Kiev and the crossing of the Dnieper River line. They were outnumbered everywhere, and knew that another Soviet strike would surely come in the new year. Faced with the reality that he could not possibly hold the line, *Feldmarschall* von Manstein realised that he needed mobile reserves.

It was at this time that the German general staff's genius for organisation and flexibility was demonstrated. Hurriedly independent units and portions of larger formations in reserve were flung together to form a *kampfgruppe*. An experienced commander, *Oberstleutnant* Dr Franz Bäke, was placed in command and soon formed a unique group of forces into a cohesive regiment. Then they were hurled into action.

Schwere Panzerregiment Bäke's first battle was at the Balabonowka pocket in January 1944, where it immediately had to halt a Soviet offensive. On the boundary between First and Fourth Panzer Armies, five Soviet Tank Corps from 1st Tank Army had broken through and headed for Vinnitsa. On the open steppe between the Bug and Dnieper Bäke's Tigers and Panthers met the Soviets. They destroyed over 250 Soviet tanks for the loss of just four. During this battle Bäke personally destroyed three Soviet tanks at close quarters with infantry weapons.

Next *Schwere Panzerregiment* Bäke led attempts to relieve the Korsun Pocket. *III Panzerkorps* and *XLVII Panzerkorps* were rapidly launched in counter-offensives to break them out. *Schwere Panzerregiment* Bäke formed the spearhead of *III Panzerkorps*. Thrusting through two Soviet Infantry corps and 5th Guards Tank Corps (including units equipped with IS-2 heavy tanks), Regiment Bäke reached the Gniloy Tikich River, just ten kilometres from the pocket. Though falling short it held the southern flank open to allow thousands of the trapped men to escape.

There was little time to recover after Korsun. In March *Schwere Panzerregiment* Bäke was itself encircled in the Kaminets-Podolsky pocket, better known as Hube's pocket. Here Zhukov and Konev launched another offensive with two huge pincers, this time cutting off the entire *1. Panzerarmee* under General Hans Hube in a huge sack between the Bug and Dniester Rivers, further to the west of Korsun. By now von Manstein and Hube were not about to sit idly by waiting for orders. While his force was mobile he would get himself out. Rather than break through the bulk of the Soviet spearhead in front of him, von Manstein ordered Hube to thrust sideways, through the weaker Soviet flanking forces. Once again *Schwere Panzerregiment* Bäke formed part of the spearhead of Brieth's *III Korps*. They were part of the force that cut through the Soviet infantry along the Zbruch River. Hube had ordered all non-fighting equipment to be ruthlessly discarded and both combat and support units moved with the panzers as transport. These desperate measures worked and led to the successful break-out. *1. Panzerarmee* reached German lines on 5 April.

With both Soviet and German forces worn out on the southern front, there was finally time to regroup and for the Germans to stabilise the front. In May 1944 *Schwere Panzerregiment* Bäke was split up and its components returned to their parent divisions.

ORGANISATION OF SCHWERE PANZER REGIMENT BÄKE

The German army had long practiced combining tactical sub-units into a *Kampfgruppe* for particular operations. As the war dragged on, and gradually turned against Germany, this virtue of flexibility became a necessity. By the Late War period many *Kampfgruppen* up to regimental or divisional size became semi-permanent, with internal logistics to sustain them. In the long term this practice was the unfortunate side effect of rendering many conventional divisions under-strength due to various missing sub-units. Nevertheless, some of the *Kampfgruppen* were highly successful.

Schwere Panzerregiment Bäke was a fully self-contained force created by combining various independent units with others withdrawn from line divisions. Although primarily a panzer unit, it had supporting infantry and artillery.

OBERSTLEUTNANT DR FRANZ BÄKE

By 1944 *Oberstleutnant* Dr Franz Bäke was one of the most capable and experienced front-line panzer commanders in the German Army. Born in 1898 in the town of Schwarzenfels, Bäke had served in World War One as an enlisted man. He fought in the infantry on the western front including the battle of Verdun, was wounded, awarded the Iron Cross Second Class and finished the war as a sergeant and officer candidate. Between the wars he studied dentistry, gained his doctorate, and established a successful professional career. Still a reservist, Bäke was called up in 1937, and was a Leutnant commanding a panzer platoon at the start of the war.

Bäke's platoon was part of the *Panzerabteilung 65*, *1. Jägerdivision*, later reorganised as *6. Panzerdivision*. Equipped with the Czech Panzer 35(t) tanks. Bäke's unit took part in the invasion of Poland, where Bäke served well and was promoted to company commander. In France 1940, *6. Panzerdivision* formed part of Guderian's strike force through the Ardennes. Bäke's Company captured a key bridge across the Meuse. Wounded twice in the campaign, Bäke was awarded the Gold Wound Badge and the Iron Cross, First Class.

In 1941 *6. Panzerdivision* was transferred to East Prussia for Operation Barbarossa. Bäke's role changed to a staff position in charge of recovery of damaged tanks. As regimental *Ordinanz-Offizier* he carried out tank recovery with his customary energy and intelligence. With Operation Typhoon reaching its climax he often led ad-hoc *kampfgruppen* forward on missions.

After the winter of 1941/42 the worn-out *6. Panzerdivision* was withdrawn to France for rebuilding. Bäke was promoted to commander of *II Abteilung/11. Panzerregiment*. By the time the rebuilt division was ready for combat Stalingrad had been encircled. *6. Panzerdivision* was transferred back east and thrown into the relief attempt. Unable to break through, it then helped encircle and destroy Soviet tanks thrusting towards Kharkov. For this action Bäke was awarded the Knights Cross in January 1943.

At Kursk *6. Panzerdivision* formed part of Hoth's *4. Panzerarmee* attacking from the south. Wounded again himself, Bäke continued and took over command of *11. Panzerregiment* when its commander was severely wounded. Bäke led the unit through the defensive battles towards the Dneiper, receiving the Oak Leaves to the Knights Cross in August 1943. He was promoted to *Oberstleutnant der Reserve* and ordered to form *Schwere Panzerregiment* Bäke in December 1943. Six months of intensive combat involving the Regiment proved Bäke's ability to command larger formations. Bäke was awarded the Swords to the Knights Cross for his efforts to relieve the Korsun pocket in February 1944.

After further defensive battles Bäke's unit was disbanded in May 1944. Bäke took command of *Panzer Brigade 106 Feldherrnhalle*, now forming in the west with Panthers and armoured infantry. By September these were fighting defensive battles against Patton's Third US Army. At first successful against US tanks, Bäke suffered his first defeat when he attacked the US 90th Infantry Division, which took the Panzer attacks in its stride and counterattacked the Panzergrenadiers. The *Feldherrnhalle* panzer brigade fell back with heavy losses.

In January 1945 Bäke completed a course in divisional command and in March led *Panzerdivision Feldehernhalle 2* in the final offensive in Hungary. Bäke led them back to Czechoslovakia where he was promoted to General in April. In May he led them in a breakout through encircling Soviet forces to surrender to American forces at the Elbe. After the war Bäke was held as a POW until 1950, and then returned to his dentistry career until his death in a car accident in 1978.

OBERSTLEUTNANT
DR FRANZ BÄKE

CHARACTERISTICS

Franz Bäke is a Warrior and Higher Command Team rated as Fearless Veteran. He may join a Kampfgruppe Bäke company mounted in a Panther tank replacing the Company Command team for +60 points.

Bäke's combat career was characterised by his intelligence, determination and energy. Bäke gained his success not by fanaticism or bravado, but through his tactical skill. Bäke always has the following abilities:

ENERGETIC COMMANDER

Bäke was an energetic commander who both led by example and pushed his men as hard as needed to achieve victory.

> *Bäke and any platoon he is attached to may make a Stormtrooper Move on a roll of 2+.*

MASTER TACTICIAN

Bäke preferred to win his battles through manoeuvre and tactics rather than bludgeoning the enemy.

> *If Bäke is present on the battlefield one Bäke Panther Platoon or Bäke Schwere Panzer Platoon may make a Reconnaissance Deployment move after deployment is completed but before the first turn begins.*

EVERY SHOT COUNTS

Bäke is a veteran of tank combat and has personally served in almost every mark of tank used by the Panzer units. He knows how to stay cool under pressure and shoot accurately.

> *Bäke's Panther tank may re-roll any failed To Hit rolls when it shoots.*

KAMPFGRUPPE BÄKE
BATTLE GROUP BÄKE

(TANK COMPANY)

You must field one platoon from each box shaded black and may field one platoon from each box shaded grey.

DIVISIONAL SUPPORT PLATOONS

COMBAT PLATOONS

ARMOUR

Bäke Panther Platoon — 23

ARMOUR

Bäke Panther Platoon — 23
Bäke Schwere Panzer Platoon — 24

ARMOUR

Bäke Panther Platoon — 23
Bäke Schwere Panzer Platoon — 24

WEAPONS PLATOONS

INFANTRY

Panzer Pioneer Platoon — 74

RECONNAISSANCE

Panzer Scout Platoon — 75

ANTI-AIRCRAFT

Bäke Anti-aircraft Gun Platoon — 24

INFANTRY

Gebirgsjäger Platoon — 25

INFANTRY

Gebirgsjäger Platoon — 25
Gepanzerte Panzerpionier Platoon — 87
Panzerpionier Platoon — 89
Schwere Panzer Armoured Scout Platoon — 25

RECONNAISSANCE

Half-tracked Panzerspäh Platoon — 91

ARTILLERY

Armoured Artillery Battery — 168
Armoured Heavy Artillery Battery — 168

ARTILLERY

Rocket Launcher Battery — 169
Armoured Rocket Launcher Battery — 170

AIRCRAFT

Air Support — 172

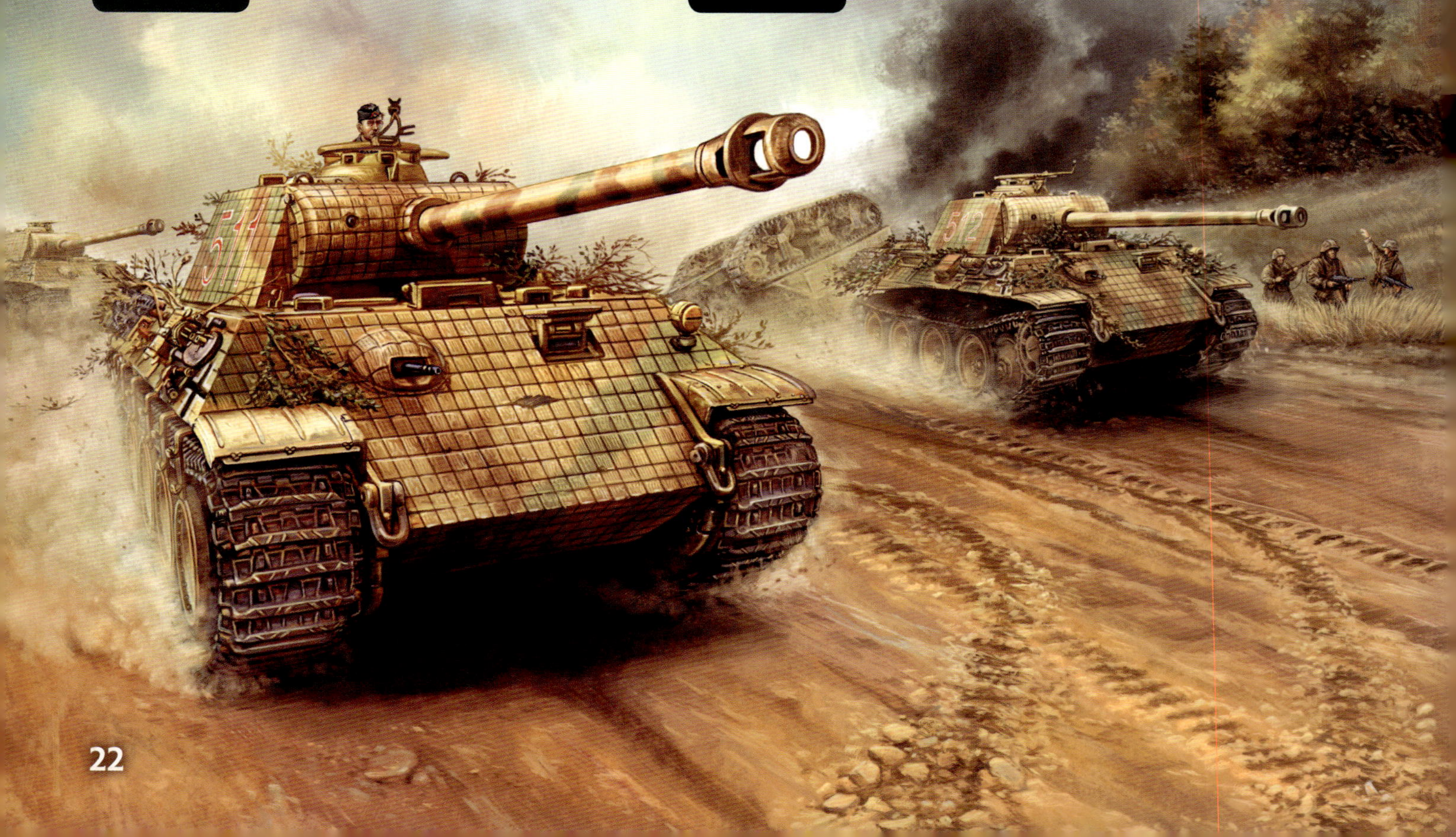

MOTIVATION AND SKILL

The companies of Schwere Panzerregiment Bäke (Heavy Tank Regiment Bäke) are led by seasoned veterans of past battles. These men teach the new recruits the tricks of tank combat and weld them into an effective fighting force. Kampfgruppe Bäke is rated as **Confident Veteran.**

RELUCTANT	CONSCRIPT
CONFIDENT	TRAINED
FEARLESS	**VETERAN**

HEADQUARTERS

KAMPFGRUPPE BÄKE HQ

HEADQUARTERS

Company HQ with:

Panther A	190 points

OPTIONS

- Add a second Panther A tank for +190 points.
- Add a Bergepanther recovery vehicle for +15 points.

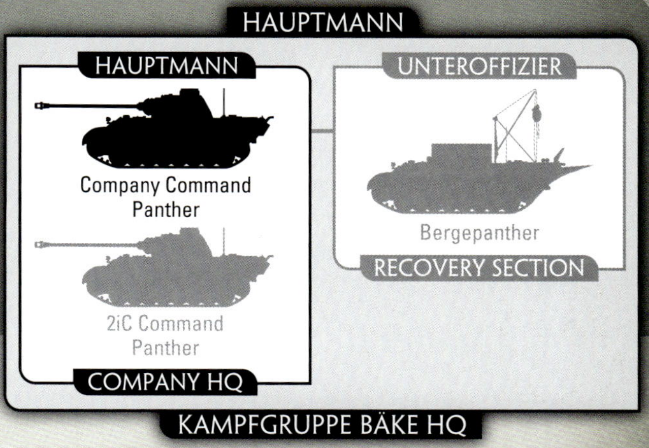

Schwere Panzerregiment Bäke was a fully self-contained force created by combining various independent units with others withdrawn from line divisions. Although primarily a panzer unit, it had supporting infantry and artillery.

Often with German *Panzer* units during intense fighting the number of operational tanks could get quite low. Consequently, a *Flames Of War* on table company may represent all the tanks available at times during the various operations of *Schwere Panzerregiment Bäke*.

HAUPTMANN

HAUPTMANN
Company Command Panther

2iC Command Panther

COMPANY HQ

UNTEROFFIZIER
Bergepanther
RECOVERY SECTION

KAMPFGRUPPE BÄKE HQ

If you form a Kampfgruppe under your 2iC Command Panther tank any Tiger I E tanks included still roll for Tiger Aces skills. However, any Tiger Ace Skill that normally effects the whole platoon will only be applied to the Tiger I E tanks in the Kampfgruppe.

COMBAT PLATOONS

BÄKE PANTHER PLATOON

PLATOON

4 Panther A	750 points
3 Panther A	560 points

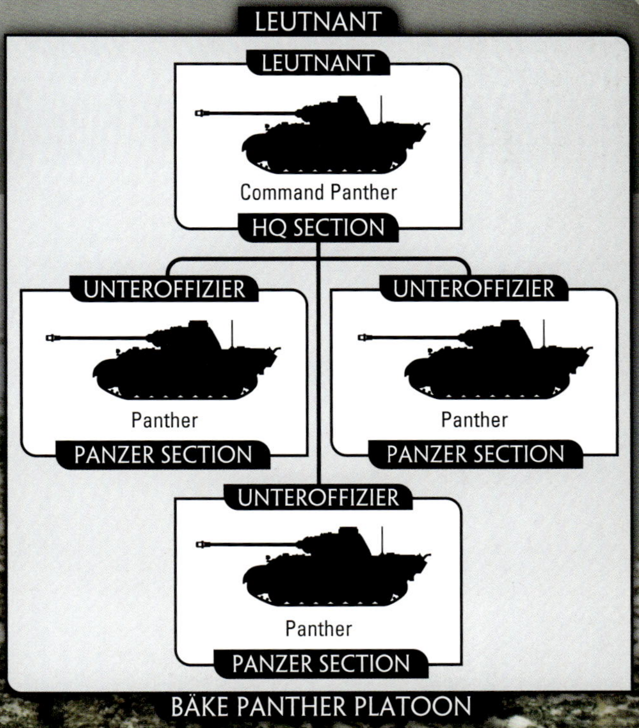

The Panthers of *Kampfgruppe Bäke* are drawn from an experienced Panzer battalion, *II Battalion, 23. Panzerregiment, 23. Panzerdivision*. Their mobility enabled them to be used on wide outflanking manoeuvres while the Tigers engaged frontally. They act as the rapier while the Tigers form the shield.

LEUTNANT

LEUTNANT
Command Panther
HQ SECTION

UNTEROFFIZIER
Panther
PANZER SECTION

UNTEROFFIZIER
Panther
PANZER SECTION

UNTEROFFIZIER
Panther
PANZER SECTION

BÄKE PANTHER PLATOON

BÄKE SCHWERE PANZER PLATOON

PLATOON

4 Tiger I E	860 points
3 Tiger I E	645 points
2 Tiger I E	430 points

Remember to roll for your Tiger Ace Skills before each game.

The Tigers of *Schwere Panzerregiment Bäke* are from *503. Schwere Panzer Abteilung*, one of the first Tiger units formed and with excellent quality crews. *509. Schwere Panzer Abteilung* also joined and fought alongside *Schwere Panzerregiment Bäke* during March 1944. Though there were never enough to go around, the German Tiger battalions proved themselves a serious problem for the Soviets. Kill ratios of three or four to one were common for Tigers across the Eastern Front.

Unleashing your Tigers upon the enemy will certainly cause their commander great concern. His armour losses will swiftly mount and his battle plan will suffer immediate reassessment. Whether counterattacking alone or in support of your other units, the Tigers can carry the battle to the enemy and reverse the fortunes of war.

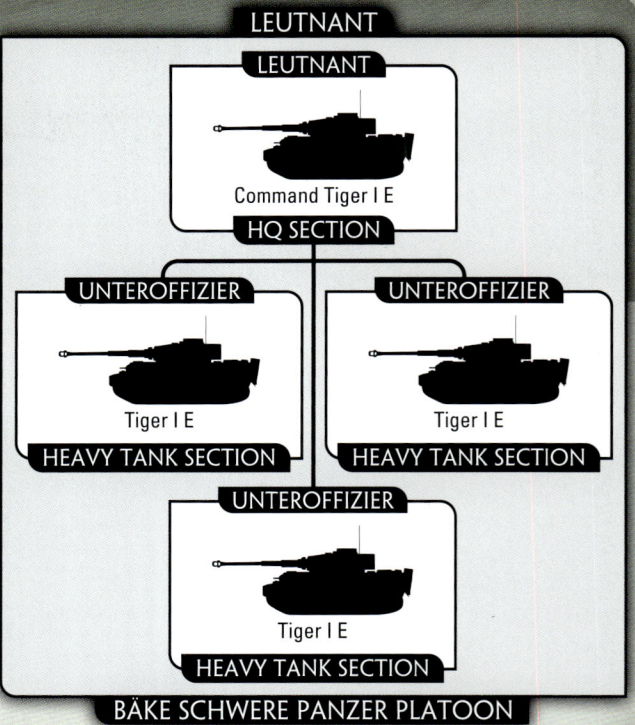

WEAPONS PLATOONS

BÄKE ANTI-AIRCRAFT GUN PLATOON

PLATOON

3 Armoured Sd Kfz 7/1 (Quad 2cm)	150 points
2 Armoured Sd Kfz 7/1 (Quad 2cm)	100 points
3 Armoured Sd Kfz 10/5 (2cm)	120 points
2 Armoured Sd Kfz 10/5 (2cm)	80 points

1944 saw the air superiority that the panzers had relied on for so long finally pass to the Soviets, as the *Luftwaffe* withdrew more and more fighters to defend Germany. Mobile anti-aircraft guns that could keep up with the panzers became essential. Both panzer battalions of the regiment have integral Anti-aircraft half-tracks, Sd Kfz 10/5 half-tracks with the panthers, and Sd Kfz 7/1 half-tracks with the Tigers.

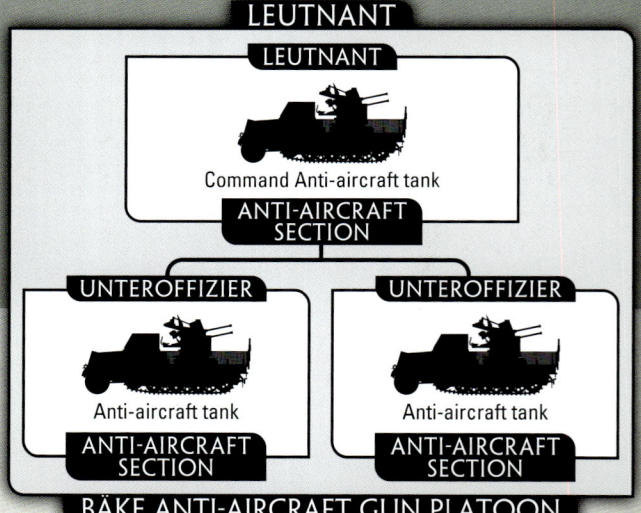

24

DIVISIONAL SUPPORT PLATOONS

GEBIRGSJÄGER PLATOON

PLATOON

HQ Section with:

3 Jäger Squads	155 points
2 Jäger Squads	110 points

OPTION

- Replace the Command Rifle/MG team with a Command Panzerknacker SMG team for +5 points or a Command Panzerfaust SMG team for +10 points.

The *Gebirgsjäger* drafted into *Schwere Panzerregiment Bäke* performed yeoman service. They were well trained and, although unaccustomed to fighting with tanks, soon learned the basics of infantry-armour cooperation.

> *Gebirgsjäger Platoons are Mountaineers.*

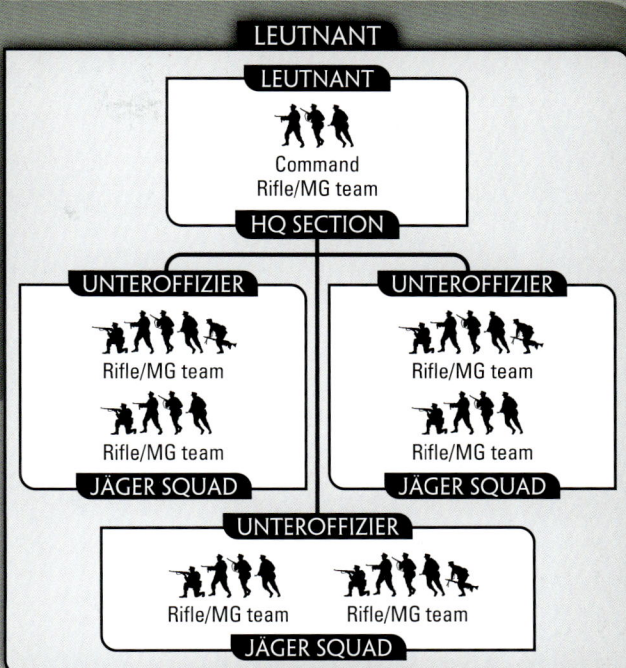

LEUTNANT

LEUTNANT

Command Rifle/MG team

HQ SECTION

UNTEROFFIZIER

Rifle/MG team
Rifle/MG team

JÄGER SQUAD

UNTEROFFIZIER

Rifle/MG team
Rifle/MG team

JÄGER SQUAD

UNTEROFFIZIER

Rifle/MG team
Rifle/MG team

JÄGER SQUAD

GEBIRGSJÄGER PLATOON

SCHWERE PANZER ARMOURED SCOUT PLATOON

PLATOON

HQ Section with:

6 Scout Squads	250 points
4 Scout Squads	175 points
2 Scout Squads	105 points

OPTION

- Replace the Command MG team with a Command Panzerfaust SMG team for +10 points.

> *Armoured Scout Platoons may use the Mounted Assault special rule.*

Armoured scouts will protect your Tigers, keep enemy infantry at bay, and allow you to exploit a hole in the enemy's defence no matter the terrain. Use terrain cover to the advance but don't get too far away from your Tigers just in case the enemy gets nasty.

LEUTNANT

LEUTNANT

Command MG team   Sd Kfz 251/1C half-track

HQ SECTION

UNTEROFFIZIER

MG team   Sd Kfz 251/1C half-track

SCOUT SQUAD

UNTEROFFIZIER

MG team   Sd Kfz 251/1C half-track

SCOUT SQUAD

UNTEROFFIZIER

MG team   Sd Kfz 251/1C half-track

SCOUT SQUAD

UNTEROFFIZIER

MG team   Sd Kfz 251/1C half-track

SCOUT SQUAD

UNTEROFFIZIER

MG team   Sd Kfz 251/1C half-track

SCOUT SQUAD

UNTEROFFIZIER

MG team   Sd Kfz 251/1C half-track

SCOUT SQUAD

SCHWERE PANZER ARMOURED SCOUT PLATOON

GRENADIERKOMPANIE
INFANTRY COMPANY

(INFANTRY COMPANY)

HEADQUARTERS

Grenadierkompanie HQ — 27

You must field one platoon from each box shaded black and may field one platoon from each box shaded grey.

COMBAT PLATOONS

INFANTRY

Grenadier Platoon — 27

INFANTRY

Grenadier Platoon — 27

INFANTRY

Grenadier Platoon — 27

WEAPONS PLATOONS

MACHINE-GUNS

Grenadier Machine-gun Platoon — 28
Field Fortifications — 155

MACHINE-GUNS

Grenadier Machine-gun Platoon — 28

ARTILLERY

Grenadier Mortar Platoon — 28

REGIMENTAL SUPPORT PLATOONS

ARMOUR

Looted Panzer Platoon — 29

ARTILLERY

Grenadier Infantry Gun Platoon — 29

ANTI-TANK

Grenadier Anti-tank Gun Platoon — 30

RECONNAISSANCE

Grenadier Scout Platoon — 31

ANTI-AIRCRAFT

Grenadier Anti-aircraft Gun Platoon — 31

INFANTRY

Pionier Platoon — 61

DIVISIONAL SUPPORT PLATOONS

ARMOUR

Panzer Platoon — 73
Schwere Panzer Platoon — 71
Radio-control Tank Platoon — 162

ARMOUR

StuG Platoon — 59
Sturmpanzer Platoon — 162
Tank-hunter Platoon — 163
Anti-tank Gun Platoon — 165

INFANTRY

Fallschirmjäger Platoon — 33

ARTILLERY

Artillery Battery — 166
Heavy Artillery Battery — 166
Armoured Train — 122

ARTILLERY

Artillery Battery — 166
Rocket Launcher Battery — 169
Armoured Rocket Launcher Battery — 170

ANTI-AIRCRAFT

Anti-aircraft Gun Platoon — 171

ANTI-TANK

Heavy Anti-aircraft Gun Platoon — 171
Luftwaffe Heavy Anti-aircraft Gun Platoon — 173
Heavy Anti-tank Gun Platoon — 165

AIRCRAFT

Air Support — 172

ALLIED PLATOONS

Luftwaffe Platoons in your force are Allies and follow the Allies rules in the rulebook.

26

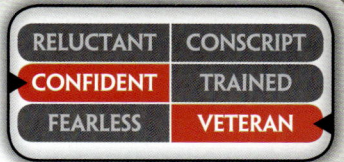
MOTIVATION AND SKILL

The Heer or German Army is well trained and has many victorious campaigns on the Eastern Front behind it. The soldiers are confident that victory lies in the near future. A Grenadierkompanie is rated as **Confident Veteran**.

RELUCTANT	CONSCRIPT
▸ CONFIDENT	TRAINED
FEARLESS	VETERAN ◂

HEADQUARTERS

GRENADIERKOMPANIE HQ

HEADQUARTERS

Company HQ	45 points

OPTIONS

- Replace either or both Command SMG teams with Command Panzerknacker SMG teams for +5 points per team or Command Panzerfaust SMG teams for +10 points per team.
- Add an Anti-tank Section for +25 points.
- Replace Panzerschreck team with a 8.8cm RW43 Püppchen rocket launcher at no cost.
- Add Mortar Section for +55 points.
- Add up to three Sniper teams for +50 points per team.

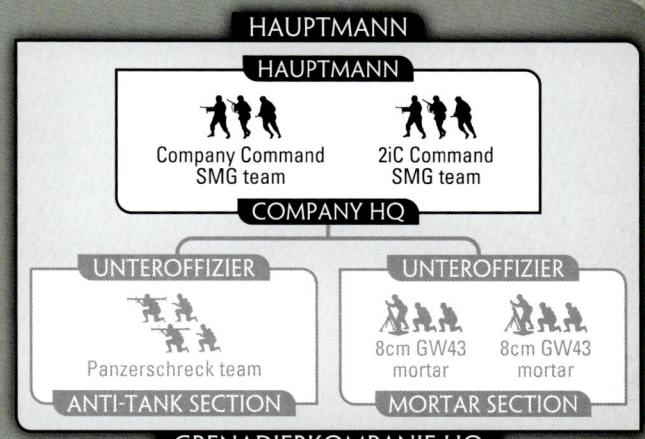

COMBAT PLATOONS

GRENADIER PLATOON

PLATOON

HQ Section with:

3 Grenadier Squads	155 points
2 Grenadier Squads	110 points

OPTION

- Replace Command Rifle/MG team with a Command Panzerknacker SMG team for +5 points or Command Panzerfaust SMG team for +10 points.

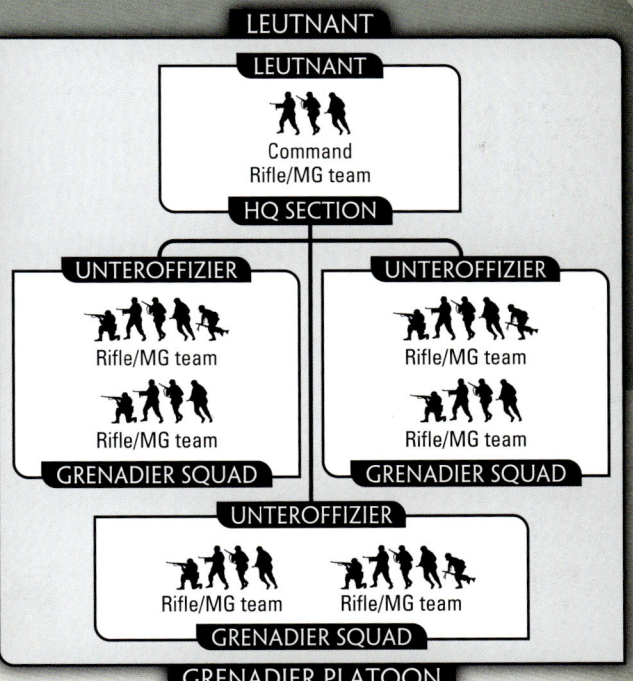

Attach machine-guns and anti-tank guns to *Grenadier* platoons forming strongpoints to anchor your defensive position. Keep other *Grenadier* platoons in reserve as a counterattack force to recover any strongpoints lost to the enemy.

The *Infanterieregiment* has considerable engineering capabilities of its own and each battalion forms a pioneer platoon from its Grenadiers as needed. These infantry pioneers are called 'white' pioneers because they wear the white piping of infantry rather than the black piping of engineers.

If your Grenadierkompanie has three Grenadier Platoons, you may upgrade the smallest Grenadier Platoon to a Grenadier Pioneer Platoon for +15 points per squad. This converts the Command team and every Rifle/MG team into Pioneer teams with the same armament. The Grenadier Pioneer Platoon may have a horse-drawn Pioneer Supply Wagon for an additional +20 points.

WEAPONS PLATOONS

GRENADIER MACHINE-GUN PLATOON

PLATOON

HQ Section with:

2 Machine-gun Sections	135 points
1 Machine-gun Section	70 points

OPTION

- Replace Command SMG team with a Command Panzerknacker SMG team for +5 points.

Grenadier Machine-gun Platoons may make Combat Attachments to Grenadier Platoons.

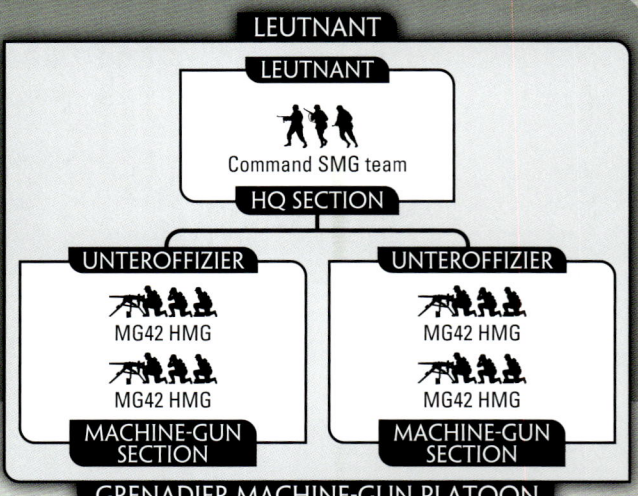

LEUTNANT
LEUTNANT
Command SMG team
HQ SECTION
UNTEROFFIZIER — MG42 HMG / MG42 HMG — MACHINE-GUN SECTION
UNTEROFFIZIER — MG42 HMG / MG42 HMG — MACHINE-GUN SECTION
GRENADIER MACHINE-GUN PLATOON

The machine-gun platoons are essential in both attack and defence. When attacking they use their speed and initiative to find covered positions from which they can engage the defences and pin them for the *Grenadier* attack. In defence they form the front line using their range and firepower to keep the enemy at bay.

GRENADIER MORTAR PLATOON

PLATOON

HQ Section with:

3 Mortar Sections	180 points
2 Mortar Sections	125 points
1 Mortar Section	65 points

OPTION

- Replace all 8cm GW34 mortars with 12cm sGW43 mortars towed by a 3-ton truck or RSO tractor for +20 points per Mortar Section.

You must upgrade the Grenadier Mortar Platoon to 12cm sGW43 mortars if you have 8cm GW34 mortars in the Company HQ.

A Grenadier Mortar Platoon upgraded to 12cm sGW43 mortars may not have more than two Mortar Sections.

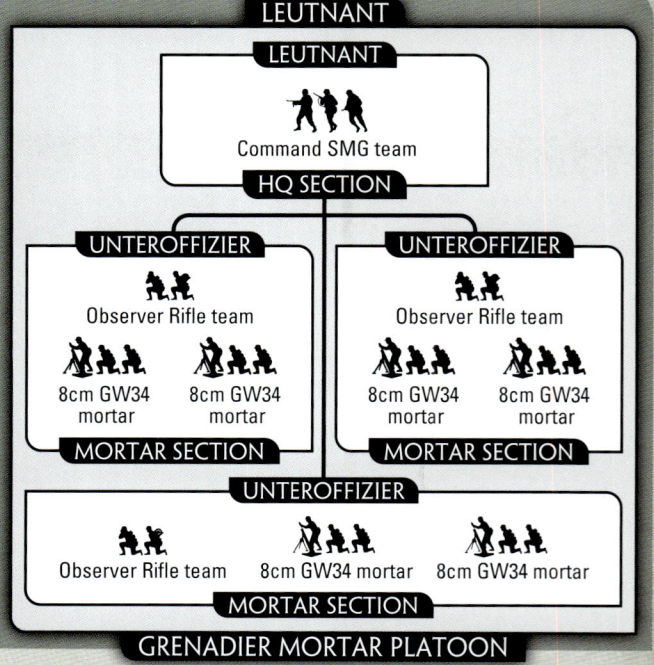

LEUTNANT
LEUTNANT
Command SMG team
HQ SECTION
UNTEROFFIZIER — Observer Rifle team — 8cm GW34 mortar / 8cm GW34 mortar — MORTAR SECTION
UNTEROFFIZIER — Observer Rifle team — 8cm GW34 mortar / 8cm GW34 mortar — MORTAR SECTION
UNTEROFFIZIER — Observer Rifle team — 8cm GW34 mortar — 8cm GW34 mortar — MORTAR SECTION
GRENADIER MORTAR PLATOON

Mortar platoons provide instant artillery support for breaking up enemy concentrations and pinning down their supporting weapons. As the heavy 12cm mortars became available, the lighter 8cm models were assigned out to the company headquarters for close support work.

REGIMENTAL SUPPORT PLATOONS

LOOTED PANZER PLATOON

PLATOON

T-70 obr 1943	30 points
M4 (M4A2 Sherman)	55 points
T-34 obr 1942	55 points
T-34/85 obr 1943	80 points

OPTION

- Upgrade T-34 obr 1942 tank to have a cupola for +5 points.

All captured tanks in a Looted Tank Platoon are rated as **Confident Trained** and are Unreliable.

CONFIDENT | TRAINED

LEUTNANT

LEUTNANT

Command tank

CAPTURED TANK

LOOTED PANZER PLATOON

In an effort to stem the tide of enemy troops, the German Grenadiers often resorted to using captured enemy tanks as a last ditch effort to plug any breaches in the line.

A *Beutepanzer* like this was used until it broke down beyond repair or was destroyed.

GRENADIER INFANTRY GUN PLATOON

PLATOON

HQ Section with:

2 7.5cm leIG18	65 points
2 15cm sIG33	145 points

OPTIONS

- Add horse-drawn limbers for +5 points for the platoon.
- Replace both horse-drawn limbers with 3-ton trucks or RSO tractors at no cost.

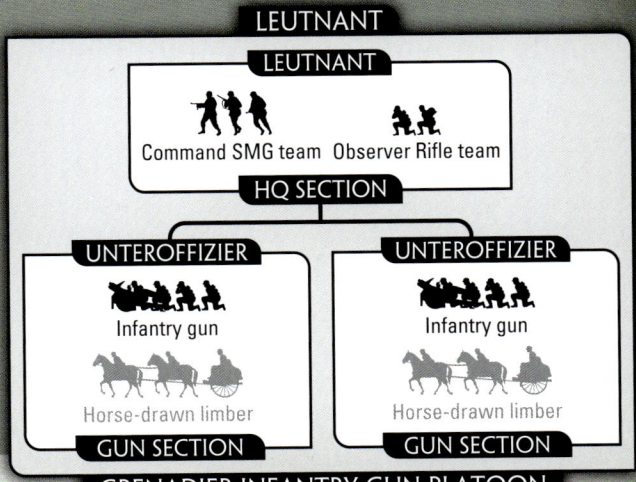

LEUTNANT

LEUTNANT

Command SMG team   Observer Rifle team

HQ SECTION

UNTEROFFIZIER

Infantry gun

Horse-drawn limber

GUN SECTION

UNTEROFFIZIER

Infantry gun

Horse-drawn limber

GUN SECTION

GRENADIER INFANTRY GUN PLATOON

Infantry gun platoons provide the *Grenadierkompanie* with close-support artillery, taking out targets such as gun positions and bunkers with direct fire. The light 7.5cm guns are useful in the forward areas firing over open sights at enemy machine gun nests. The heavy 15cm guns are more suited to sitting back and firing as heavy artillery.

GRENADIER ANTI-TANK GUN PLATOON

PLATOON

HQ Section with:

3 3.7cm PaK36	80 points
2 3.7cm PaK36	55 points

- All 3.7cm PaK36 guns are equipped with Stielgranate ammunition at no cost.

3 5cm PaK38	90 points
2 5cm PaK38	60 points
3 7.5cm PaK97/38	95 points
2 7.5cm PaK97/38	65 points
3 7.62cm PaK36(r)	125 points
2 7.62cm PaK36(r)	85 points
3 7.5cm PaK40	155 points
2 7.5cm PaK40	105 points

OPTIONS

- Add Kfz 15 field car and either Kfz 70 trucks in platoons equipped with 3.7cm PaK36, 5cm PaK38, or 7.5cm PaK97/38 guns or 3-ton trucks in platoons equipped with 7.62cm PaK36(r) or 7.5cm PaK40 guns for +5 points for the platoon.
- Replace all trucks with RSO tractors at no cost.

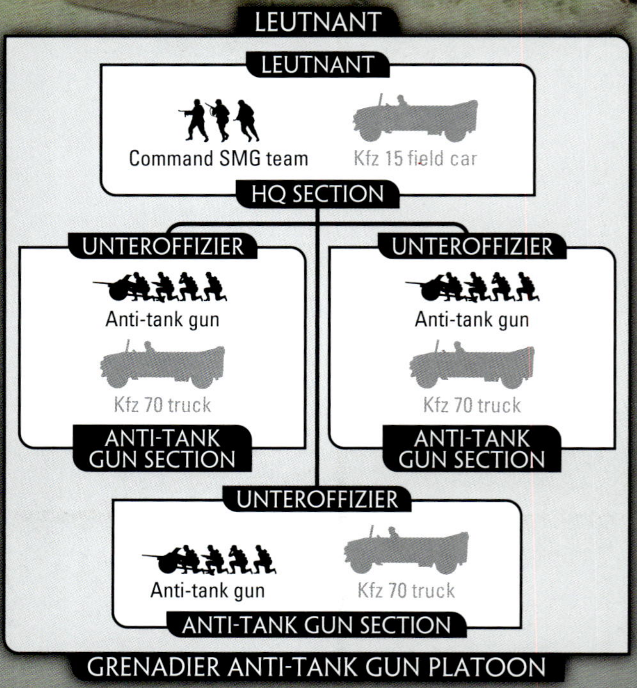

LEUTNANT

LEUTNANT

Command SMG team — Kfz 15 field car

HQ SECTION

UNTEROFFIZIER

Anti-tank gun

Kfz 70 truck

ANTI-TANK GUN SECTION

UNTEROFFIZIER

Anti-tank gun

Kfz 70 truck

ANTI-TANK GUN SECTION

UNTEROFFIZIER

Anti-tank gun

Kfz 70 truck

ANTI-TANK GUN SECTION

GRENADIER ANTI-TANK GUN PLATOON

The Grenadiers' regimental anti-tank guns are few and far between, but are often the only thing they have to keep enemy tanks at bay. Even the older models can provide good service if carefully positioned so that they cannot be seen until the enemy is at point-blank range.

GRENADIER SCOUT PLATOON

PLATOON

HQ Section with:

2 Scout Squads	115 points
1 Scout Squad	70 points

OPTIONS

- Replace Command Rifle team with a Command Panzerknacker SMG team for +15 points.
- Replace all Rifle teams with Assault Rifle teams for +15 points per team.
- Replace Command Assault Rifle team with Command Panzerknacker Assault Rifle team for +5 points.

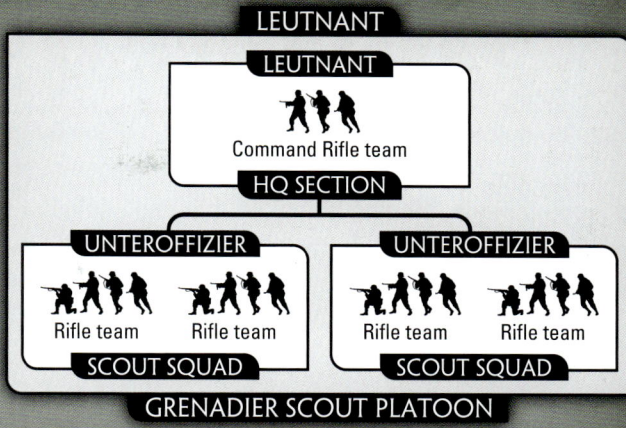

Grenadier Scout Platoons are Reconnaissance Platoons.

The regimental scouts are the only reconnaissance troops left in the *Infanteriedivision*. Their main role is scouting the flanks of an advance to prevent the Grenadiers from being ambushed. They are also a useful combat reserve for last-ditch counterattacks to regain lost positions.

GRENADIER ANTI-AIRCRAFT GUN PLATOON

PLATOON

HQ Section with:

3 2cm FlaK38	75 points

OPTIONS

- Add Kfz 15 field car and 3-ton trucks for +5 points for the platoon.
- Replace all trucks with RSO tractors at no cost.
- Mount 2cm FlaK38 guns on 3-ton trucks as Portees at no cost.

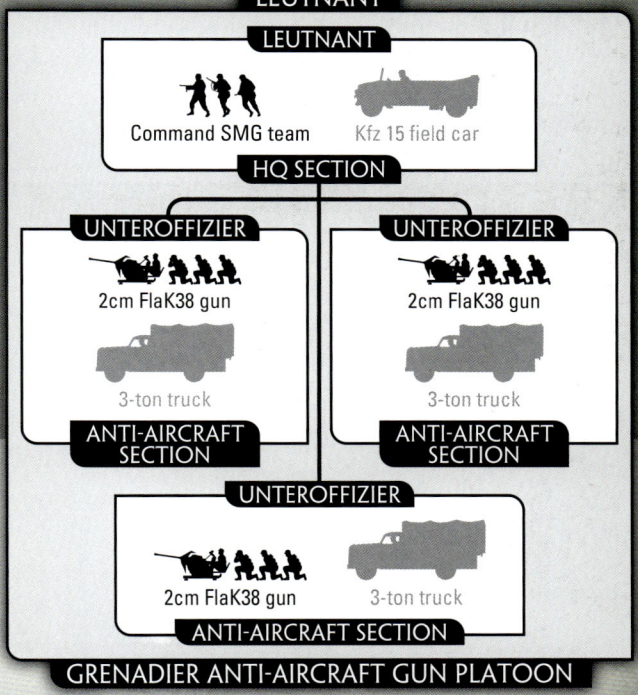

Unfortunately, the Grenadiers have very little in the way of anti-aircraft gun support, sporting just one company within the anti-tank battalion.

However, these light anti-aircraft guns can move and hide from marauding aircraft along with the infantry they are supporting, much better than the tanks and half-tracks of the armoured divisions.

FALLSCHIRMJÄGERKOMPANIE
PARACHUTE INFANTRY COMPANY

(INFANTRY COMPANY)

HEADQUARTERS

Fallschirmjägerkompanie HQ · 33

You must field one platoon from each box shaded black and may field one platoon from each box shaded grey.

DIVISIONAL SUPPORT PLATOONS

COMBAT PLATOONS

INFANTRY

Fallschirmjäger Platoon · 33

INFANTRY

Fallschirmjäger Platoon · 33

INFANTRY

Fallschirmjäger Platoon · 33

ALLIED PLATOONS

Heer Platoons in your force are Allies and follow the Allies rules in the rulebook.

WEAPONS PLATOONS

MACHINE-GUNS

Fallschirmjäger Machine-gun Platoon · 34

MACHINE-GUNS

Fallschirmjäger Machine-gun Platoon · 34

ARTILLERY

Fallschirmjäger Mortar Platoon · 34

ANTI-TANK

Fallschirmjäger Light Gun Platoon · 34

REGIMENTAL SUPPORT PLATOONS

INFANTRY

Fallschirmjäger Heavy Mortar Platoon · 35

ANTI-TANK

Fallschirmjäger Anti-tank Gun Platoon · 35

ARMOUR

Panzer Platoon · 73
Schwere Panzer Platoon · 71
Radio-control Tank Platoon · 162

ARMOUR

StuG Platoon · 59
Tank-hunter Platoon · 163
Anti-tank Gun Platoon · 165
Heavy Anti-tank Gun Platoon · 166

INFANTRY

Fallschirmpionier Platoon · 37

INFANTRY

Panzergrenadier Platoon · 81
Grenadier Platoon · 27

ARTILLERY

Fallschirmjäger Artillery Battery · 36
Artillery Battery · 166
Heavy Artillery Battery · 166
Motorised Artillery Battery · 167
Motorised Heavy Artillery Battery · 167

ARTILLERY

Rocket Launcher Battery · 169
Armoured Rocket Launcher Battery · 170
Artillery Battery · 166
Motorised Artillery Battery · 167

ANTI-AIRCRAFT

Fallschirmjäger Anti-aircraft Gun Platoon · 37
Anti-aircraft Gun Platoon · 171

ANTI-AIRCRAFT

Luftwaffe Heavy Anti-aircraft Gun Platoon · 173

AIRCRAFT

Air Support · 172

MOTIVATION AND SKILL

All Fallschirmjäger *(paratroopers) are volunteers. They are put through rigorous selection examinations and hard training before they win their wings. A Fallschirmjägerkompanie is rated as* **Fearless Veteran.**

RELUCTANT	CONSCRIPT
CONFIDENT	TRAINED
FEARLESS	**VETERAN**

HEADQUARTERS

FALLSCHIRMJÄGERKOMPANIE HQ

HEADQUARTERS

Company HQ	55 points

OPTIONS

- Replace Command SMG teams with Command Panzerknacker SMG teams for +5 points per team or Command Panzerfaust SMG teams for +10 points per team.
- Add an Anti-tank Section for +30 points.
- Replace Panzerschreck team with 8.8cm RW43 (Püppchen) launcher at no cost.
- Add a Mortar Section of up to three 8cm GW42 (Stummelwerfer) mortars for +25 points per mortar.
- Add up to three Sniper teams for +50 points per team.

HAUPTMANN

HAUPTMANN

Company Command SMG team

2iC Command SMG team

COMPANY HQ

OBERJÄGER

Panzerschreck team

ANTI-TANK SECTION

OBERJÄGER

8cm GW42 mortar 8cm GW42 mortar

8cm GW42 mortar

MORTAR SECTION

FALLSCHIRMJÄGERKOMPANIE HQ

The *Fallschirmjäger* are Germany's elite paratroops. They fall under the operational control of the army, but are part of the *Luftwaffe* or air force. As such they wear air force uniforms and rank insignia, and of course, consider themselves far better than the army!

While most of the *Fallschirmjäger* are trained for parachute operations, they have not conducted large-scale parachute operations in three years. Instead they fight as elite light infantry wherever the army needs assistance.

COMBAT PLATOONS

FALLSCHIRMJÄGER PLATOON

PLATOON

HQ Section with:

3 Fallschirmjäger Squads	265 points
2 Fallschirmjäger Squads	185 points

OPTION

- Replace Command Rifle/MG team with a Command Panzerknacker SMG team for +5 points or with a Command Panzerfaust team for +10 points.

LEUTNANT OR OBERFELDWEBEL

LEUTNANT

Command Rifle/MG team

HQ SECTION

OBERJÄGER

Rifle/MG team Rifle/MG team

Rifle/MG team

FALLSCHIRMJÄGER SQUAD

OBERJÄGER

Rifle/MG team Rifle/MG team

Rifle/MG team

FALLSCHIRMJÄGER SQUAD

OBERJÄGER

Rifle/MG team Rifle/MG team Rifle/MG team

FALLSCHIRMJÄGER SQUAD

FALLSCHIRMJÄGER PLATOON

The *Fallschirmjäger* platoon provides the manpower to hold off the staunchest assaults. These platoons were made larger than normal rifle platoons to allow for expected parachuting casualties on landing. This added manpower gives them greater resilience in prolonged ground operations, helping them retain their reputation for holding at all costs.

The *esprit de corps* of the *Fallschirmjäger* together with their extra team per squad makes them the toughest and most feared German light infantry, capable of truly heroic operations.

FALLSCHIRMJÄGER MACHINE-GUN PLATOON

PLATOON

HQ Section with:

2 Machine-gun Sections	150 points
1 Machine-gun Section	80 points

OPTION

- Replace Command SMG team with a Command Panzerknacker SMG team for +5 points.

Fallschirmjäger Machine-gun Platoons may make Combat Attachments to Fallschirmjäger Platoons.

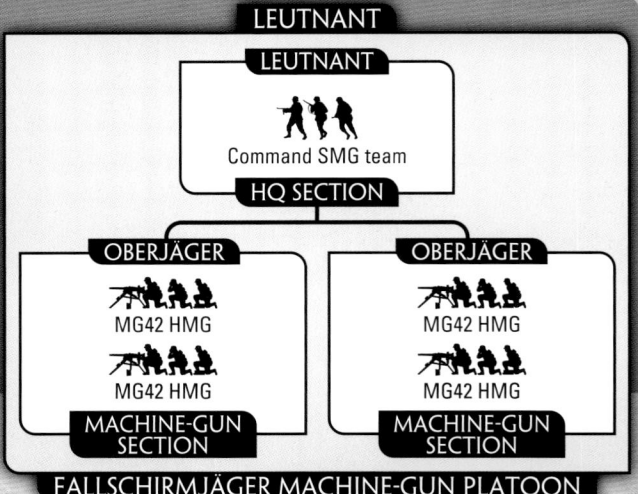

FALLSCHIRMJÄGER MORTAR PLATOON

PLATOON

HQ Section with:

2 Mortar Sections	120 points
1 Mortar Section	65 points

OPTION

- Replace Command SMG team with a Command Panzerknacker SMG team for +5 points.

The *Fallschirmjäger* use light 8cm GW42 mortars. These are nicknamed *Stummelwerfer* or 'Stump mortar' for their short barrels. Between the mortars in the company HQ and the mortar platoon, they have plenty of firepower to break up enemy assaults.

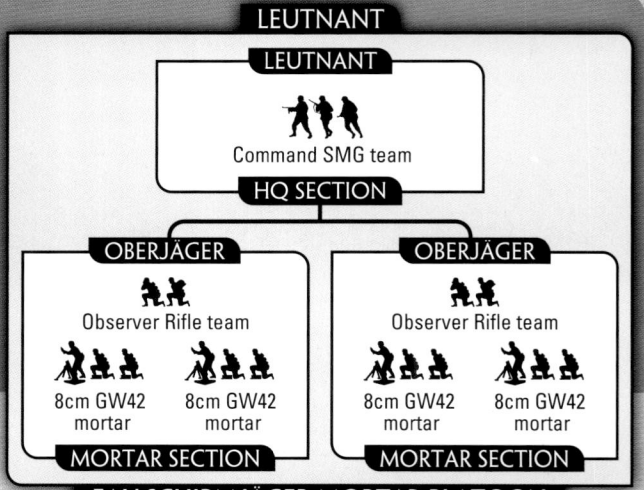

FALLSCHIRMJÄGER LIGHT GUN PLATOON

PLATOON

HQ Section with:

2 Gun Sections	55 points

OPTION

- Replace Command SMG team with a Command Panzerknacker SMG team for +5 points.

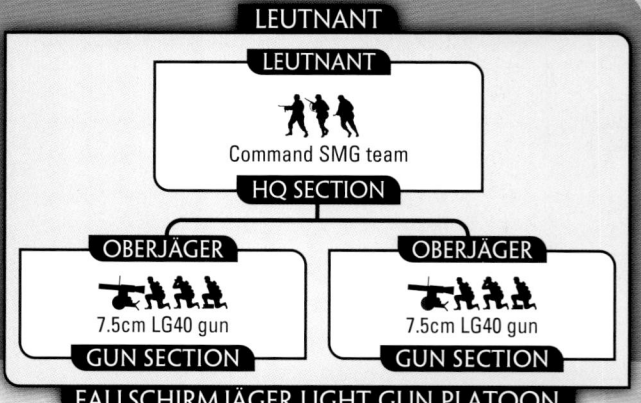

The 7.5cm light guns combine the role of a light anti-tank weapon and an infantry gun. They are small, mobile, and deadly at short range. These weapons can be devastating if used properly. Use light anti-tank guns to prevent the enemy from driving their armour over a critical position. Use them at close range and wait for the flanking shot so that their maximum effectiveness can be used in destroying the attacking enemy armour.

Keep them dug-in and well hidden until the opportunity to strike presents itself. Well-placed anti-tank weapons supported by infantry can hold a static position for quite some time.

FALLSCHIRMJÄGER HEAVY MORTAR PLATOON

PLATOON

HQ Section with:

4 10.5cm NbW35	165 points
2 10.5cm NbW35	85 points

OPTIONS

- Add Kfz 15 field car and 3-ton trucks to the platoon for +5 points.
- Replace 10.5cm NbW35 mortars with 12cm sGW43 mortars for +10 points per Mortar Section.

The Fallschirmjäger recognised the need for heavier fire support early, initially the only weapons they were able to secure were old 10.5cm NbW35 *Nebelwerfer* chemical mortars that had been made redundant by the new rocket launchers issued to the chemical troops. Later they were able to obtain 12cm sGW43 heavy mortars.

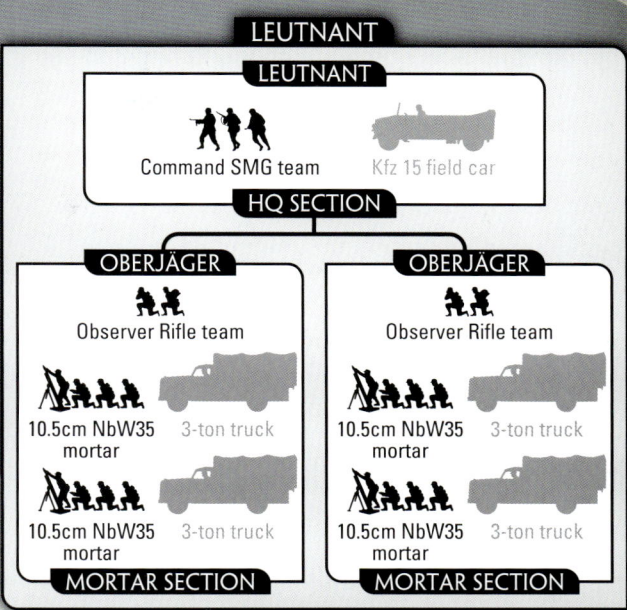

FALLSCHIRMJÄGER HEAVY MORTAR PLATOON

FALLSCHIRMJÄGER ANTI-TANK GUN PLATOON

PLATOON

HQ Section with:

4 3.7cm PaK36	115 points
3 3.7cm PaK36	90 points
2 3.7cm PaK36	65 points

- All 3.7cm PaK36 guns are equipped with Stielgranate ammunition at no cost.

4 5cm PaK38	140 points
3 5cm PaK38	100 points
2 5cm PaK38	75 points

4 7.5cm PaK40	240 points
3 7.5cm PaK40	180 points
2 7.5cm PaK40	120 points

OPTION

- Add Kfz 15 field car and either Kfz 70 trucks in platoons equipped with 3.7cm PaK36 or 5cm PaK38, or 3-ton trucks in platoons equipped with 7.5cm PaK40 guns for +5 points for the platoon.

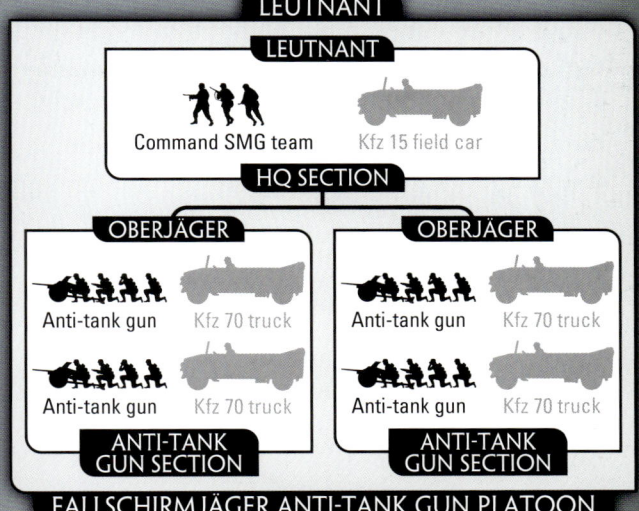

FALLSCHIRMJÄGER ANTI-TANK GUN PLATOON

FALLSCHIRMJÄGER ARTILLERY BATTERY

PLATOON

HQ Section with:

4 7.5cm GebG36	175	points
2 7.5cm GebG36	95	points
4 10.5cm LG40	245	points
2 10.5cm LG40	130	points
4 10.5cm leFH18	230	points
2 10.5cm leFH18	120	points
4 10.5cm leFH18/40	240	points
2 10.5cm leFH18/40	125	points

OPTION

- Add Kfz 15 field car, Kfz 68 radio truck and 3-ton trucks for +5 points for the battery.

You may replace all 10.5cm LG40 recoilless guns with 10.5cm leFH18 howitzers at the start of any game before deployment.

HAUPTMANN

HAUPTMANN

Command SMG team · Kfz 15 field car · Staff team · Kfz 68 radio truck

HQ SECTION

LEUTNANT

Observer Rifle team
Gun or howitzer
3-ton truck
Gun or howitzer
3-ton truck

GUN SECTION

LEUTNANT

Observer Rifle team
Gun or howitzer
3-ton truck
Gun or howitzer
3-ton truck

GUN SECTION

FALLSCHIRMJÄGER ARTILLERY BATTERY

The flexibility of the *Fallschirmjäger* light artillery battery is showcased with the ability to use two different types of artillery pieces. They have both a conventional 7.5cm mountain gun and the newly-developed 10.5cm recoilless gun. Both offer solid artillery support without the hindrance of immobile guns. Before each operation the *Fallschirmjäger* commander may select the most appropriate weapon for the current mission.

With the end of airborne operations, many *Fallschirmjäger* units replaced their light 7.5cm guns with heavier 10.5cm leFH18 howitzers. This increased their range and effectiveness in providing artillery support for their light infantry. Many units still retained their light 10.5cm recoilless guns for any airborne operations that might eventuate.

Artillery support can be critical in defending important positions. Providing smoke and targeting enemy troops at the proper moment can render their attacks useless and provide the cover needed to unleash a devastating counterattack with anti-tank and infantry assets.

FALLSCHIRMJÄGER ANTI-AIRCRAFT GUN PLATOON

PLATOON

HQ Section with:

3 2cm FlaK38	80 points

OPTIONS

- Add Kfz 15 field car and 3-ton trucks for +5 points for the platoon.
- Mount 2cm FlaK38 guns on 3-ton trucks as Portees at no cost.

As part of the *Luftwaffe*, the *Fallschirmjäger* have always been aware of the need for protection from enemy air attack. The lightweight 2cm FlaK38 gun is easy to move and hide while its good rate of fire provides adequate protection without the need for heavy anti-aircraft support.

This platoon can also provide additional fire support against attacking infantry tipping the scales in favour of the defender.

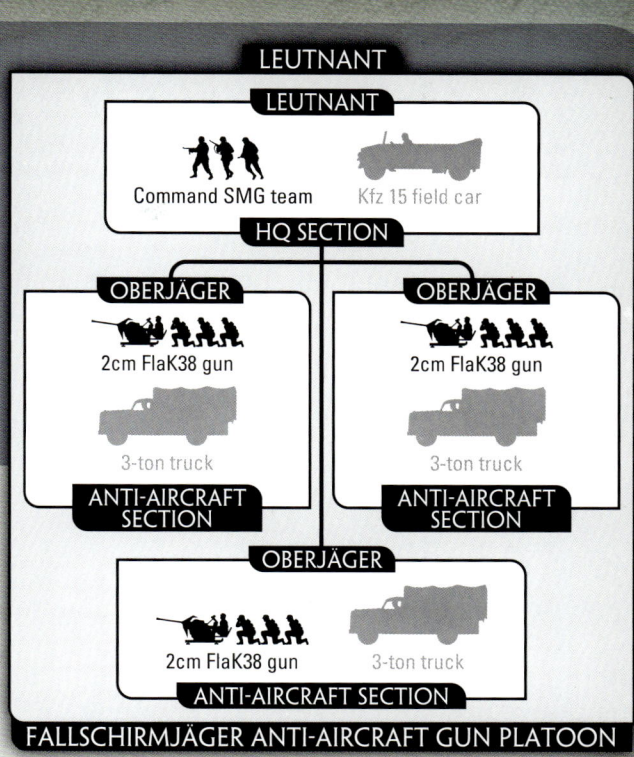

FALLSCHIRMPIONIER PLATOON

PLATOON

HQ Section with:

3 Pioneer Squads	345 points
2 Pioneer Squads	240 points
1 Pioneer Squad	135 points

OPTION

- Add a Pioneer Supply truck for +25 points.

You may replace up to one Pioneer Rifle/MG team per Pioneer Squad with a Flame-thrower team at the start of the game before deployment.

The *Fallschirmpionier* have a history as long and illustrious as the *Fallschirmjäger*. Their role is both field engineering, laying and clearing minefields and other defences, and leading attacks as assault engineers.

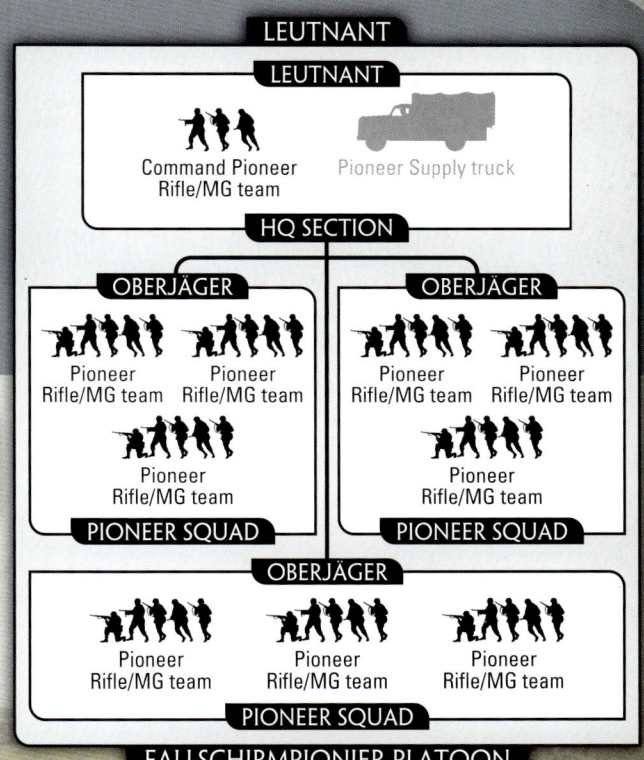

SS-FREIWILLIGEN-PANZERGRENADIERKOMPANIE
SS MOTORISED INFANTRY COMPANY

(INFANTRY COMPANY)

HEADQUARTERS

SS-Freiwilligen-Panzergrenadierkompanie HQ — 39

You must field one platoon from each box shaded black and may field one platoon from each box shaded grey.

Your Company HQ, Combat and Support platoons must be all either from 4. SS-Freiwilligen-Panzergrenadier Brigade 'Nederland' (Marked ⚔) or 11. SS-Freiwilligen-Panzergrenadierdivision 'Nordland' (Marked ✠), unless otherwise noted.

COMBAT PLATOONS

INFANTRY

SS-Freiwilligen-Panzergrenadier Platoon — 39

INFANTRY

SS-Freiwilligen-Panzergrenadier Platoon — 39

INFANTRY

SS-Freiwilligen-Panzergrenadier Platoon — 39

ALLIED PLATOONS

Heer and Luftwaffe Platoons in your force are Allies and follow the Allies rules in the rulebook.

WEAPONS PLATOONS

MACHINE-GUNS

SS-Freiwilligen-Heavy Platoon — 40

Field Fortifications — 155

ARTILLERY

SS-Freiwilligen-Mortar Platoon — 40

ANTI-AIRCRAFT

SS-Freiwilligen-Anti-aircraft Gun Platoon — 41

ARTILLERY

SS-Freiwilligen-Light Infantry Gun Platoon — 42

ANTI-TANK

SS-Freiwilligen-Anti-tank Gun Platoon — 41

REGIMENTAL SUPPORT PLATOONS

ARTILLERY

SS-Freiwilligen-Heavy Infantry Gun Platoon — 42

DIVISIONAL SUPPORT PLATOONS

ARMOUR

Schwere Panzer Platoon — 71

Veteran Tank-hunter Platoon — 164

ARMOUR

Nordland SS-Panzer Platoon — 42

INFANTRY

SS-Freiwilligen-Aufklärüngs Platoon — 43

SS-Panzergrenadier Pioneer Platoon — 177

Luftwaffe Jäger Platoon — 172

RECONNAISSANCE

SS-Panzerspäh Platoon — 176

Heavy SS-Panzerspäh Platoon — 176

ARTILLERY

Motorised SS-Artillery Battery — 178

Motorised Heavy SS-Artillery Battery — 178

ROCKET ARTILLERY

SS-Rocket Launcher Battery — 180

SS-Vielfachwerfer Battery — 181

Motorised SS-Artillery Battery — 178

ANTI-AIRCRAFT

Heavy SS-Anti-aircraft Gun Platoon — 180

Self-propelled SS-Anti-aircraft Gun Platoon — 181

Luftwaffe Heavy Anti-aircraft Gun Platoon — 173

AIRCRAFT

Air Support — 172

An SS-Freiwilligen-Panzergrenadierkompanie from 11. SS-Freiwilligen-Panzergrenadierdivision Nordland uses the Nordland SS special rules on page 189 as well as all the normal German special rules on pages 183 to 187.

An SS-Freiwilligen-Panzergrenadierkompanie from 4. SS-Freiwilligen-Panzergrenadier Brigade Nederland uses the Nederland SS special rules on page 189 as well as all the normal German special rules on pages 183 to 187.

MOTIVATION AND SKILL

The 4. SS-Freiwilligen-Panzergrenadier Brigade Nederland *retains a core of combat veterans who had been fighting on the Russian Front since 1941. An SS-Panzergrenadierkompanie from* 4. SS-Freiwilligen-Panzergrenadier Brigade Nederland *is rated* **Fearless Veteran**.

The 11. SS-Freiwilligen-Panzergrenadierdivision Nordland *was formed around a core of veterans from the* Wiking *SS-Division,* Freikorps Danmark *and* Legion Norwegen, *though many of the recruits are* Volksdeutsch *Germans from other parts of Europe. An SS-Panzergrenadierkompanie from* 11. SS-Freiwilligen-Panzergrenadierdivision Nordland *is rated* **Fearless Trained**.

RELUCTANT	CONSCRIPT
CONFIDENT	TRAINED
FEARLESS	**VETERAN**

SS-FREIWILLIGEN-PANZERGRENADIER BRIGADE 'NEDERLAND'

RELUCTANT	CONSCRIPT
CONFIDENT	**TRAINED**
FEARLESS	VETERAN

SS-FREIWILLIGEN-PANZERGRENADIER-DIVISION 'NORDLAND'

HEADQUARTERS

SS-FREIWILLIGEN-PANZERGRENADIERKOMPANIE HQ

PANZERGRENADIER HQ

Company HQ	55 points	40 points

- Replace either or both Command SMG teams with Command Panzerfaust SMG teams for +10 points per team.

OPTION

Anti-tank Section with:

2 Panzerschreck teams	+65 points	+50 points
1 Panzerschreck team	+35 points	+25 points

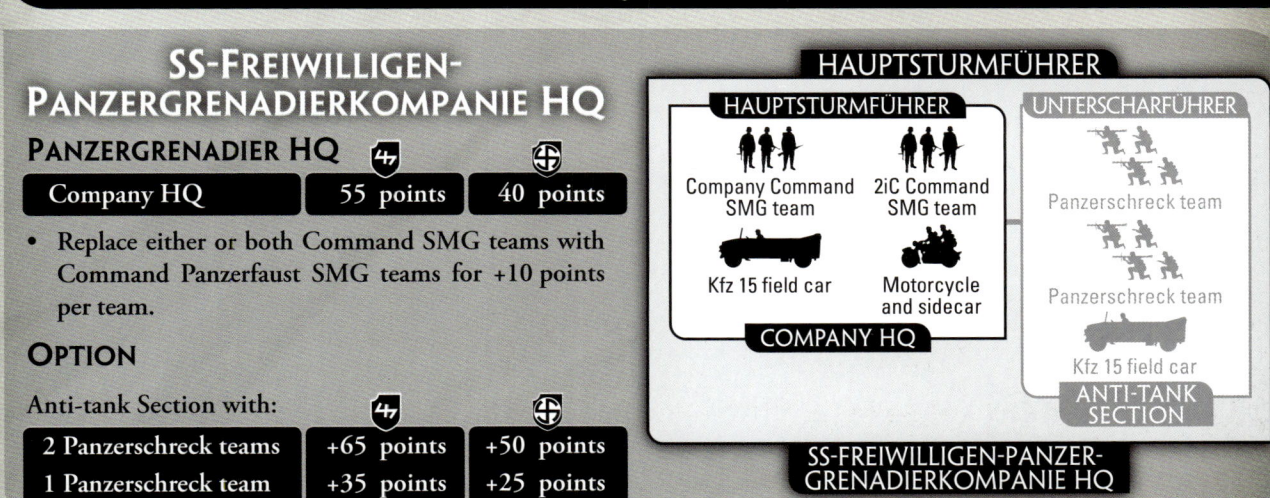

HAUPTSTURMFÜHRER

HAUPTSTURMFÜHRER

Company Command SMG team

2iC Command SMG team

Kfz 15 field car

Motorcycle and sidecar

COMPANY HQ

UNTERSCHARFÜHRER

Panzerschreck team

Panzerschreck team

Kfz 15 field car

ANTI-TANK SECTION

SS-FREIWILLIGEN-PANZER-GRENADIERKOMPANIE HQ

COMBAT PLATOONS

SS-FREIWILLIGEN-PANZERGRENADIER PLATOON

PLATOON

HQ Section with:

3 Panzergrenadier Squads	220 points	165 points
2 Panzergrenadier Squads	155 points	115 points

OPTIONS

- Replace the Command MG team with a Command Panzerfaust SMG team for +10 points.
- Replace up to one MG team per squad with a Panzerfaust MG team for +10 points per team.
- Add 3-ton trucks for +5 points for the platoon.

If MG teams, other than the Command MG team, are replaced with Panzerfaust MG teams your force may not contain a Nordland SS-Panzer Platoon armed with Panther D tanks.

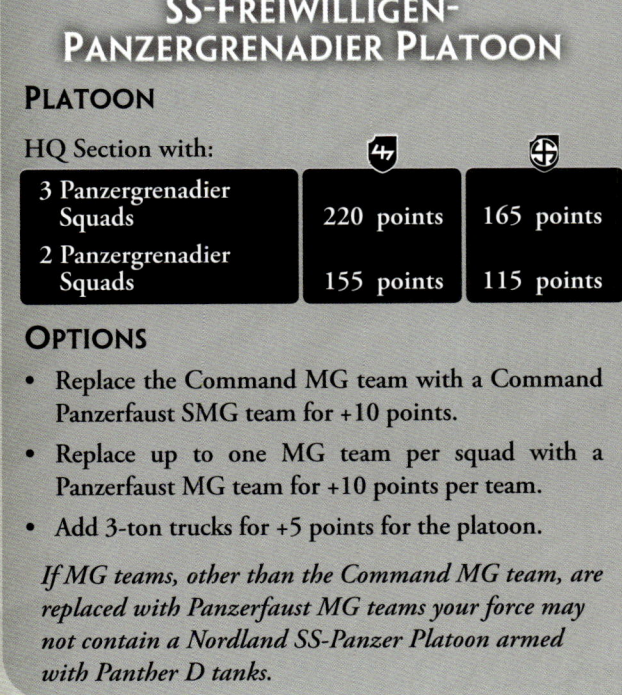
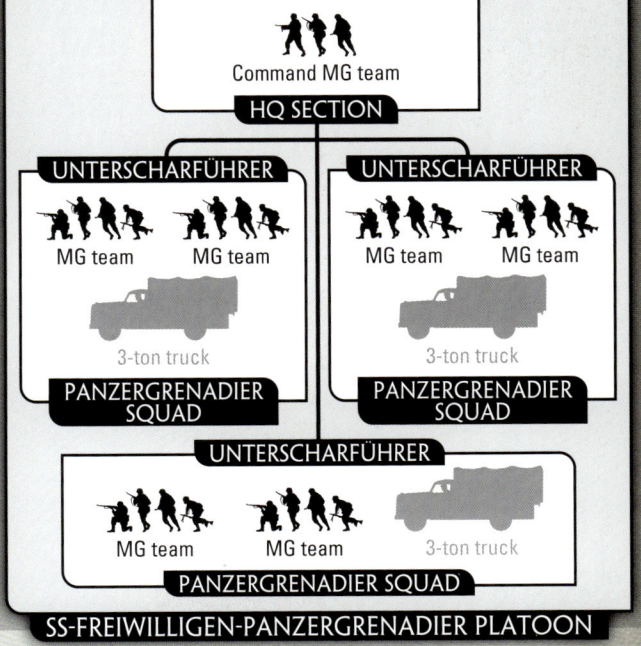

UNTERSTURMFÜHRER

UNTERSTURMFÜHRER

Command MG team

HQ SECTION

UNTERSCHARFÜHRER

MG team MG team

3-ton truck

PANZERGRENADIER SQUAD

UNTERSCHARFÜHRER

MG team MG team

3-ton truck

PANZERGRENADIER SQUAD

UNTERSCHARFÜHRER

MG team MG team 3-ton truck

PANZERGRENADIER SQUAD

SS-FREIWILLIGEN-PANZERGRENADIER PLATOON

The SS-Panzergrenadiers are motorised infantry and fight their battles on foot. They don't have the advantage of armoured half-tracks like their comrades from a SS-Panzerdivision. Use the trucks to move them quickly around the battlefield, but be sure to dismount before entering the combat zone.

SS-FREIWILLIGEN-HEAVY PLATOON

PLATOON

HQ Section with:

	47	4/
2 Machine-gun Sections	155 points	120 points
1 Machine-gun Section	85 points	65 points
No Machine-gun Sections	10 points	10 points

ADD

	47	4/
1 Mortar Section	+70 points	+55 points

OPTION

- Add additional Kfz 70 trucks at no cost.

A Heavy Platoon must have a Mortar Section if it has no Machine-gun Sections.

SS-Freiwlligen-Heavy Platoons may make Combat Attachments to SS-Freiwlligen-Panzergrenadier Platoons.

In one platoon you can consolidate all the heavy weapons support you may need to complement your Panzergrenadiers whether they be on offence, defence or in counterattack. This one platoon can provide the necessary firepower to halt most any infantry assaults against you.

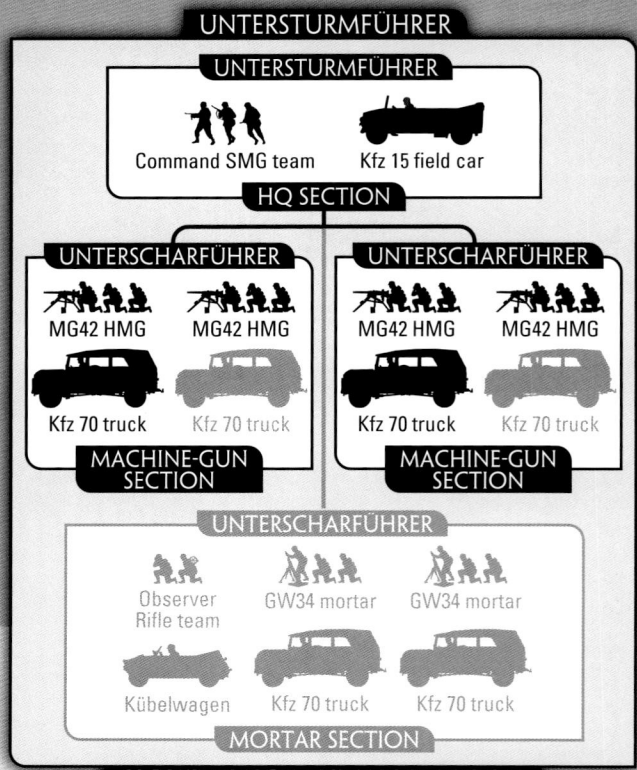

SS-FREIWILLIGEN-HEAVY PLATOON

Alternately, if you assault or counterattack, a heavy weapons platoon can provide the required edge to ensure your assaulting forces reach the enemy lines with minimal casualties.

SS-FREIWILLIGEN-MORTAR PLATOON

MOTORISED PLATOON

HQ Section with:

	47	4/
3 Mortar Sections	210 points	160 points
2 Mortar Sections	150 points	115 points
1 Mortar Section	80 points	60 points

OPTIONS

- Add Kfz 15 field car and 3-ton trucks for +5 points for the platoon.
- Replace 3-ton trucks with Opel Maultier half-tracks for +5 points for the platoon.

Good light artillery is critical in pinning or blinding enemy positions or strongpoints with smoke. SS-Mortar platoons are excellent in this role due to the speed with which they respond to calls for fire.

With plenty of observer teams they can engage any target across the whole battlefield much faster than the big guns of the artillery. The 8cm GW34 mortar can pin down enemy attacks as well as deliver covering smoke just as well as any artillery battery.

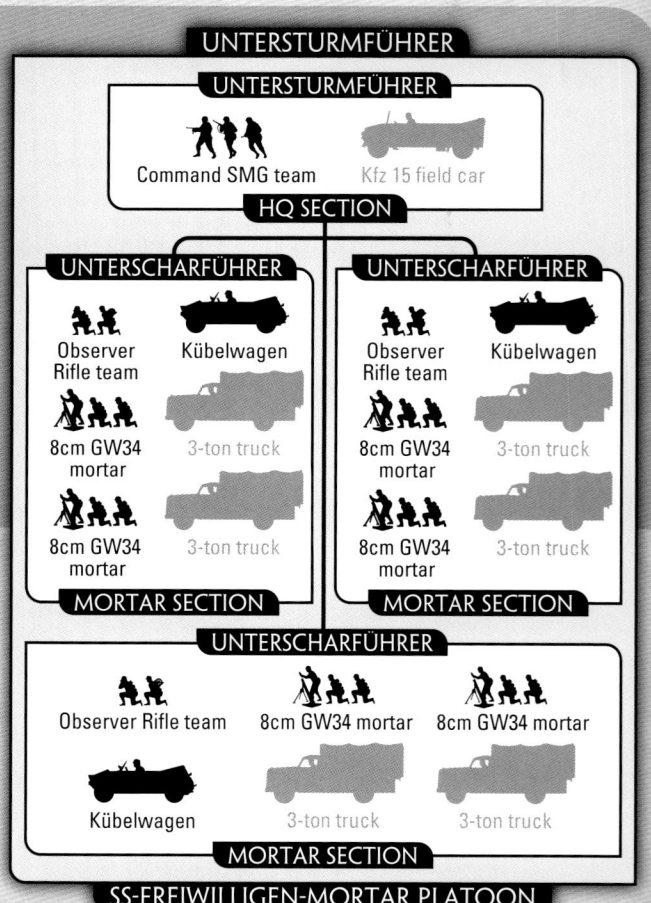

SS-FREIWILLIGEN-MORTAR PLATOON

SS-FREIWILLIGEN-ANTI-AIRCRAFT GUN PLATOON

PLATOON

HQ Section with:

	47	SS
3 2cm FlaK38	80 points	60 points
3 3.7cm FlaK43	110 points	85 points

OPTION

- Add Kfz 15 Field Car and 3-ton trucks for +5 points for the platoon.

Protection from air attacks is a vital part of any division's arsenal. The 2cm Flak38 guns are light and easy to move allowing them to cover the SS-Panzergrenadiers even during a counterattack. It has a good rate of fire to provide adequate protection without the need for heavy anti-aircraft support.

The platoon can also provide additional fire support against attacking infantry tipping the scales in favour of the defenders.

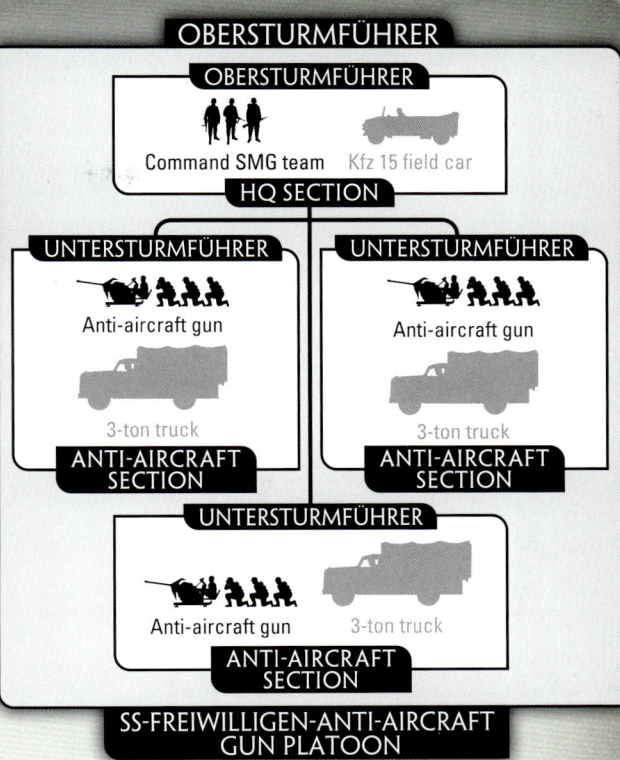

OBERSTURMFÜHRER

OBERSTURMFÜHRER
Command SMG team — Kfz 15 field car

HQ SECTION

UNTERSTURMFÜHRER — Anti-aircraft gun — 3-ton truck — ANTI-AIRCRAFT SECTION

UNTERSTURMFÜHRER — Anti-aircraft gun — 3-ton truck — ANTI-AIRCRAFT SECTION

UNTERSTURMFÜHRER — Anti-aircraft gun — 3-ton truck — ANTI-AIRCRAFT SECTION

SS-FREIWILLIGEN-ANTI-AIRCRAFT GUN PLATOON

SS-FREIWILLIGEN-ANTI-TANK GUN PLATOON

PLATOON

HQ Section with:

	47	SS
3 7.5cm PaK40	180 points	135 points
2 7.5cm PaK40	120 points	90 points

OPTION

- Add Kübelwagen jeep and 3-ton trucks for +5 points for the platoon.

SS-Freiwilligen-Anti-tank Gun Platoons may make Combat Attachments to SS-Freiwilligen-Panzergrenadier Platoons.

The 7.5cm PaK40 gun has become the standard anti-tank gun of the SS-Panzergrenadier divisions. The hard hitting PaK40 will destroy almost any tank the enemy cares to put in range of them.

Place your anti-tank assets in good cover behind your front lines, then spring your ambush when the enemy tanks close to overrun your position. Your concentrated fire will stop the assault in its tracks.

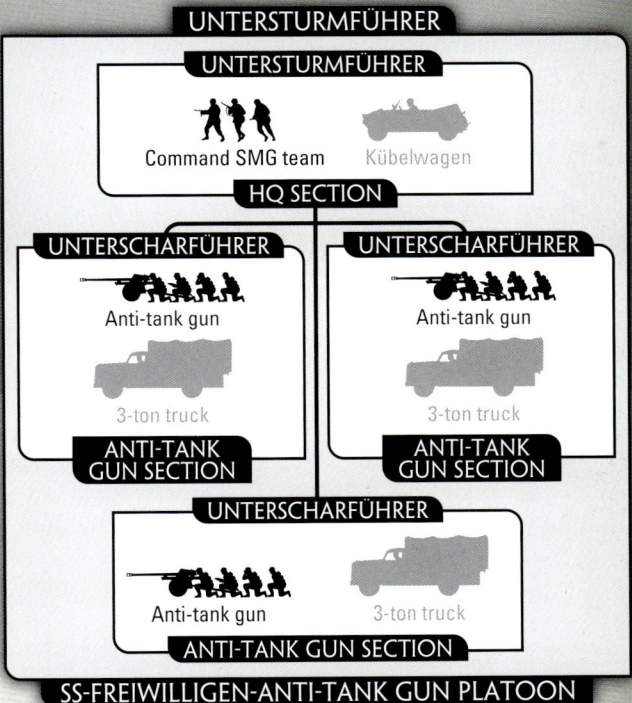

UNTERSTURMFÜHRER

UNTERSTURMFÜHRER
Command SMG team — Kübelwagen

HQ SECTION

UNTERSCHARFÜHRER — Anti-tank gun — 3-ton truck — ANTI-TANK GUN SECTION

UNTERSCHARFÜHRER — Anti-tank gun — 3-ton truck — ANTI-TANK GUN SECTION

UNTERSCHARFÜHRER — Anti-tank gun — 3-ton truck — ANTI-TANK GUN SECTION

SS-FREIWILLIGEN-ANTI-TANK GUN PLATOON

SS-FREIWILLIGEN-LIGHT INFANTRY GUN PLATOON

PLATOON

HQ Section with:

2 7.5cm leIG18	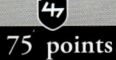 75 points	 60 points

OPTIONS

- Add Kfz 15 field car and 3-ton trucks for +5 points for the platoon.
- Replace 3-ton trucks with RSO tractors at no cost.

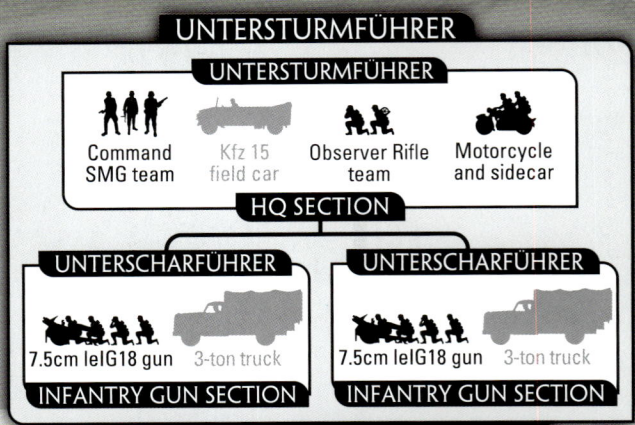

REGIMENTAL SUPPORT PLATOON

SS-FREIWILLIGEN-HEAVY INFANTRY GUN PLATOON

PLATOON

HQ Section with:

2 15cm sIG33	165 points	125 points

OPTION

- Add Kfz 15 field car and Sd Kfz 11 half-tracks for +5 points for the platoon.

Heavy infantry guns can and will destroy any enemy resistance nest. Even bunkers are not immune.

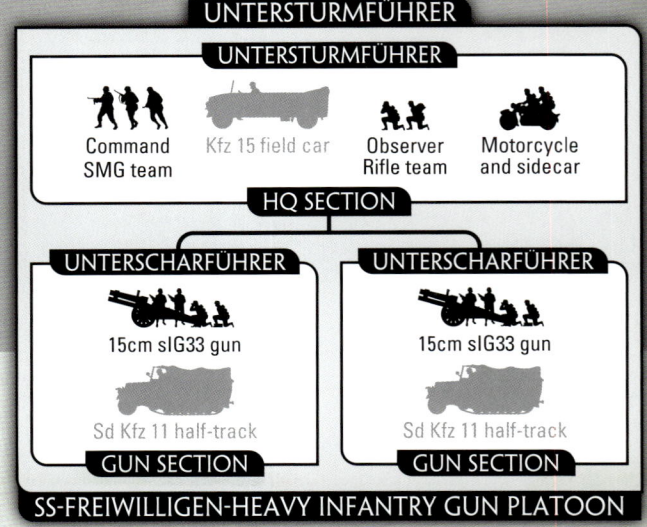

DIVISIONAL SUPPORT PLATOONS

NORDLAND SS-PANZER PLATOON

PLATOON

5 StuG G	545 points
4 StuG G	435 points
3 StuG G	325 points
3 Panther D	610 points

The Panzerabteilung *of the* Nordland *SS-Division also supported the* Nederland *SS-Brigade. They use the* Nordland Waffen-SS *special rules and are rated* **Fearless Veteran**.

| FEARLESS | VETERAN |

If a company with Nordland SS-Panzer Platoon armed with Panther D tanks is defending in a Mission with the Prepared Positions special rule they may be deployed in Tank Pits (see the rulebook).

The 'Hermann von Salza' Panzerabteilung of the 11. SS-Freiwilligen-Panzergrenadierdivision Nordland was armed with StuG III G assault guns. On a number of occasions these assault guns also supported the troops of the 4. SS-Freiwilligen-Panzergrenadier Brigade Nederland'

SS-FREIWILLIGEN-AUFKLÄRUNGS PLATOON

PLATOON

HQ Section with:

3 Aufklärungs Squads	305 points	235 points
2 Aufklärungs Squads	220 points	170 points

OPTIONS

- Replace Command Motorcycle MG team with a Command Motorcycle Panzerfaust SMG team for +10 points.
- Replace all Motorcycle teams with the equivalent Schwimmwagen teams for +5 points for the platoon.

While still capable of deep penetration behind enemy lines, the prevalence of enemy armour requires the intelligent use of this highly valued unit to ensure its maximum effectiveness.

SS-Freiwilligen-Aufklärungs Platoons use the Motorcycle Reconnaissance rules on page 187 and are Reconnaissance Platoons while mounted.

You may model your Motorcycle MG teams with Kübelwagen jeeps instead of motorcycles, they are based the same way as the Motorcycle MG teams and use the same rules.

UNTERSTURMFÜHRER

UNTERSTURMFÜHRER

Command Motorcycle MG team

HQ SECTION

UNTERSCHARFÜHRER

Motorcycle MG team

Motorcycle MG team

AUFKLÄRUNGS SQUAD

UNTERSCHARFÜHRER

Motorcycle MG team

Motorcycle MG team

AUFKLÄRUNGS SQUAD

UNTERSCHARFÜHRER

Motorcycle MG team

Motorcycle MG team

AUFKLÄRUNGS SQUAD

SS-FREIWILLIGEN-AUFKLÄRUNGS PLATOON

Having swept through the German trench lines, the Red Army advances at lightning speed...

Ambushing PaK40 anti-tank guns stop Soviet attempts to encircle the StuGs...

...only to have their advance slowed at the river by the retreating German defenders.

...which gives the Germans time to re-establish their defences.

OPERATION BAGRATION

BUILD UP TO BAGRATION

From the Soviet *Stavka* (High Command) perspective, the Byelorussian offensive was a result of three years of learning by ordeal at the hands of the German *Wehrmacht* (armed forces). The miracle at Moscow, where the first German strategic offensive was halted, and the epic struggle at Stalingrad, gave *Stavka* the confidence to plan their offensive.

Geographically, the operation dwarfed all other campaigns to date. They constructed a plan to crush the German armies, and drive the war into the German homeland and on to victory. The Red Army's furious blitzkrieg was timed to launch on the third anniversary of the German invasion of the Soviet Union.

THE GERMAN ARMY

The German forces of Army Group Centre had been stripped of the majority of their *Panzer* elements, which had been sent to reinforce Army Group Northern Ukraine, where Hitler and the majority of the high command were convinced the main thrust of the Soviet summer campaign would fall.

German forces in the region consisted of, from north to south, *3. Panzerarmee* (3rd Armoured Army) in Vitebsk, *4. Armee* (4th Army) centred near Orsha and Mogilev and the *9. Armee* (9th Army) based out of Bobruisk.

Hitler had ordered that German forces maintain a static line of defence, while also holding several fortified cities, or *Festerplatz*. The major communications and supply centres of Vitebsk, Bobruisk, Orsha, and Mogilev all became *Festerplatz* towns. Each received a division of infantry to garrison the city and were ordered to defend at all costs. Retreat was simply not an option!

VITEBSK

The Soviet war machine launched Operation Bagration with the assault on *Festerplatz* Vitebsk on 22 June. The Soviet First Baltic and Third Byelorussian Fronts attacked the German Third Panzer Army and the Fourth Army defending the region with overwhelming force pushing the Germans back.

The Third Panzer Army counterattacked in an attempt to hold open an escape corridor for the forces defending Vitebsk. This attack was defeated by the Soviet infantry, assisted by support from the Soviet First Air Army.

On the morning of 27 June, the remains of the German garrison evacuated the city and began their break out. They managed to push twelve miles to the south-west of the city where they quickly became encircled once more.

The Soviet juggernaut rapidly overtook the isolated German soldiers. As they retreated west, lack of ammunition, fuel and supplies led to the majority of the Third Panzer Army being killed or captured. Its battered remnants withdrew towards Minsk.

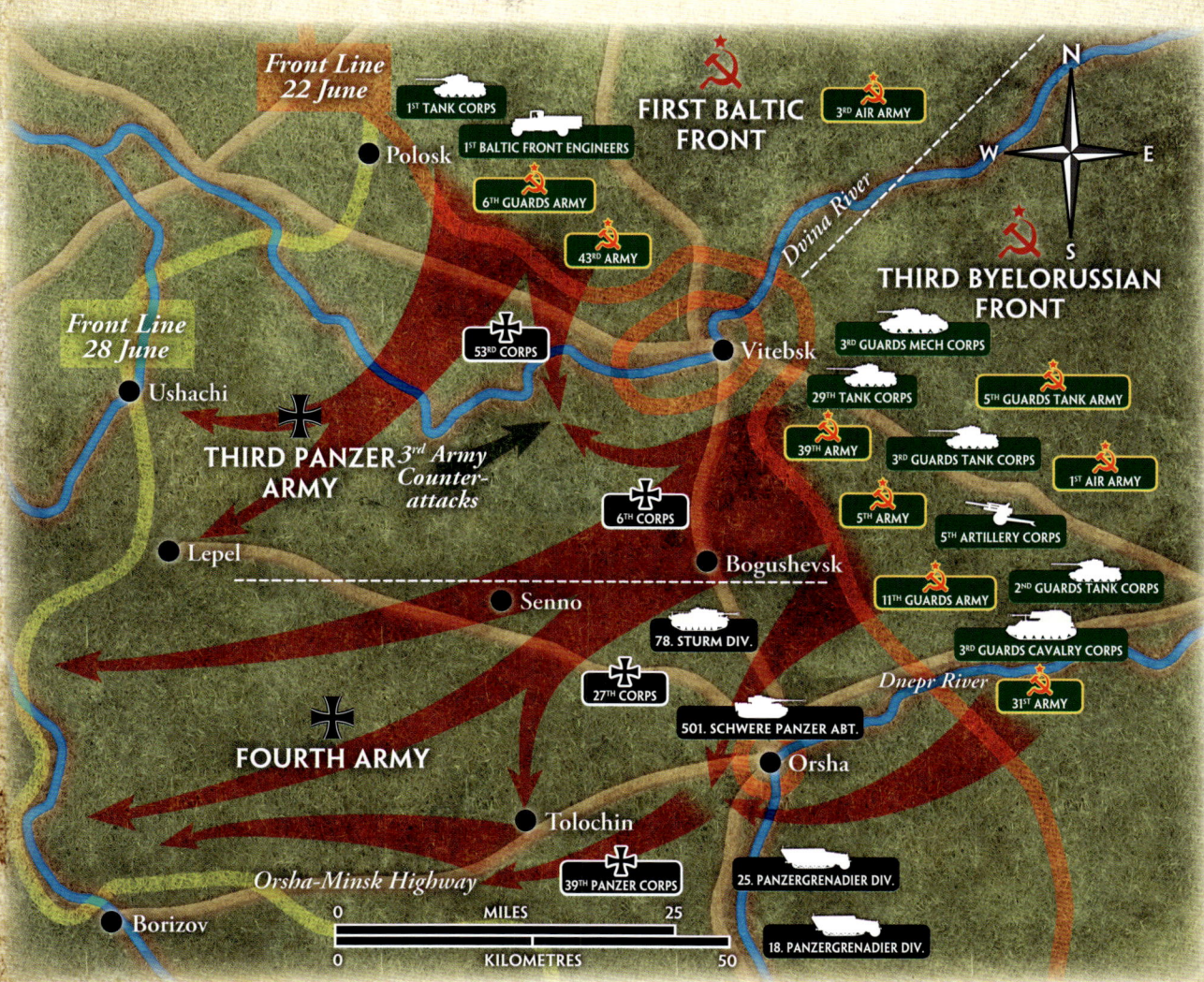

ORSHA

Immediately south of Vitebsk the Soviet 11th Guards Army launched a brutal assault against *Festerplatz* Orsha. The German *78. Sturmdivision* (78th Assault Division) and *501. Schwere Panzerabteilung* (501st Heavy Tank Battalion) defended the region using a vast network of trenches and strongpoints. The town itself was fortified by troops from the 78th and other divisions.

The battle for Orsha was fierce and the well-placed fortifications initially stalled the Soviet assault. The Soviet 2nd Guards

Tank Corps found a weak point and broke through the line on the left flank of the *Sturmdivision*. The Germans moved to counter, but the flank was lost. The 2nd Tank Corps bolted through the gap and took off towards Minsk, leaving a tank brigade behind to complete the encirclement of Orsha.

Despite the heroic and desperate defence of the city by *78. Sturmdivision* and the heavy Tiger tanks of *501. Schwere Panzerabteilung*, *Festerplatz* Orsha fell on 27 June.

BOBRUISK

The Soviet onslaught in the south began on 24 June. The Soviet First Byelorussian Front attack against the German Ninth Army near Bobruisk made some progress. The German *20. Panzerdivision* (20th Armoured Division) quickly counterattacked and bought time for the German infantry to re-establish the front line.

The Soviet 9th Tank Corps and the 1st Guards Tank Corps combined their efforts against *20. Panzerdivision*. The renewed Soviet attack proved too much for the German defenders who reluctantly retreated out of the city.

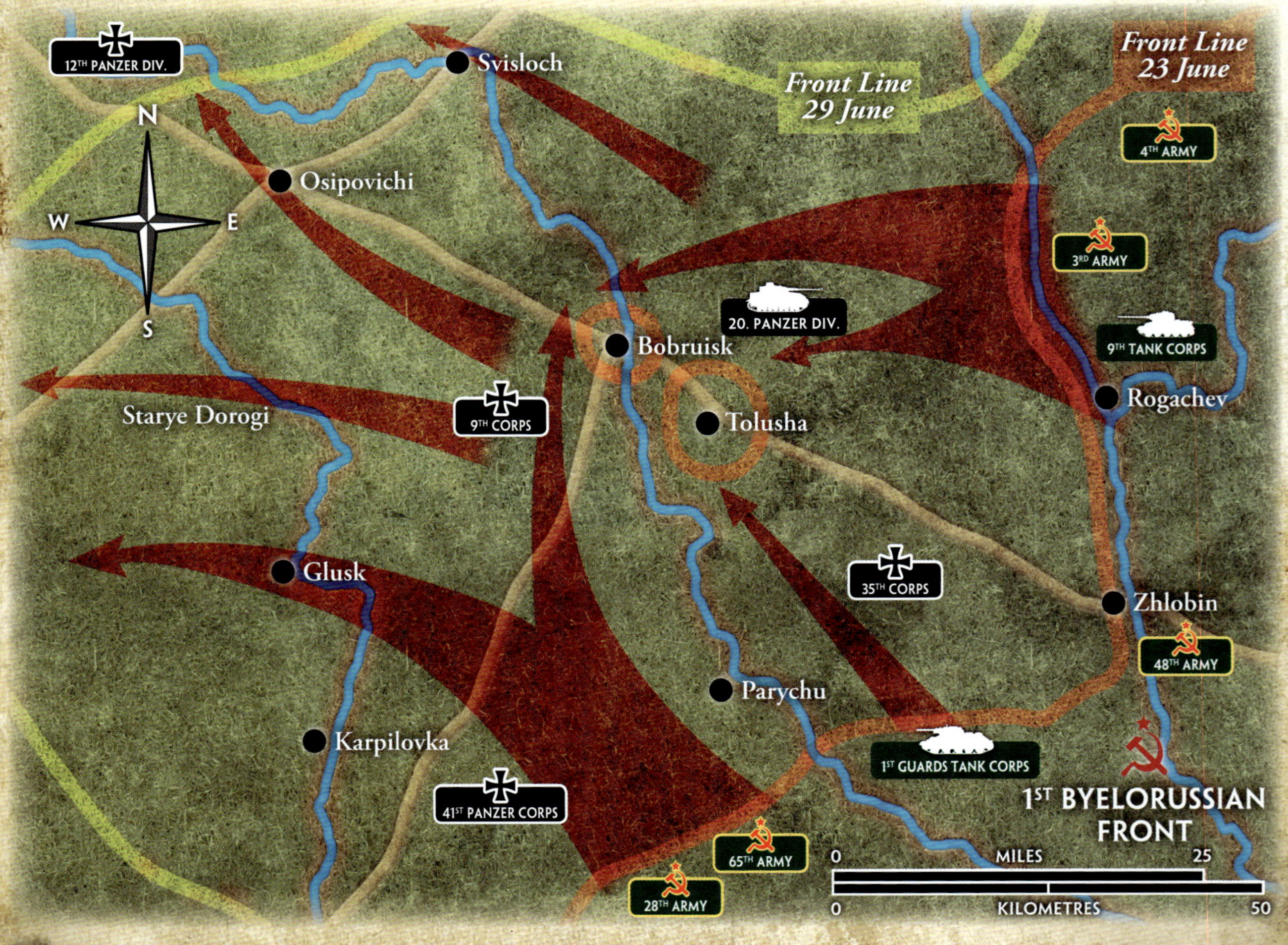

Front Line 29 June

Front Line 23 June

12TH PANZER DIV.

Svisloch

4TH ARMY

N
W E
S

Osipovichi

3RD ARMY

20. PANZER DIV.

9TH TANK CORPS

Starye Dorogi

9TH CORPS

Bobruisk

Tolusha

Rogachev

35TH CORPS

Glusk

Zhlobin

48TH ARMY

Parychu

1ST GUARDS TANK CORPS

Karpilovka

41ST PANZER CORPS

1ST BYELORUSSIAN FRONT

65TH ARMY

0   MILES   25

28TH ARMY

0   KILOMETRES   50

78. STURMDIVISION

The *78. Sturmdivision* was originally raised in 1939 as *78. Grenadierdivision* in Stuttgart, Germany. The division participated in the invasion of Russia and advanced to the gates of Moscow, before being forced back.

In January 1943, *78. Grenadierdivision* reformed as the *78. Sturmdivision*. The motive behind these changes was to create a division with many weapons and few men. Each *Sturmkompanie* was authorised enlarged infantry platoons and significant increases in heavy weapons. Additionally, each battalion had nine integral 7.5cm PaK40 anti-tank guns, a number usually held by an entire Grenadier regiment. *Sturmgeschütz-Abteilung 189* was also attached to the division, along with *Panzerjäger-Abteilung 178*.

Soon after its reorganisation *78. Sturmdivision* was committed to Operation *Zitadelle* (Citadel) where the unit fought with distinction at Kursk.

By October 1943, the veteran *78. Sturmdivision* was placed along the Panther Line near Orsha. It was now part of the XXVII Army Corps along with *25. Panzergrenadierdivision* and *260. Grenadierdivision*, and defended the critical junction along the Moscow–Minsk highway.

Three years after the German invasion of the Soviet Union, on 22 June 1944, the Soviets launched Operation Bagration with a huge preparatory bombardment. The initial attacks failed with heavy losses, despite the use of special armoured assault groups by the 11th Guards Army.

By the evening of 23 June *78. Sturmdivision* had withdrawn back in good order, about 5km to its second defensive line.

Problems arose when a Soviet reconnaissance patrol discovered a disused narrow-gauge rail line through the swampy forest between the *Sturmdivision* and *256. Grenadierdivision*. The 2nd Guards Tank Corps pushed through along the rail line, and the *Sturmdivision* was outflanked.

On 26 June the division withdrew again into Orsha, and fell under attack by two Soviet divisions. Meanwhile, *5. Panzerdivision* and the Tiger tanks of *505. Schwere Panzer Abteilung* arrived to try to stem the Soviet advance. However, by the morning of the 27 June, Orsha had fallen, and the road to Minsk was wide open.

The *Sturmdivision* continued to retreat, abandoning its heavy equipment at the Berezina River. Any stragglers risked attack by units of partisans lurking in the forests.

By 5 July the *Sturmdivision* had fought from encirclement to encirclement, having conducted a 200km-long fighting withdrawal. The division managed to get clear of the encircling enemy, despite losing the majority of *78. Sturmdivision* and its commander, General Traut. They were not alone, as Army Group Centre lost 17 other divisions completely, with many more severely depleted.

Those few survivors that managed to escape the collapse of Army Group Centre were absorbed by *565. Volksgrenadierdivision*, and the *Sturmdivision* would be reformed as *78. Volksgrenadierdivision* a few weeks later. The reformed division would fight through Poland, Silesia and Czechoslovakia.

Of all the German divisions in Byelorussia, the *78. Sturmdivision* is the most prepared to receive the impending Soviet attack. The *Sturmdivision* has been kept at high strength with a trench-strength of 5,700 troops compared to the average 2,500 troops of the other German divisions. In addition to its high strength, the assault division has been issued plenty of heavy weapons to support its troops! The *78.Sturmdivision* is fully prepared for a Soviet assault!

StuG Assault guns

Since the Battle of Kursk, the *Sturmdivision* has had its own *Sturmgeschützabteilung* (Assault gun battalion). The StuG G Assault guns of *189. Sturmgeschützabteilung* (189th Assault Gun Battalion) offer an immediate tank response to any breakthroughs, helping shore up the line and restore defensive positions or follow up a *Sturmgrenadier* assault. The StuG assault guns of the *Batterie* offer a huge morale boost to the men and are a vital asset to the *Sturmdivision!*

THE MG42 LIGHT MACHINE-GUN

The *Sturmdivision* has been issued two MG42 Light machine-guns per squad, rather than the usual *Grenadierkompanie* organisation of one per squad. This sort of additional firepower will hold the line against the Soviet onslaught!

StG44 Assault Rifle

BEGLIET RIDERS

The *189. Sturmgeschützabteilung* pioneered the use of *Begleit*, or Escort, troops at Kursk. These *Grenadiers* ride atop the StuGs when the battalion is manoeuvring, offering the assault guns their own infantry protection. When in combat, the *Begleit* riders often dismount and fight alongside their tanks, working together to mount successful counterattacks.

PANZERFAUST ANTI-TANK LAUNCHER

By June 1944, Panzerfaust anti-tank launchers are a welcome addition to the Grenadier's arsenal. These weapons are deadly at short range, roughly 30 meters (98 feet) and will penetrate armour up to 200mm (8 inches) thick! By June these inexpensive tank-busters can be found in large quantities all along the 78th Assault Division's fortifications.

StG44 Assault Rifle

The *Begleit* riders in the *189. Sturmgeschützabteilung* are among the first to receive the new StG44 (or, *Sturmgewehr 44*) Assault rifle, also known as the MP44 (or, *Maschinenpistole 44*). The StG44 assault rifle combines the firepower of a submachinegun in close combat with the accuracy of a rifle at medium ranges. It fires a shortened version of the 7.92mm round, also used in the standard Karabiner 98k rifle. The gun's ammunition supply also increases the standard five-round capacity of the Karabiner 98k to an impressive 30-round magazine in the StG44 assault rifle!

Panzerfaust Anti-tank Launcher

THE 7.5CM PAK40 ANTI-TANK GUN

Each *Sturmkompanie* has its own *Sturm* Anti-tank platoon with two 7.5cm PaK40 anti-tank guns. This gives each company its own anti-tank guns in addition to the battalion and regimental anti-tank guns. No other German division can boast having 99 PaK40 anti-tank guns at its disposal!

This fully automatic rifle will help the *Begleit* riders to assault the enemy infantry with their Assault guns in support!

FELDWEBEL
DIETRICH UTHOFF

Dietrich Uthoff joined the army after a series of unsuccessful careers. In the military, Uthoff discovered his calling. He rapidly gained the rank of *Unteroffizier*, or corporal.

He transferred to the *78. Infanteriedivision* just as the division was sent to Russia, where he was wounded while manning a machine-gun after its crew had been wiped out by a 76mm shell. This earned him an Iron Cross and a promotion to *Feldwebel*, or sergeant.

In 1943, Uthoff and the *78. Infantriedivision* reorganised as a *Sturmdivision*. During the conversion, Uthoff acquired the latest weapons, such as his own personal *Sturmgewehr 44* assault rifle and a pile of *Panzerfaust* launchers for his men.

In late June 1944, Operation Bagration erupted in front of Uthoff and his men and by the afternoon of the first day the platoon had been reduced to just six men. He asked for two volunteers and ordered the remainder of the men to rejoin the division in Orsha.

Uthoff and the volunteers were last seen dashing between mortar craters and upturned earth, each with three *Panzerfaust* launchers. Several Soviet T-34 and IS-2 tanks exploded as they made their way through no-man's-land.

For his bravery, Dietrich Uthoff was posthumously recommended for the Knight's Cross.

CHARACTERISTICS

Feldwebel Dietrich Uthoff is a Warrior and a Panzerfaust Assault Rifle team. He is an Independent team and rated as **Fearless Veteran**.

Uthoff may join a Sturmkompanie from the *78. Sturmdivision* for +50 points.

AMBUSH!

Feldwebel Uthoff used craters, destroyed tanks and bombed-out trenches to surprise hapless Soviet tanks.

> *Uthoff may be deployed using the Ambush special rule in the rulebook in addition to any other platoons that would normally be deployed in Ambush.*
>
> *While defending, Uthoff may deploy in Ambush even in missions that do not normally use the Ambush special rule.*

TANK HUNTER

Uthoff knew that there was no time to waste in stemming the Red tide!

> *Feldwebel Uthoff may move and shoot with his Panzerfaust anti-tank launcher.*

FAUST EXPERT

After several years of hunting tanks with Panzerknackers, the new Panzerfaust launchers make tank-hunting even easier!

> *Feldwebel Uthoff may re-roll failed To Hit rolls with his Panzerfaust anti-tank launcher.*

STURMKOMPANIE
STORM COMPANY

(INFANTRY COMPANY)

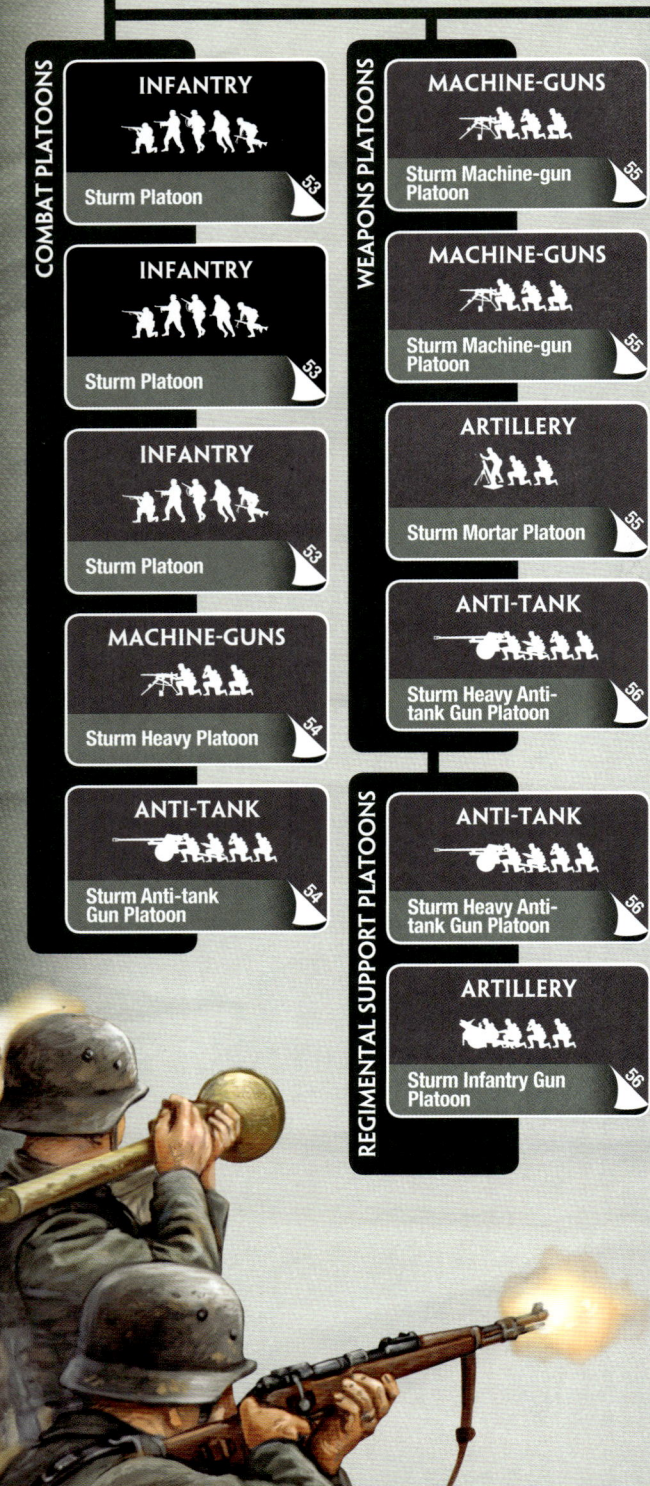

HEADQUARTERS

HEADQUARTERS

Sturmkompanie HQ — 53

You must field one platoon from each box shaded black and may field one platoon from each box shaded grey.

COMBAT PLATOONS

INFANTRY

Sturm Platoon — 53

INFANTRY

Sturm Platoon — 53

INFANTRY

Sturm Platoon — 53

MACHINE-GUNS

Sturm Heavy Platoon — 54

ANTI-TANK

Sturm Anti-tank Gun Platoon — 54

WEAPONS PLATOONS

MACHINE-GUNS

Sturm Machine-gun Platoon — 55

MACHINE-GUNS

Sturm Machine-gun Platoon — 55

ARTILLERY

Sturm Mortar Platoon — 55

ANTI-TANK

Sturm Heavy Anti-tank Gun Platoon — 56

REGIMENTAL SUPPORT PLATOONS

ANTI-TANK

Sturm Heavy Anti-tank Gun Platoon — 56

ARTILLERY

Sturm Infantry Gun Platoon — 56

DIVISIONAL SUPPORT PLATOONS

ARMOUR

Bäke Schwere Panzer Platoon — 24

ARMOUR

StuG Platoon — 59

Tank-hunter Platoon — 163

INFANTRY

Pionier Platoon — 61

INFANTRY

Sturm Scout Platoon — 57

ARTILLERY

Sturm Heavy Mortar Platoon — 57

ARTILLERY

Artillery Battery — 166

ARTILLERY

Heavy Artillery Battery — 166

ARTILLERY

Rocket Launcher Battery — 169

Armoured Rocket Launcher Battery — 170

ANTI-AIRCRAFT

Heavy Anti-aircraft Gun Platoon — 171

ANTI-AIRCRAFT

Sturm Anti-aircraft Gun Platoon — 57

Luftwaffe Anti-aircraft Gun Platoon — 173

AIRCRAFT

Air Support — 172

FORTIFICATIONS

Field Fortifications — 155

ALLIED PLATOONS

Luftwaffe Platoons in your force are Allies and follow the Allies rules in the rulebook.

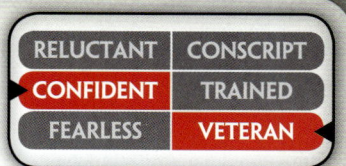

MOTIVATION AND SKILL

The illustrious 78. Sturmdivision has seen combat from the gates of Moscow to Kursk, and now guards the critical point at Orsha. The high concentration of integrated infantry weapons and assault guns makes the unit a very powerful formation. The 78. Sturmdivision will keep the Soviets in check, should they decide to assault! A Sturmkompanie is rated **Confident Veteran.**

RELUCTANT	CONSCRIPT
CONFIDENT	TRAINED
FEARLESS	**VETERAN**

HEADQUARTERS

STURMKOMPANIE HQ

HEADQUARTERS

Company HQ	45 points

OPTIONS

- Replace either or both Command SMG teams with Command Panzerfaust SMG teams for +10 points per team.
- Add an Anti-tank section with up to three Panzerschreck teams for +25 points per team.
- Add up to three Sniper teams for +50 points per team.

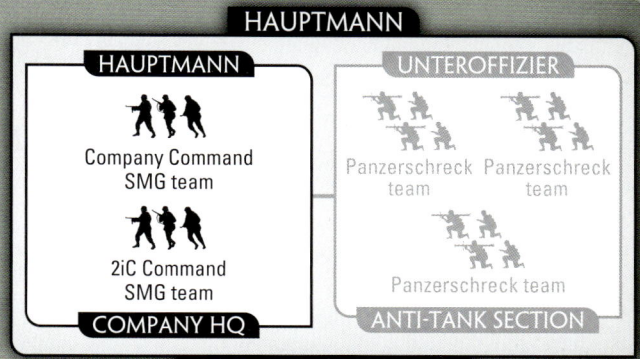

Now is the time to dig in and prepare for the expected Soviet offensive. We cannot be certain where this blow will fall, but we can be assured that the mighty *78. Sturmdivision* is ready.

Your *Sturmkompanie* is well supplied with the new *Panzerfaust* and *Panzerschreck* anti-tank rocket launchers able to repel any Bolshevik attack!

COMBAT PLATOONS

STURM PLATOON

PLATOON

HQ Section with:

3 Grenadier Squads	180 points
2 Grenadier Squads	130 points

OPTION

- Replace Command MG team with a Command Panzerfaust SMG team for +10 points. If you do this, you may replace all remaining MG teams with Panzerfaust MG teams for +20 points per squad.

Not only is the *Sturmkompanie* well equipped with machine-guns and *Panzerfaust* anti-tank launchers, but it also has a higher unit strength compared to other Grenadier units in the front line. This combination of defensive weaponry and manpower should make it easier to build a formidable defensive line.

On the other hand, your *Sturmkompanie* has not lost its offensive potential. The company is prepared to leave its trenches and counterattack against whatever small gains the Soviets might manage to secure.

If your Sturmkompanie has three Sturm platoons, you may upgrade the smallest Sturm Platoon to a Sturm Pionier Platoon for +15 points per squad. This converts the Command team and every MG team into Pioneer teams with the same armament. The Sturm Pionier Platoon may have a Pioneer Supply Truck for an additional +25 points.

Sturm Heavy Platoon

Platoon

HQ Section with:

1 Machine-gun Section	70 points
No Machine-gun Section	10 points

Options

- Replace Command SMG team with a Command Panzerfaust SMG team for +10 points.
- Add a Mortar Section for +60 points.

A Heavy Platoon must have a Mortar Section if it has no Machine-gun Section.

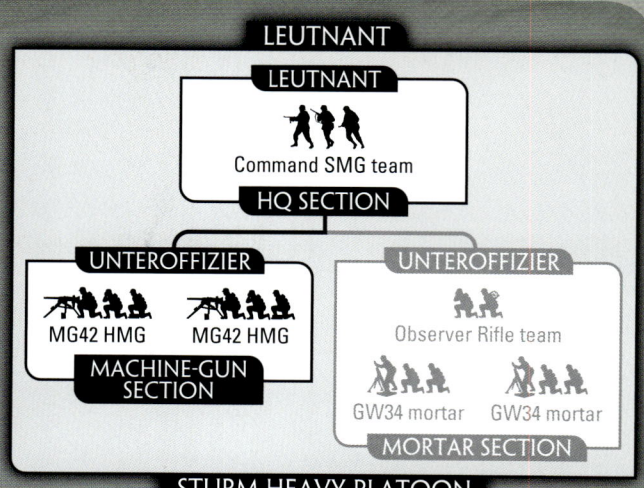

Sturm Heavy Platoons may make Combat Attachments to Sturm Platoons.

As a Grenadier force, your company is unique in having a heavy platoon available to it. The company's fourth, heavy platoon is a great asset to use along the defensive line and lend its fire against the *Bolshevik* assault.

Assigning a heavy machine-gun to each of your *Sturmgrenadier* platoons will boost their defensive capability.

While the fourth platoon's heavy machine-guns are not much use against waves of Soviet tanks, they are vitally needed to cut down their infantry support. The platoon's 8cm mortars pin down attacking formations and lay smoke to conceal a counterattack.

Sturm Anti-tank gun Platoon

Platoon

HQ Section with:

2 Anti-tank Gun Sections	110 points

Option

- Replace Command SMG team with a Command Panzerfaust SMG team for +10 points.

Sturm Anti-tank gun Platoons may make Combat Attachments to Sturm Platoons.

Pakfront!

German defensive doctrine used the *Pakfront* to efficiently devastate Soviet tank assaults. Guns are deployed carefully with overlapping fields of fire, inflicting maximum damage. The *78. Sturmdivision* made expert use of the *Pakfront* defending *Festerplatz* Orsha.

You may choose your Sturm Anti-tank gun platoon or one of your Sturm Heavy Anti-tank gun platoons to form a Pakfront. Do not deploy the platoon when you would normally. Instead, it is deployed at the same time as Independent Teams.

The 7.5cm PaK40 anti-tank guns of your fifth, anti-tank platoon give the *Sturmkompanie* its own guns. If positioned carefully, these can destroy even heavy tanks. They are also useful as infantry guns, to knock out enemy machine-gun nests.

WEAPONS PLATOONS

STURM MACHINE-GUN PLATOON

COMPANY

HQ Section with:

2 Machine-gun Sections	135 points
1 Machine-gun Section	70 points

OPTION

- Replace Command SMG team with a Command Panzerfaust SMG team for +10 points.

Sturm Machine-gun Platoons may make Combat Attachments to Sturm Platoons.

The battalion's assault machine-gun platoons strengthen the line, forming strong points to stop Soviet infantry and allow the Grenadiers to counterattack. Position them carefully to cover objectives and dig them in quickly. They will be the rock upon which the waves of Soviet infantry break!

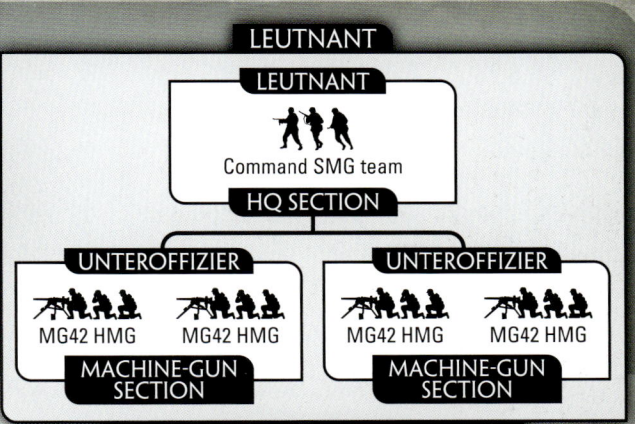

LEUTNANT

LEUTNANT

Command SMG team

HQ SECTION

UNTEROFFIZIER

MG42 HMG   MG42 HMG

MACHINE-GUN SECTION

UNTEROFFIZIER

MG42 HMG   MG42 HMG

MACHINE-GUN SECTION

STURM MACHINE-GUN PLATOON

STURM MORTAR PLATOON

PLATOON

HQ Section with:

3 Mortar Sections	180 points
2 Mortar Sections	125 points
1 Mortar Section	65 points

OPTION

- Replace Command SMG team with a Command Panzerfaust SMG team for +10 points.

The battalion's integral mortar platoon delivers a quick response to calls for fire. It is excellent at pinning the enemy as they advance, or as your troops close to assault.

A well-placed mortar platoon can stop a Soviet advance cold with six mortar tubes in the platoon. A mortar barrage can be devastating as enemy troops approach in the open.

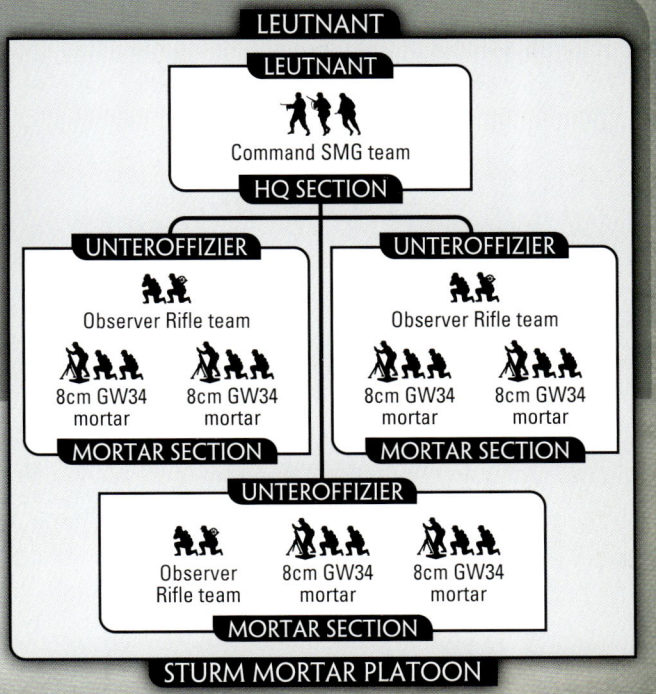

LEUTNANT

LEUTNANT

Command SMG team

HQ SECTION

UNTEROFFIZIER

Observer Rifle team

8cm GW34 mortar   8cm GW34 mortar

MORTAR SECTION

UNTEROFFIZIER

Observer Rifle team

8cm GW34 mortar   8cm GW34 mortar

MORTAR SECTION

UNTEROFFIZIER

Observer Rifle team   8cm GW34 mortar   8cm GW34 mortar

MORTAR SECTION

STURM MORTAR PLATOON

STURM HEAVY ANTI-TANK GUN PLATOON

PLATOON

HQ Section with:

3 Anti-tank Gun Sections	155 points
2 Anti-tank Gun Sections	105 points

OPTIONS

- Replace Command SMG team with a Command Panzerfaust SMG team for +10 points.
- Add RSO tractors for +5 points for the platoon.

Each regiment of the *78. Sturmdivision* has a *14. Kompanie* holding twelve heavy anti-tank guns. The *Sturmdivision* is well stocked with anti-tank guns, far exceeding the allocation for a *Grenadierdivision*.

Sight your guns so you can maximise their fields of fire and inflict devastating losses to your opponent's tanks! The 7.5cm PaK40 anti-tank guns are ideal for ambushing enemy platoons from concealing terrain.

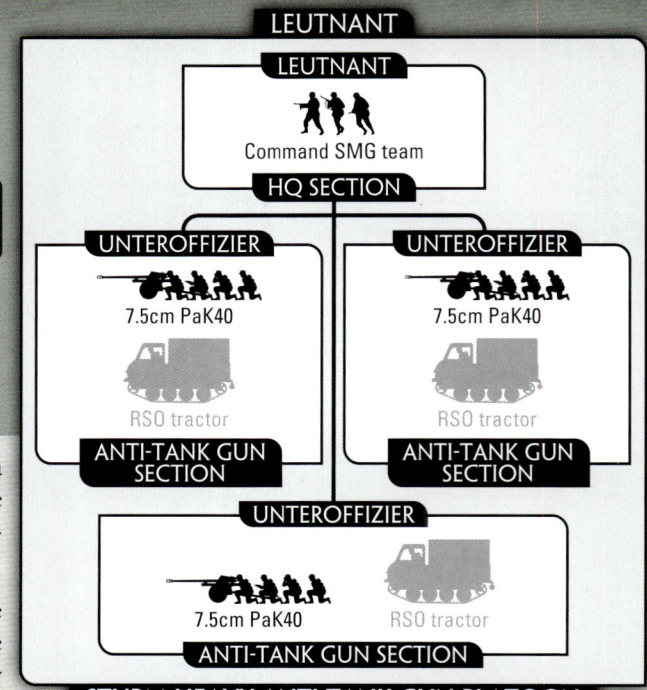

LEUTNANT
LEUTNANT
Command SMG team

HQ SECTION

UNTEROFFIZIER
7.5cm PaK40
RSO tractor
ANTI-TANK GUN SECTION

UNTEROFFIZIER
7.5cm PaK40
RSO tractor
ANTI-TANK GUN SECTION

UNTEROFFIZIER
7.5cm PaK40
RSO tractor
ANTI-TANK GUN SECTION

STURM HEAVY ANTI-TANK GUN PLATOON

REGIMENTAL SUPPORT PLATOONS

STURM INFANTRY GUN PLATOON

PLATOON

HQ Section with:

2 7.5cm leIG18	65 points
2 15cm sIG33	145 points

OPTIONS

- Replace Command SMG team with a Command Panzerfaust SMG team for +10 points.
- Add horse-drawn limbers for +5 points for the platoon.
- Replace both horse-drawn limbers with 3-ton trucks or RSO tractors at no cost.

The 7.5cm leIG18 and 15cm sIG33 infantry guns of the Sturm Regiment's *13. Kompanie* may be few, but they are versatile weapons on defence or when counterattacking. Move them up to pin down the enemy, or take out enemy gun pits, or dig them in to help defend against Soviet armour.

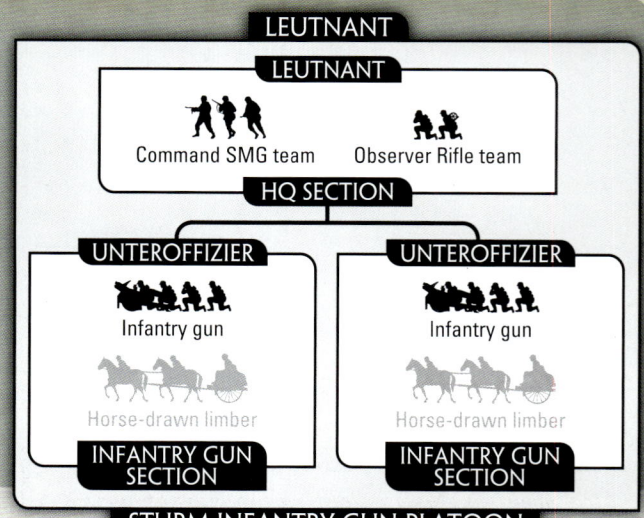

LEUTNANT
LEUTNANT
Command SMG team   Observer Rifle team

HQ SECTION

UNTEROFFIZIER
Infantry gun
Horse-drawn limber
INFANTRY GUN SECTION

UNTEROFFIZIER
Infantry gun
Horse-drawn limber
INFANTRY GUN SECTION

STURM INFANTRY GUN PLATOON

STURM HEAVY MORTAR PLATOON

PLATOON

HQ Section with:

4 12cm sGW43	160 points
2 12cm sGW43	80 points

OPTION

- Add Kfz 15 field car and 3-ton trucks for +5 points for the platoon.

5. Schwere Granatwerfer Bataillon (5th Heavy Mortar Battalion) is armed with thirty-six of the brilliant new 12cm heavy mortars. These are able to destroy even dug-in targets.

The mortar's inability to maintain their fire over a prolonged period is more than overcome by their ability to provide rapid-fire support to the assault troops.

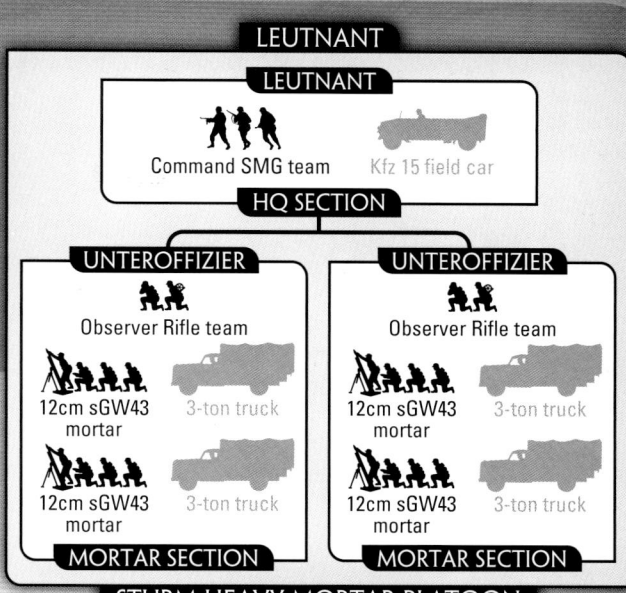

STURM ANTI-AIRCRAFT GUN PLATOON

PLATOON

HQ Section with

3 2cm FlaK38	75 points

OPTIONS

- Replace Command SMG team with a Command Panzerfaust SMG team for +10 points.
- Add Kfz 15 field car and 3-ton trucks for +5 points for the platoon.
- Replace all trucks with RSO tractors at no cost.
- Mount 2cm FlaK38 guns on 3-ton trucks as Portees at no cost.

The *293. Heeres Flakabteilung* is the dedicated anti-aircraft battalion for our *78. Sturmdivision*.

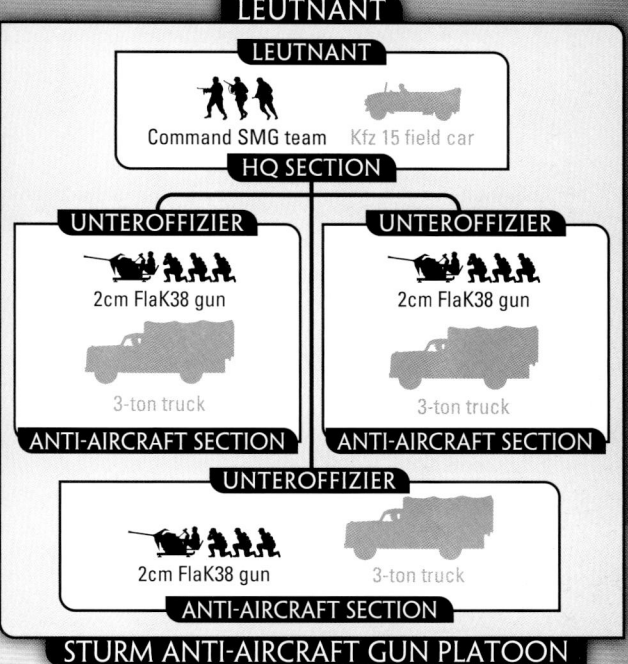

STURM SCOUT PLATOON

PLATOON

HQ Section with:

2 Scout Squads	180 points
1 Scout Squad	110 points

OPTION

- Replace Command Motorcycle MG team with a Command Panzerfaust Motorcycle SMG team for +10 points.

You may model your Motorcycle MG teams with Kübelwagen jeeps instead of motorcycles, they are based the same way as the Motorcycle MG teams and use the same rules.

The *178. Aufklarungsabteilung* will provide you with necessary field reconnaissance.

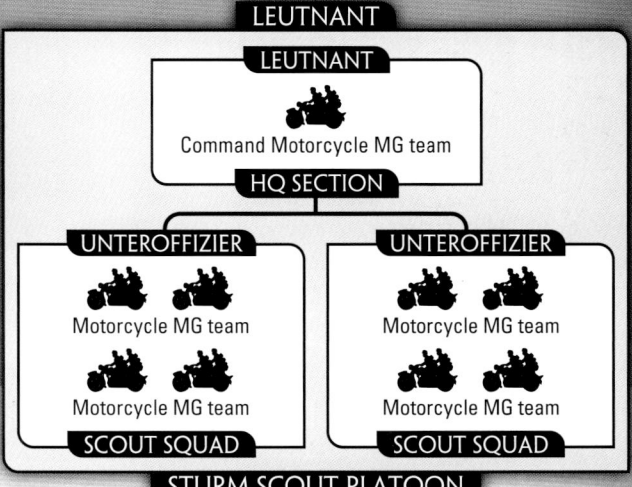

Sturm Scout Platoons use the Motorcycle Reconnaissance rules on page 187 and are Reconnaissance Platoons while mounted.

STUG BATTERIE
Assault Gun Battery

(Tank Company)

HEADQUARTERS

StuG Batterie HQ — 59

You must field one platoon from each box shaded black and may field one platoon from each box shaded grey.

DIVISIONAL SUPPORT PLATOONS

COMBAT PLATOONS

ARMOUR — StuG Platoon — 59

ARMOUR — StuG Platoon — 59

ARMOUR — StuG Platoon — 59

WEAPONS PLATOONS

INFANTRY — Begleit Platoon — 59

INFANTRY — Pionier Platoon — 61

ARMOUR — Schwere Panzer Platoon — 71

ARMOUR — StuG Platoon — 59

ARMOUR — Tank-hunter Platoon — 163

INFANTRY — Sturm Platoon — 53

INFANTRY — Sturm Scout Platoon — 57 — Sturm Platoon — 53

ARTILLERY — Sturm Heavy Mortar Platoon — 57 — Artillery Battery — 166

ARTILLERY — Heavy Artillery Battery — 166

ARTILLERY — Rocket Launcher Battery — 169 — Armoured Rocket Launcher Battery — 170

ANTI-AIRCRAFT — Heavy Anti-aircraft Gun Platoon — 171

ANTI-AIRCRAFT — Sturm Anti-aircraft Gun Platoon — 57

AIRCRAFT — Air Support — 172

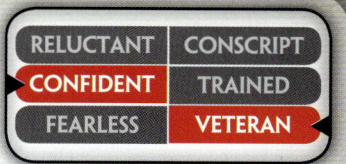

MOTIVATION AND SKILL

The *189. Sturmgeschützabteilung* has a long history working with the *78. Sturmdivision* and has earned recognition as an elite assault gun force. The StuG assault guns and their *Begleit escort* troops are ready to be deployed wherever needed to help the Sturmgrenadiers repel Soviet attacks. A StuG Batterie is rated **Confident Veteran**.

RELUCTANT	CONSCRIPT
CONFIDENT	TRAINED
FEARLESS	VETERAN

HEADQUARTERS

StuG Batterie HQ

HEADQUARTERS

StuG G	95 points

OPTION

- Mount Assault Rifle Tank Escortson StuG G for +15 points.

HAUPTMANN
HAUPTMANN
Company Command StuG G
COMPANY HQ
STUG BATTERIE HQ

COMBAT PLATOONS

StuG Platoon

PLATOON

3 StuG G	285 points

OPTIONS

- Replace one StuG G assault gun with a StuH42 assault gun at no cost.
- Mount Assault Rifle Tank Escorts on each StuG G or StuH42 assault gun for +45 points for the platoon.

The *189. Sturmgeschützabteilung*, the 189th Assault Gun Battalion, has been attached to *78. Sturmdivision* since December 1942. It fought with the *Sturmdivision* during the tough battles at Kursk.

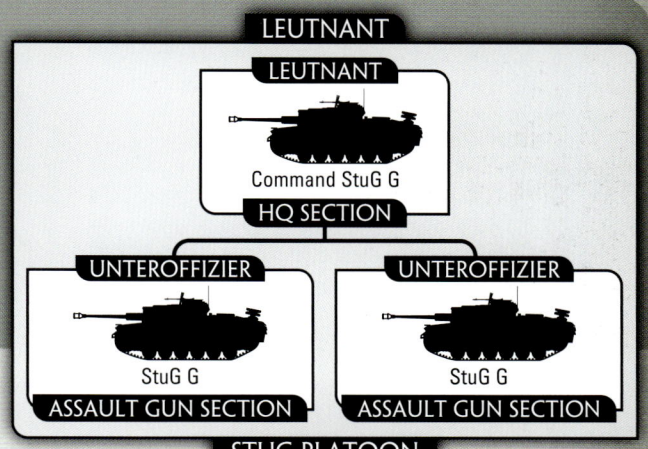

LEUTNANT
LEUTNANT
Command StuG G
HQ SECTION
UNTEROFFIZIER
StuG G
ASSAULT GUN SECTION
UNTEROFFIZIER
StuG G
ASSAULT GUN SECTION
STUG PLATOON

WEAPONS PLATOONS

BEGLEIT Platoon

PLATOON

3 Begleit Squads	310 points
2 Begleit Squads	220 points

Your force may not contain a Begliet Platoon if any of your StuG Platoons have Tank Escort teams.

At the start of the game before deployment you may select not to use your Begleit Platoon during the game. Instead you may give your StuG Batterie HQ and all your StuG Platoons Assault Rifle Tank Escorts.

189. Sturmgeschützabteilung was the first unit to use *Begleit* (pronounced be-glite), or escort, tank riders to escort their assault guns. Sometimes they fought dismounted away from the assault gun as the battle conditions dictated.

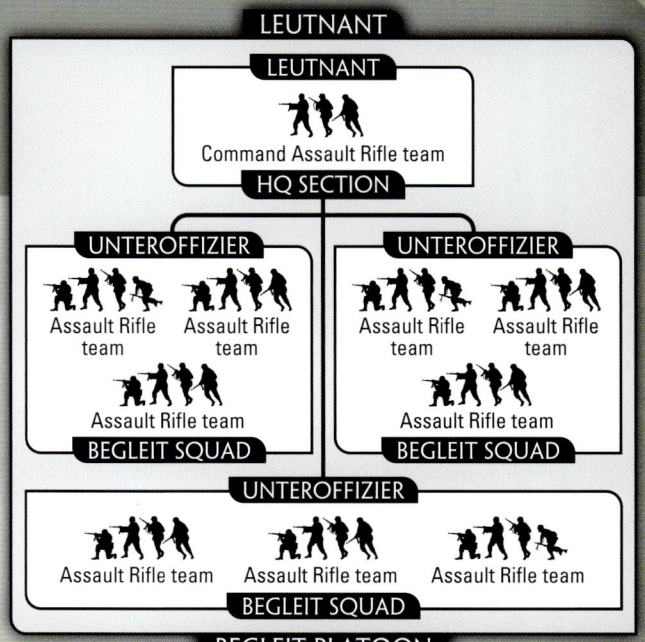

LEUTNANT
LEUTNANT
Command Assault Rifle team
HQ SECTION
UNTEROFFIZIER
Assault Rifle team / Assault Rifle team / Assault Rifle team
BEGLEIT SQUAD
UNTEROFFIZIER
Assault Rifle team / Assault Rifle team / Assault Rifle team
BEGLEIT SQUAD
UNTEROFFIZIER
Assault Rifle team / Assault Rifle team / Assault Rifle team
BEGLEIT SQUAD
BEGLEIT PLATOON

PIONIERKOMPANIE
ENGINEER COMPANY

(INFANTRY COMPANY)

HEADQUARTERS

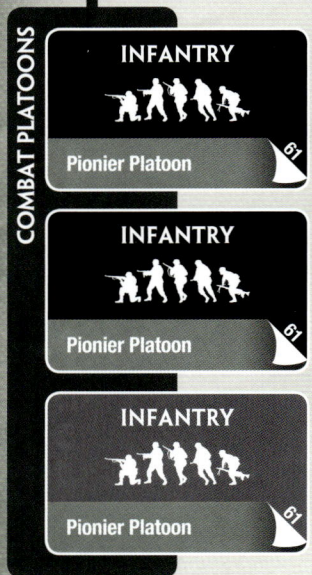

Pionierkompanie HQ *61*

You must field one platoon from each box shaded black and may field one platoon from each box shaded grey.

COMBAT PLATOONS

INFANTRY

Pionier Platoon *61*

INFANTRY

Pionier Platoon *61*

INFANTRY

Pionier Platoon *61*

ALLIED PLATOONS

Luftwaffe Platoons in your force are Allies and follow the Allies rules in the rulebook.

DIVISIONAL SUPPORT PLATOONS

ARMOUR

Panzer Platoon *73*

Schwere Panzer Platoon *71*

Radio-control Tank Platoon *162*

ARMOUR

StuG Platoon *59*

Sturmpanzer Platoon *162*

Tank-hunter Platoon *163*

Anti-tank Gun Platoon *165*

INFANTRY

Grenadier Platoon *27*

Fallschirmjäger Platoon *33*

Sturm Platoon *53*

ARTILLERY

Artillery Battery *166*

Heavy Artillery Battery *166*

ARTILLERY

Rocket Launcher Battery *169*

Armoured Rocket Launcher Battery *170*

Artillery Battery *166*

ANTI-AIRCRAFT

Anti-aircraft Gun Platoon *171*

ANTI-TANK

Heavy Anti-aircraft Gun Platoon *171*

Luftwaffe Heavy Anti-aircraft Gun Platoon *173*

Heavy Anti-tank Gun Platoon *165*

AIRCRAFT

Air Support *172*

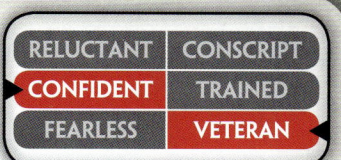

MOTIVATION AND SKILL

*Pioniers are tough fighters, as well trained and equipped to lead assaults as they are for other hazardous tasks like laying and clearing minefields. They know what they are about and are confident of their ability to do it. A Pionierkompanie is rated as **Confident Veteran**.*

RELUCTANT	CONSCRIPT
CONFIDENT	TRAINED
FEARLESS	**VETERAN**

HEADQUARTERS

PIONIERKOMPANIE HQ

HEADQUARTERS

Company HQ	45 points

OPTIONS

- Replace either or both Command SMG teams with Command Panzerknacker SMG teams for +5 points per team or Command Panzerfaust SMG teams for +10 points per team.
- Add Mortar Section for +55 points.
- Add Machine-gun Section for + 60 points.

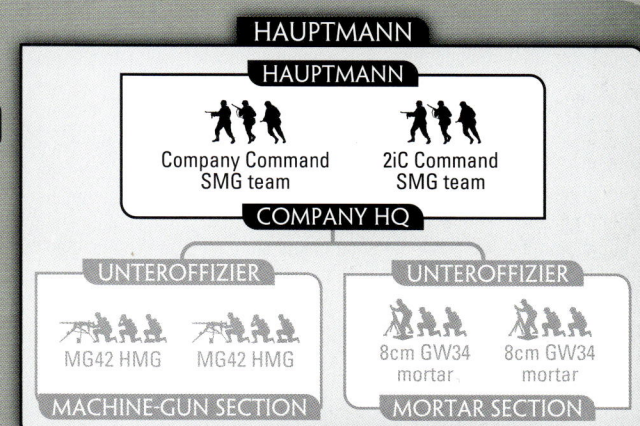

The *Pionierkompanie* is composed of tough well-trained combat engineers ready to do two things: undertake dangerous engineering assignments while under fire, and storm enemy positions by close assault.

The 'black' pioneers of the *Infanteriedivision* perform all the specialist engineering tasks. They lay minefields, prepare bunkers, and create other defensive positions.

When the Grenadiers counterattack to regain lost positions, the pioneers lead the way with their flame-throwers.

COMBAT PLATOONS

PIONIER PLATOON

PLATOON

HQ Section with:

3 Pioneer Squads	235 points
2 Pioneer Squads	165 points

OPTIONS

- Replace Command Pioneer Rifle team with Command Pioneer Panzerknacker SMG team for +5 points or Command Pioneer Panzerfaust SMG team for +10 points.
- Equip one Pioneer Rifle team with a Goliath demolition carrier in addition to its normal weapons for +30 points.
- Add Pioneer Supply horse-drawn wagon for +20 points, or Pioneer Supply RSO tractor for +25 points.

You may replace up to one Pioneer Rifle team per Pioneer Squad with a Flame-thrower team at the start of the game before deployment.

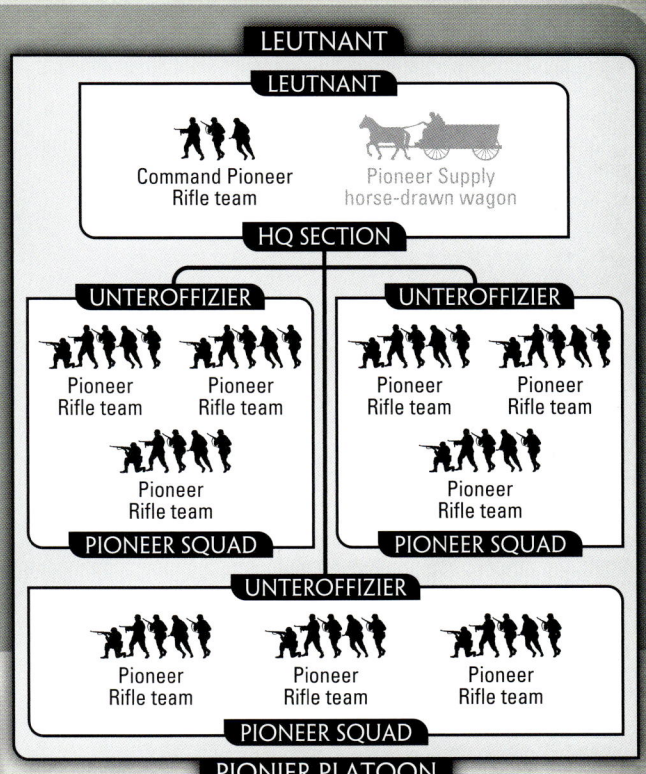

General Der Panzertruppen Von Saucken directs his Panzers to the front line...

... while RSO auf PaK40 tank-hunters ambush and slow down the enemy advance.

...giving the infantry time to prepare their defences...

Feldwebel Windgruber leads the counter-assault to drive the communists back.

BAGRATION BREAKTHROUGH

THE PANZER DIVISIONS FACING OPERATION BAGRATION

When *Feldmarschal* Model took command of Army Group Centre during the chaos of the Soviet breakthrough those panzer divisions that could be quickly pushed into action were sent against the Soviet onslaught. The panzer divisions were used to reinforce ad hoc blocking forces (*Sperrverbände*) and battle groups *(kampfgruppen)*. The first to see action was *20. Panzerdivison* who were already deployed near Bobruisk as the front reserve. As they were pushed back their men joined the blocking forces holding against the Soviet breakthrough.

They were joined at the front by *5. Panzerdivision* under *General der Panzertruppen* von Saucken and the *505. Schwere*

Panzerabteilung (505th Heavy Tank Battalion). They fought around Borisov and Minsk, delaying the breakthrough forces of the Red Army as they sped towards Minsk.

To the south *4. Panzerdivision* and *12. Panzerdivision* arrived to reinforce various *Sperrverband* defending the road between Bobruisk and Baranovichi. They were further reinforced by Hungarian and German cavalry units who arrived from anti-partisan duties to be thrown in the line against Soviet tank, mechanised and cavalry corps rampaging westwards.

THE FALL OF MINSK

On 2 July, fierce fighting erupted to the immediate north of Minsk, the Byelorussian capital. Elements of *5. Panzerdivision* (5th Armoured Division) dutifully contained various threats along the northern flank of the city. From the south, the Soviet 2nd Guards Tank Corps captured Smolevichi before crashing into Minsk. Along the way the Soviets only encountered sporadic German resistance.

Realising the flaws in the *Festerplatz* strategy, Hitler authorised the evacuation of Minsk on 3 July. German troops

chaotically abandoned the city under the protection of *5. Panzerdivision*.

The remnants of the German Fourth Army were trapped in a large pocket east of Minsk, resisting desperately against Soviet efforts to reduce it. By 5 July, the Fourth Army concluded that they would have to attempt a break out. Despite achieving initial successes, on 6 July, the majority of the army's soldiers were either killed or captured.

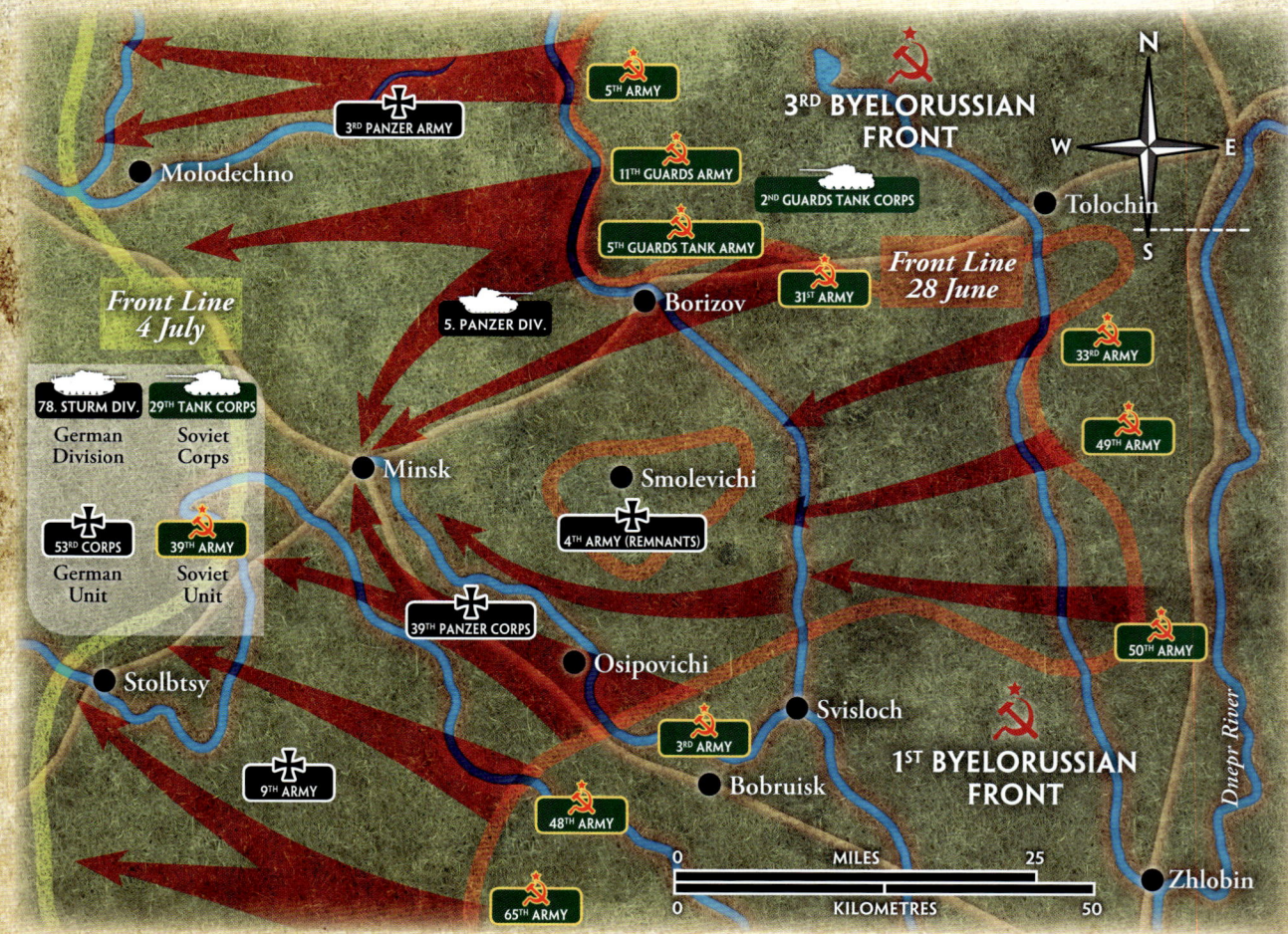

THE AFTERMATH

Hitler's insistence on constructing a rigid defence along a static front using *Festerplatz* tied down the defence and played perfectly into Soviet strategy. The Red Army simply bypassed the fortified cities, trapping the defending units within and then reducing the pockets at will.

Soviet forces advanced extremely rapidly making it almost impossible for the few units breaking out of encirclement to reach the safety of their own lines.

With Minsk liberated and the German Fourth Army reduced, Soviet forces paused to reorganise and prepare to push into Poland.

The operation was an unqualified Soviet success, not only liberating the last areas of Soviet territory held by the Germans, but also bringing about the total collapse of Army Group Centre. Of the thirty four German divisions in the field, twenty eight were destroyed or rendered ineffective.

Total casualties for the Germans reached over 300,000 troops killed or captured, while the Soviet casualty count on all fronts numbered 765,000. In less than two weeks the Red Army managed to break open the front, leaving the road to Berlin wide open.

TIGERS

As well as *505. Schwere Panzerabteilung* (505th Heavy Tank Battalion), which fought to halt the Soviet Operation Bagration breakthrough, other *Heer* Tiger battalions played an important role in the fighting on the Eastern Front in 1944 and 1945.

501. SCHWERE PANZERABTEILUNG

The first battalion to receive Tiger I tanks was also the first battalion to receive the Königstiger or Tiger II tanks. From 1942 to 1944 *501. Schwere Panzerabteilung* fought in Africa, the Eastern Front, and finally in Poland.

Part of the battalion received its first Tiger II tanks while another part was still fighting in Minsk during Operation Bagration. Initially equipped with 45 Tiger II, the battalion was attached to *16. Panzerdivision* in the counterattack against the Sandomierz bridgehead on the Vistula River.

In December 1944 the battalion was redesignated the *424. Schwere Panzerabteilung* and moved to the Western Front.

502. SCHWERE PANZERABTEILUNG

502. Schwere Panzerabteilung was formed in 1943 dispatched to the Leningrad front in July 1943. It supported the troops

of Army Group Narva during the battles for Narva and the Tannenberg Line. It spent the war fighting in the north of the Eastern Front in support of Army Group North.

The most famous of its Tiger I E tank commanders was Otto Carius, who was awarded a Knight's Cross on 4 May 1944 for his efforts during the Narva fighting.

503. SCHWERE PANZERABTEILUNG

503. Schwere Panzerabteilung (later *Feldherrnhalle*) was initially equipped with Tiger I tanks and Panzer IIIs. In 1944, it was re-equipped with the new Tiger II. The battalion saw action on the Eastern and Western Fronts.

During early November 1944 *13. Panzerdivision* fought alongside *503. Schwere Panzerabteilung* around Hatvan and Jakohalma during the battles to halt the Soviet advance on Budapest.

503. Schwere Panzerabteilung was heavily involved in the fighting around Debrecen in October 1944 leading the counterattack with the *24. Panzerdivision* and *4. Polizei SS-Panzergrenadierdivision*.

Later the battalion became part of the newly formed *Panzerkorps Feldherrnhalle*.

KAMPFGRUPPEN

A *Kampfgruppe* (battle group) was an ad-hoc combined arms formation, usually employing a combination of tanks, infantry, anti-tank and artillery elements. It was generally organised for a particular task or operation.

A *Kampfgruppe* was usually named after its commanding officer or its parent division. The battle group could range in size from a corps to a company, but the most common was an *Abteilung* (battalion) sized formation.

During Operation Bagration, the Germans employed these combined arms formations to stymy the rapid advance of Soviet forces. For the first few weeks of July 1944, these formations formed the backbone of the German defences.

The *Kampfgruppen* formed during Operation Bagration were built around infantry and security divisions, and even headquarters units that survived the initial Soviet onslaught. These were then reinforced by the 5th and 12th Panzer Divisions brought up from reserve.

SPERRVERBAND KAMPFGRUPPEN

KAMPFGRUPPE LINDIG, SPERRVERBAND BERGEN, SPERRVERBAND MEINECKE, KAMPFGRUPPE BERCKEN, KAMPFGRUPPE VON VORMANN, KAMPFGRUPPE BIRKEL, PANZERGRUPPE HOPPE

The *9. Armee Sperrverbände* (Blocking Forces) consisted of *Kampfgruppen* from the *XXXXI* and *LV Armeekorps*. Primarily constituted from the remnants of the forward infantry divisions, these *Sperrverbände* were supported by a varied selection of troops and vehicles.

These included the retreating mobile assets of the infantry divisions, independent anti-tank battalions and *Luftwaffe* Flak battalions. Eventually the *Kampfgruppen* were supported by elements of the newly arrived *12. Panzerdivision*, as well as by retreating elements of *20. Panzerdivision* after the fall of Bobruisk.

Additional *Kampfgruppen* were formed to block the routes into Minsk using regiments from *9. Armee* Weapons School and *390. Feldausbildungs* (field replacement) Division. This replacement unit was made from veteran leaders of field units, returning wounded and new recruits. They fought like demons possessed.

Another *Sperrverband Kampfgruppe* known as *Panzergruppe* Hoppe was formed in the area occupied by *3. Panzerarmee*. It contained units from *252. Infanteriedivision* (infantry), *391. Sicherungsdivision* (Security), a battalion of StuG assault guns, and an independent battalion of *Hornisse* tank-hunters.

KAMPFGRUPPE VON SAUCKEN

KAMPFGRUPPE LENDIL, KAMPFGRUPPE METZ, KAMPFGRUPPE I, KAMPFGRUPPE II, KAMPFGRUPPE III

General der Panzertruppen Dietrich von Saucken had gained the reputation as a field commander who could stop any Soviet offensive. When *Generalfeldmarschall* Model took over Army Group centre, *General* von Saucken was immediately tasked to develop a mobile counterattacking force.

Taking over the reserve forces available to keep the bridges over the Berezina River and the railroad to Minsk open, von Saucken built *Kampfgruppen* to meet the oncoming spearheads of the Soviet mechanised and armoured corps.

All told, he created five separate *Kampfgruppen*, mixing the assets of *5. Panzerdivision* with various other units to take the brunt of Soviet armour. His *Kampfgruppen* took enormous casualties. However, they inflicted many more and ultimately allowed the remaining divisions of *4. Armee* to escape the closing pincers of the ever-advancing Soviet forward detachments. *Kampfgruppe von Saucken* later became the *XXXIX Panzer Korps*.

SICHERUNGS AND WALKÜRE

Until reserve infantry divisions could be brought up to form *Sperrverbände*, the Germans utilized their *Sicherungs* (Security) Divisions as stopgap forces. Although lacking the full support options of a German infantry division, small *Kampfgruppen* were organized from these as emergency forces.

The *Walküre* or German homeguard infantry battalions were brought up to form *Sperrverband Kampfgruppen*. These infantry regiments had their full complement of soldiers but had never faced combat.

German Karabiner K98k Rifle with Bayonet

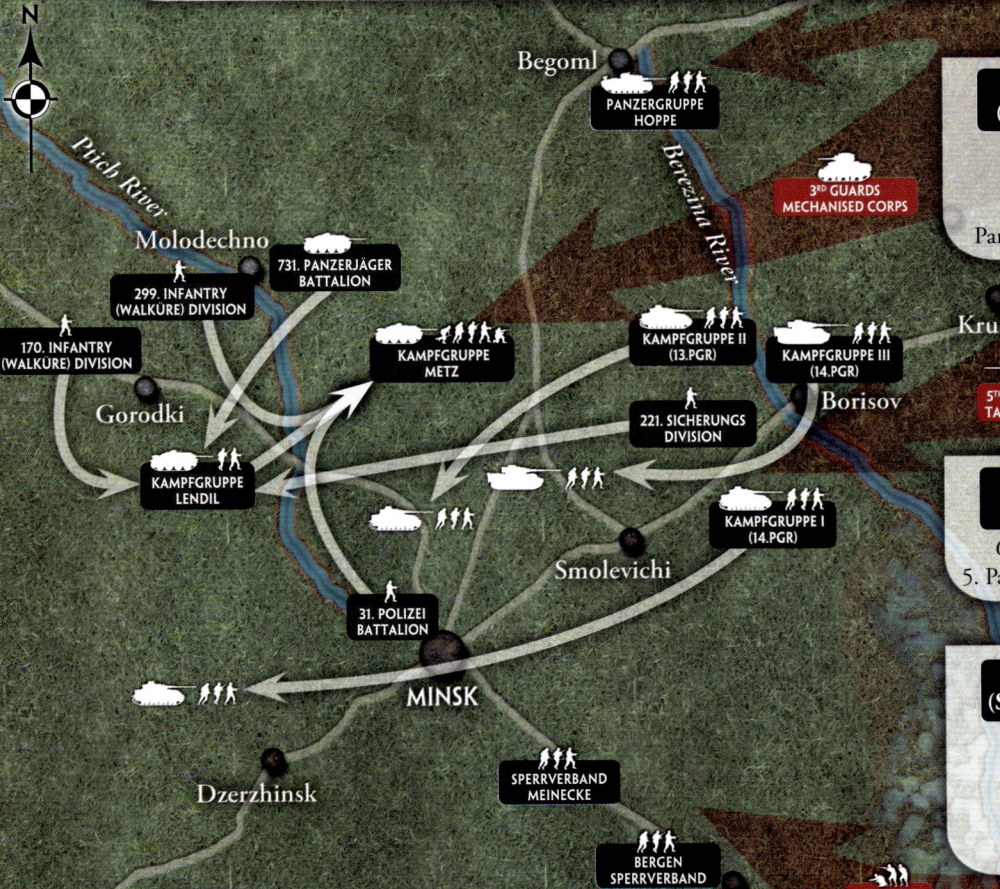

KAMPFGRUPPE VON SAUCKEN KAMPFGRUPPEN

Kampfgruppe Metz
(Walküre or Sicherungs)
170. Infantry Division
14. Infantry Division
299. Infantry Division
31. Polizei Battalion
221. Sicherungs Division
18. FlaK Battalion

Kampfgruppe Lendil
(Walküre or Sicherungs)
170. Infantry Division
731. Panzerjäger Battalion
221. Sicherungs Division

Kampfgruppe I
(14. Panzergrenadierregiment, 5. PD)
II 14. Panzergrenadier Regiment
5. Ersatz Pionierabteilung
elements of 31. Panzer Regiment
elements of 85. Pionierabteilung

Kampfgruppe II
(13. Panzergrenadierregiment, 5. PD)
13. Panzergrenadier Regiment
89. Panzerpionier Battalion
elements of 31. Panzerregiment

Kampfgruppe III
(31. Panzerregiment, 5. PD)
I 14. Panzergrenadier Regiment
505. Schwere Panzerabteilung
elements of 31. Panzer Regiment

Dvina River

VITEBSK

Begoml

PANZERGRUPPE HOPPE

3RD GUARDS MECHANISED CORPS

Panzergruppe Hoppe
(Sperrverband/Sicherungs)
252. Infantry Division
391. Sicherungs Division
Sturmgeschütz Battery
Panzerjäger Company (Hornisse)

Ptich River

Molodechno

731. PANZERJÄGER BATTALION

299. INFANTRY (WALKÜRE) DIVISION

170. INFANTRY (WALKÜRE) DIVISION

Gorodki

KAMPFGRUPPE LENDIL

KAMPFGRUPPE METZ

221. SICHERUNGS DIVISION

KAMPFGRUPPE II (13.PGR)

KAMPFGRUPPE III (14.PGR)

Borisov

Krupki

Berezina River

5TH GUARDS TANK ARMY

KAMPFGRUPPE I (14.PGR)

Smolevichi

Sperrverband Meinecke
(Sperrverband)
60. Panzergrenadier Division
5. Panzergrenadier Regiment, 12. PD

31. POLIZEI BATTALION

MINSK

Dzerzhinsk

SPERRVERBAND MEINECKE

Sperrverband Bergen
(Sperrverband or Sicherungs)
390 Feldausbildungs HQ
791. Sicherungs Battalion
915. Sicherungs Battalion
2/603. Sicherungs Regiment
3/22. Artillerie Regiment

BERGEN SPERRVERBAND

Marina Gorka

65TH ARMY

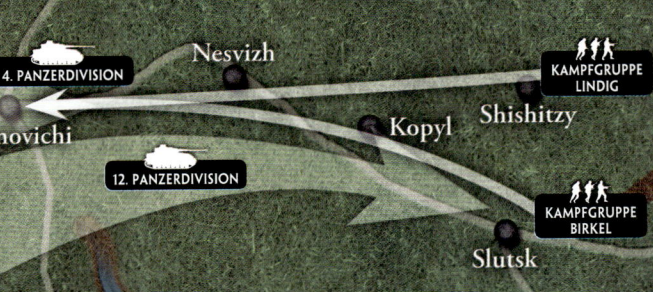

4. PANZERDIVISION

Nesvizh

KAMPFGRUPPE LINDIG

Bobruisk

28TH ARMY

Baranovichi

Kopyl

Shishitzy

1ST GUARDS MECHANISED CORPS

12. PANZERDIVISION

KAMPFGRUPPE VON VORMANN

KAMPFGRUPPE BIRKEL

Slutsk

Kampfgruppe von Vormann
(Sperrverband)
102. Infantry Division
129. Infantry Division
292. Infantry Division

KAMPFGRUPPE VON BERCKEN

Glusk

Kampfgruppe Birkel
(Sperrverband)
35. Infantry Division
9. Armee Weapons School

Kampfgruppe Lindig
(Sperrverband)
9. Armee Weapons School
390 Feldausbildungs Division
12. Panzerdivision incl.
I/25 Panzergrenadier Regiment
(armoured)

Kampfgruppe von Bercken
(Sperrverband)
102. Infantry Division
35. Infantry Division
129. Infantry Division

| 0 | Miles | 20 | 40 | 60 | 80 | 100 | 120 |
| 0 | Kilometres | 40 | 60 | 80 | 100 | 120 | 140 | 160 | 180 | 200 |

FELDWEBEL LUDWIG WINDGRUBER

Born in Oppeln, Upper Silesia in 1922, Windgruber joined the *Wehrmacht* in 1940 just before the invasion of France. He was initially assigned to a rifle company of the 84th Infantry Regiment, 8th Infantry Division. Ludwig received his baptism under fire in Flanders when the 84th Regiment marched through Belgium on its way around the Maginot Line as part of *16. Armee*.

It was here, around the towns of Verdun and Metz that Ludwig began to show his leadership abilities, helping his fellow soldiers tie down the French divisions in the Maginot Line while the rest of the German forces marched on. With the fall of Paris and the capitulation of France, Ludwig found himself polishing his newly acquired *Obergrenadier* pip and witnessing the *Führer's* acceptance of the French surrender. For the next six months the 84th regiment did garrison duty in France.

In December 1940 the division was sent to Poland. As a veteran of Bryansk, Vyazma, and the bitter battle around Moscow, *Obergrenadier* Windgruber earned his first medals for courage and wounds. During the battle of Moscow in 1941 he was wounded by shrapnel and sent back to

Germany to recover. Prior to returning to the division he earned his black epaulettes at the Pioneer Training Battalion. When he was ready to return to the front he had found that his old regiment had been transferred to the 102nd Infantry Division, and subsequently Windgruber found himself in the *2. Kompanie* of the 102nd Pioneer Battalion now wearing the black epaulettes of a Pioneer.

For the next two years Windgruber fought across the steppes and marshes of Russia and the Ukraine. Through attrition, good fortune and veteran instinct, *Feldwebel* Windgruber found himself, at the age of 22, the *Alter Häse* (Old Hare, a seasoned and crafty survivor) of *2. Kompanie*. His combat experience was known throughout the battalion and any officer wishing to succeed would first seek out the *Alter Grüber* (Old Digger) before attempting any combat patrols. Windgruber's ability to lead an attack and assault Soviet tanks was legend within the division.

The Iron Cross and *Panzerknacker* badge on his uniform told but a small part of the close combat he had survived.

CHARACTERISTICS

Windgruber replaces the 2iC in a *Sperrverband* or *Grenadierkompanie* for +30 points. He is a Warrior and a Pioneer Panzerfaust SMG team rated **Confident Veteran**.

GREEN HELL (GRÜN HÖLLE)

Grün Hölle, literally *Green Hell*, referred to the constant close combat in the green fields of Russia. *Feldwebel* Windgruber learned his lessons well instilling the ferocity needed to turn back Soviet attacks while keeping his men alive to fight again.

> All teams from a Sperr, Sperr Pionier, or Grenadier Platoon joined by, or in a Kampfgruppe formed by, Feldwebel Windgruber hit on 2+ in assaults.

OLD HARE (ALTER HÄSE)

Feldwebel Windgruber was an 'Old Hare' who could inspire his troops to attack when the odds favoured the opponent. This can destroy your enemy's will to fight and disrupt their plans before they can execute them. *Feldwebel* Windgruber's leadership enabled him to take the fight to the enemy before they could overrun the defences.

> Feldwebel Windgruber may launch an Assault from 6"/15cm away from enemy teams. He moves up to 6"/15cm when Charging into Contact or Counterattacking. Any platoon he has joined still move 4"/10cm in assaults, often meaning that they don't get into the fight until the platoon Counterattacks.

GENERAL DER PANZERTRUPPEN
DIETRICH VON SAUCKEN

In 1910 at the age of eighteen, Dietrich von Saucken entered the Imperial Army as an officer cadet. After two years he was appointed *Leutnant* serving as a cavalry officer of the 'Kaiser Wilhelm I' Cavalry Regiment. Wounded seven times in the First World War von Saucken rose to the rank of *Rittmeister* (Captain). He remained with the army between the wars as a tactics teacher at the Hannover War School.

Oberst von Saucken led *2. Reiterregiment* (2nd Cavalry Regiment) throughout the fighting in Poland and France. In November 1940 he was posted to *4. Schützenbrigade* (4th Rifle Brigade) of *4. Panzerdivision* until he took command of the division in December 1941. Promoted to *Generalmajor* (Major General) on 1 January 1942, von Saucken was severely wounded the next day. He returned to duty as commander of the Cavalry School in August 1942.

After promotion to *Generalleutnant* (Lieutenant General) in 1943, von Saucken returned to the front to command *4. Panzerdivision* once again. For the next year, von Saucken's division distinguished itself in a number of decisive engagements, culminating with the battle of Kovel. Becoming

known as a defensive and counterattack specialist, after halting numerous Red Army offensives, von Saucken gained the Oak Leaves and Swords to his Knight's Cross.

Von Saucken's mobile counterattack expertise was once again called upon during the Soviet summer offensive of 1944. He led *Kampfgruppe von Saucken* against Marshal Rotmistrov's 5th Guards Tank Army. Though casualties were high, and while stopping the overwhelming force of the Soviet advance proved impossible, von Saucken's blocking actions allowed many German units to escape westward, avoiding capture.

An aristocratic officer, the monocle-wearing General von Saucken came from Prussian nobility and had little time for the Nazis political agenda. He was one of the few high-ranking members of the *Wehrmacht* who was neither intimidated by Hitler's intense ravings nor hypnotized by the charisma of the *Führer*.

In early 1945 when the Vistula front collapsed, von Saucken once more successfully smashed his way through the Russian lines and led his corps, consisting of the *Grossdeutschland* and *Hermann Göring* divisions, back to the Oder and safety.

CHARACTERISTICS

General der Panzertruppen von Saucken is mounted in a Sd Kfz 250 half-track. He is a Warrior and a Higher Command SMG team rated as **Fearless Veteran**. Von Saucken may join a German company for +65 points.

COUNTERATTACK EXPERT

Von Saucken was repeatedly called upon by *Der Führer* to prevent Soviet breakthroughs with thrown together forces. Von Saucken's proven ability to organize and launch mobile counterattack became his forte, having demonstrated his tactical skill many times on the Eastern Front.

> *All platoons within Command Distance of von Saucken may reroll any failed Skill Tests to make Stormtrooper moves.*

PANZERS MARCH

Von Saucken's keen tactical abilities and strong leadership skills provided his forces the opportunities to provide decisive force when needed to stop Soviet spearheads. Time and again, across the Eastern Front, von Saucken's units managed to apply the right force at the right point in meeting Soviet advancing formations.

> *Each turn von Saucken may re-roll one die rolled to receive Reserves for his company.*

SCHWERE PANZERKOMPANIE
HEAVY TANK COMPANY

(TANK COMPANY)

HEADQUARTERS

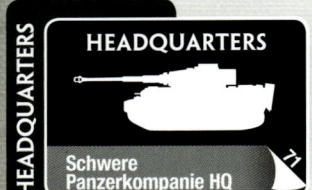

Schwere Panzerkompanie HQ — 71

You must field one platoon from each box shaded black and may field one platoon from each box shaded grey.

COMBAT PLATOONS

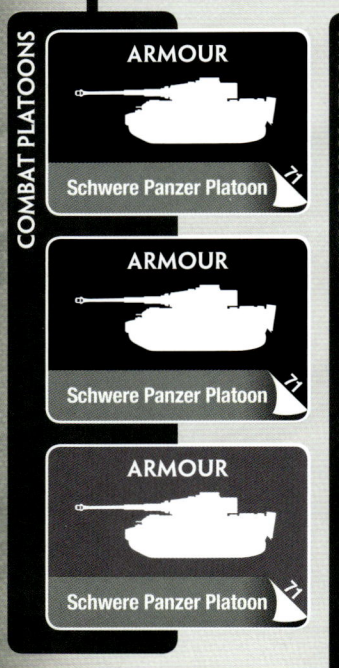

ARMOUR
Schwere Panzer Platoon — 71

ARMOUR
Schwere Panzer Platoon — 71

ARMOUR
Schwere Panzer Platoon — 71

WEAPONS PLATOONS

INFANTRY
Panzer Pioneer Platoon — 74

ANTI-AIRCRAFT
Panzer Anti-aircraft Gun Platoon — 75

RECONNAISSANCE
Panzer Scout Platoon — 75
Schwere Panzer Armoured Scout Platoon — 25

DIVISIONAL SUPPORT PLATOONS

ARMOUR
Panzer Platoon — 73
Panther Platoon — 74

ARMOUR
Veteran Tank-hunter Platoon — 164
Tank-hunter Platoon — 163

INFANTRY
Gepanzete Panzergrenadier Platoon — 77
Panzergrenadier Platoon — 81
SS-Freiwilligen-Panzergrenadier Platoon — 39
Gepanzerte Panzerpionier Platoon — 87
Panzerpionier Platoon — 89
Grenadier Platoon — 27

ARTILLERY
Motorised Artillery Battery — 167
Motorised Heavy Artillery Battery — 167
Armoured Artillery Battery — 168
Armoured Heavy Artillery Battery — 168

ARTILLERY

Motorised Artillery Battery — 167
Rocket Launcher Battery — 169
Armoured Rocket Launcher Battery — 170

ANTI-AIRCRAFT

Luftwaffe Heavy Anti-aircraft Gun Platoon — 173
Luftwaffe Light Anti-aircraft Gun Platoon — 173

ANTI-AIRCRAFT

Luftwaffe Heavy Anti-aircraft Gun Platoon — 173
Luftwaffe Light Anti-aircraft Gun Platoon — 173

AIRCRAFT

Air Support — 172

---

505. SCHWERE PANZERABTEILUNG

A company of Tigers would do wonders for morale.
— Feldwebel Windgruber, 102. Infanteriedivision.

Though there were never enough to go around, the Tiger battalions proved themselves a serious problem for Soviet forward units. Kill ratios of three or four to one were common for Tiger units across the Eastern Front. Even a platoon of four could hold the enemy at bay or stop an enemy column.

The *505. Schwere Panzerabteilung* spent the Winter and Spring of 1943-44 on the Eastern Front countering small Soviet incursions into the German lines. In early April it found itself in Orsha preparing for deployment south of Pinsk to face a suspected Soviet offensive into the Ukraine. The offensive never materialised, so after a few skirmishes the battalion found itself relocated west towards the Polish border to refit. Here the battalion was brought back up to full strength and by 17 June sported 51 new Tiger I heavy tanks.

On 24 June, the battalion joined *Kampfgruppe* von Saucken and 5. *Panzerdivision*. Deploying on 26 June under air attack, the battalion fought continuously for the next eight days destroying nearly 130 Soviet tanks. However, by 5 July this constant combat reduced the 505. *Schwere Panzerabteilung* to only 15 operational Tiger tanks.

ALLIED PLATOONS

SS and Luftwaffe Platoons in your force are Allies and follow the Allies rules in the rulebook.

MOTIVATION AND SKILL

The 505. Schwere Panzerabteilung was attached to the 5. Panzerdivision in defence of Byelorussia during the summer of 1944. The 501. Schwere Panzerabteilung was attached to the 16. Panzerdivision in defence of Poland during the summer of 1944. A Schwere Panzerkompanie is rated as **Confident Veteran.**

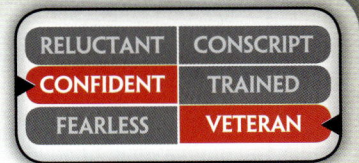

RELUCTANT	CONSCRIPT
CONFIDENT	TRAINED
FEARLESS	**VETERAN**

HEADQUARTERS

SCHWERE PANZERKOMPANIE HQ

HEADQUARTERS

2 Tiger I E	430	points
1 Tiger I E	225	points
2 Königstiger (Henschel)	695	points
1 Königstiger (Henschel)	350	points

OPTION

- Add a Bergepanther recovery vehicle for +15 points.

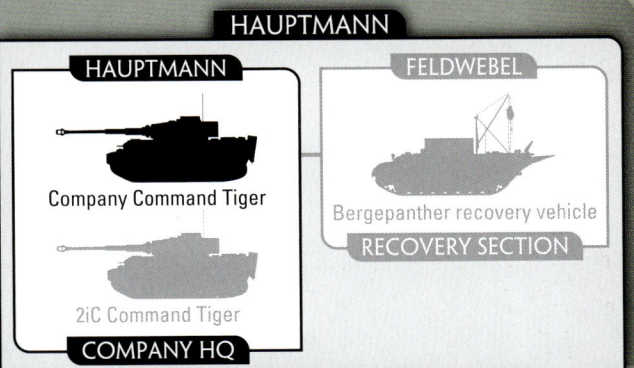

SCHWERE PANZERKOMPANIE HQ

The Company Command tank always has two Tiger Ace Skills (see page 184). Roll 2 dice and any roll of a 6 (or ♠) allows you to choose your Tiger Ace skill.

A *Schwere Panzerkompanie* (heavy tank company—pronounced shvair-rer pant-serr kom-pan-ee) can easily spoil the plans of even the most confident enemy tank commander. Their firepower can wreak havoc way beyond their numbers. They excel in either an offensive or defensive role. And though slow, can turn the tide in their favour in any tank battle.

COMBAT PLATOONS

SCHWERE PANZER PLATOON

PLATOON

4 Tiger I E	860	points
3 Tiger I E	645	points
2 Tiger I E	430	points
4 Königstiger (Henschel)	1380	points
3 Königstiger (Henschel)	1035	points
2 Königstiger (Henschel)	690	points
1 Königstiger (Henschel)	345	points

Remember to roll for your Tiger Ace Skills before each game.

All the Schwere Panzer Platoons in your force must be entirely equipped with the same type of Tiger as your Schwere Panzerkompanie Command tank.

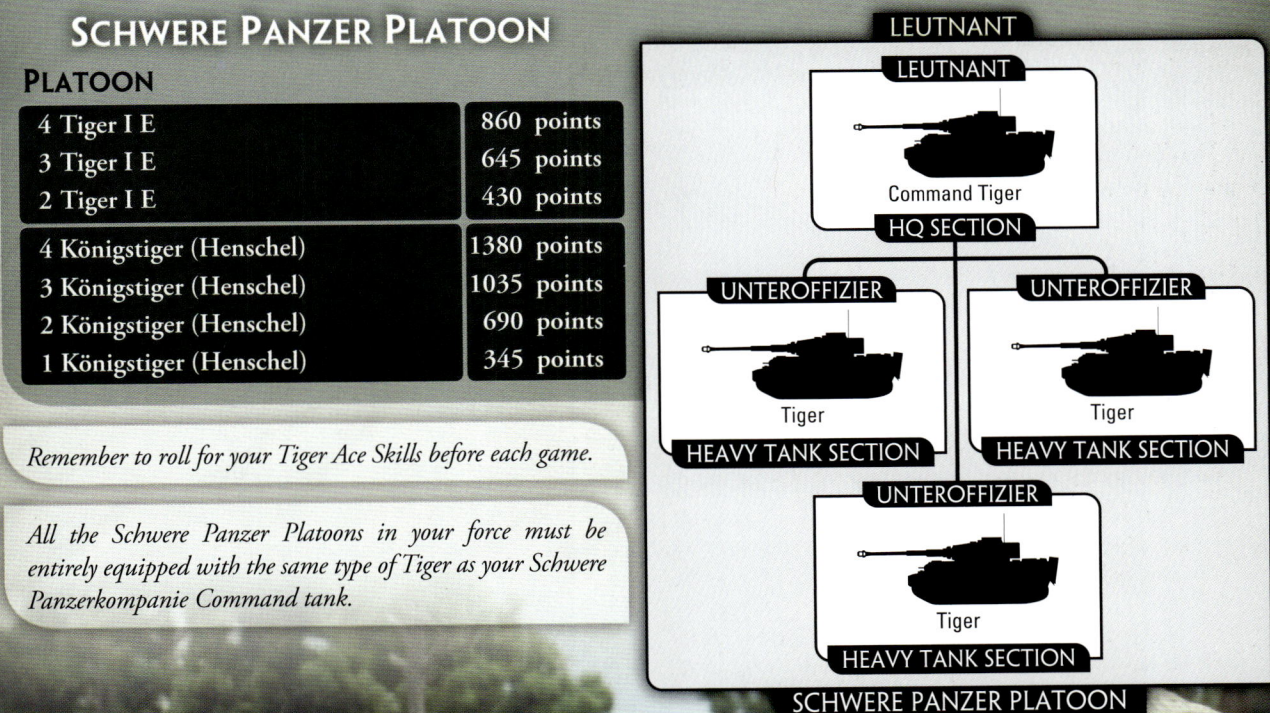

SCHWERE PANZER PLATOON

PANZERKOMPANIE
TANK COMPANY

(TANK COMPANY)

HEADQUARTERS

Panzerkompanie HQ *73*

You must field one platoon from each box shaded black and may field one platoon from each box shaded grey.

DIVISIONAL SUPPORT PLATOONS

COMBAT PLATOONS

ARMOUR

Panzer Platoon *73*
Panther Platoon *74*

ARMOUR
Panzer Platoon *73*
Panther Platoon *74*

ARMOUR
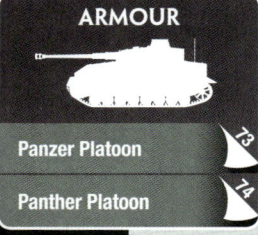
Panzer Platoon *73*
Panther Platoon *74*

ARMOUR

Panzer Platoon *73*

WEAPONS PLATOONS

INFANTRY

Panzer Pioneer Platoon *74*

RECONNAISSANCE
Panzer Scout Platoon *75*

ANTI-AIRCRAFT
Panzer Anti-aircraft Gun Platoon *75*

RECONNAISSANCE

Light Panzerspäh Platoon *91*
Half-tracked Panzerspäh Platoon *91*
Tracked Panzerspäh Platoon *92*
Puma Panzerspäh Platoon *92*
Heavy Panzerspäh Platoon *92*

ALLIED PLATOONS

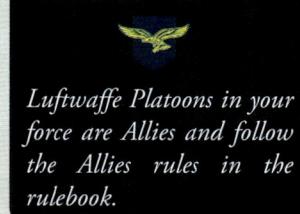

Luftwaffe Platoons in your force are Allies and follow the Allies rules in the rulebook.

ARMOUR

Schwere Panzer Platoon *71*
Radio-control Tank Platoon *162*

ARMOUR

Sturmpanzer Platoon *162*
Tank-hunter Platoon *163*
Anti-tank Gun Platoon *165*

INFANTRY

Gepanzerte Panzergrenadier Platoon *77*
Panzergrenadier Platoon *81*
Fallschirmjäger Platoon *33*
Aufklärungs Platoon *93*
Gepanzerte Aufklärungs Platoon *93*
Walküre Platoon *169*

INFANTRY

Gepanzerte Panzerpionier Platoon *87*
Panzerpionier Platoon *89*

ARTILLERY

Motorised Artillery Battery *167*
Motorised Heavy Artillery Battery *167*
Armoured Artillery Battery *168*
Armoured Heavy Artillery Battery *168*
Armoured Train *122*

ARTILLERY

Rocket Launcher Battery *169*
Armoured Rocket Launcher Battery *170*
Motorised Artillery Battery *167*

ANTI-AIRCRAFT
Anti-aircraft Gun Platoon *171*

ANTI-TANK

Heavy Anti-aircraft Gun Platoon *171*
Luftwaffe Heavy Anti-aircraft Gun Platoon *173*
Heavy Anti-tank Gun Platoon *165*

AIRCRAFT
Air Support *172*

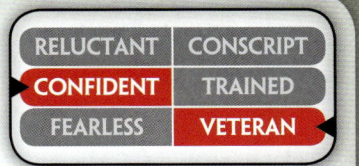
MOTIVATION AND SKILL

The Panzertruppen *have years of combat behind them and are confident of their ability to utilise their superior tactics and equipment to defeat their enemies. A Panzerkompanie is rated as* **Confident Veteran.**

RELUCTANT	CONSCRIPT
CONFIDENT	TRAINED
FEARLESS	**VETERAN**

HEADQUARTERS

PANZERKOMPANIE HQ

HEADQUARTERS

Company HQ with:

2 Panzer IV H	180 points
2 StuG G or StuG IV	190 points
2 Panther A or G	375 points

OPTION

- Add an Sd Kfz 9 (18t) recovery half-track for +5 points, a Bergepanzer III recovery vehicle for +10 points, or a Bergepanther recovery vehicle for +15 points.

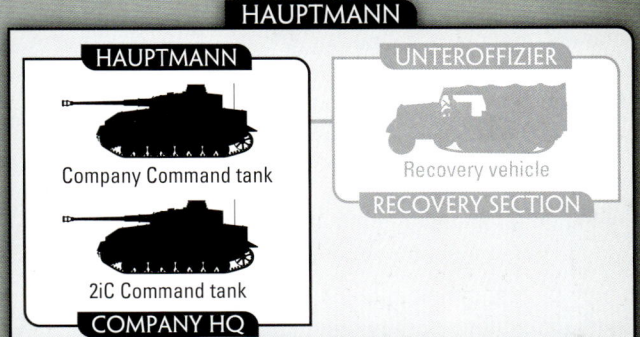

You must field at least one Panzer Platoon or Panther Platoon entirely equipped with the same model of tank as the Company HQ.

COMBAT PLATOONS

PANZER PLATOON

PLATOON

5 Panzer IV H	450 points
4 Panzer IV H	360 points
3 Panzer IV H	270 points
5 StuG G or StuG IV	475 points
4 StuG G or StuG IV	380 points
3 StuG G or StuG IV	285 points

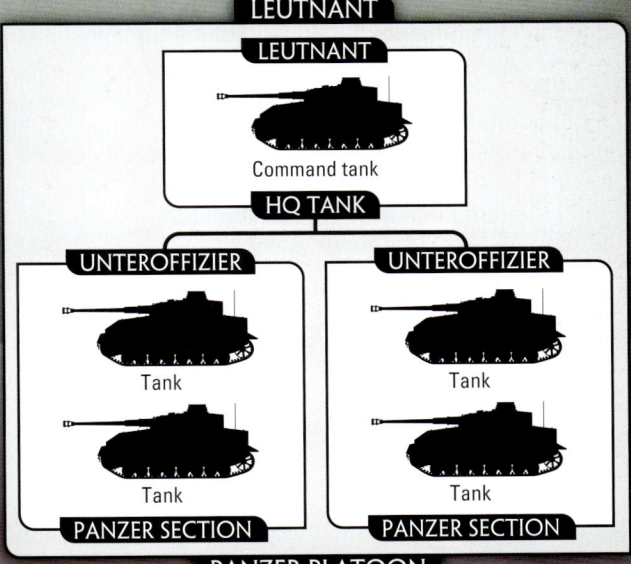

PANTHER PLATOON

PLATOON

5 Panther A or G	940	points
4 Panther A or G	750	points
3 Panther A or G	560	points

German doctrine requires tanks to be used as a concentrated strike force at the decisive point. Massed tanks attacking across good ground are almost impossible to stop. As a tank commander, you must be decisive. Choose your objective, then overwhelm it with everything you have. Do not take unnecessary risks, but by the same token, do not give the enemy time to counter your plan.

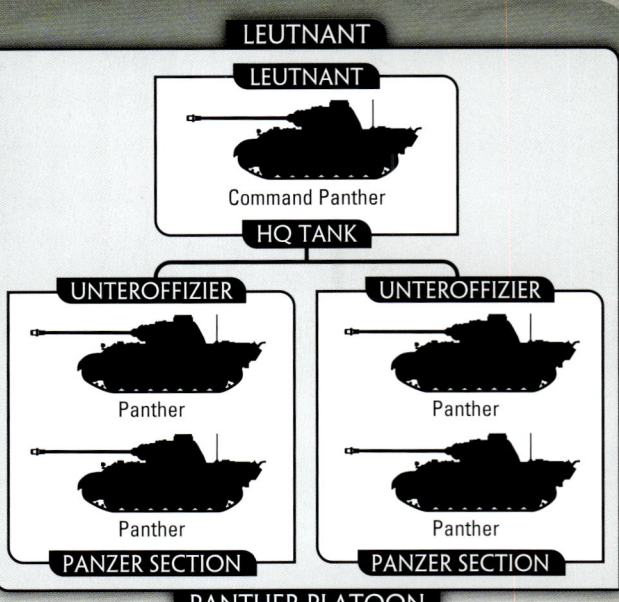

LEUTNANT

LEUTNANT

Command Panther

HQ TANK

UNTEROFFIZIER

Panther

Panther

PANZER SECTION

UNTEROFFIZIER

Panther

Panther

PANZER SECTION

PANTHER PLATOON

WEAPONS PLATOONS

PANZER PIONEER PLATOON

PLATOON

HQ Section with:

3 Pioneer Squads	135	points
2 Pioneer Squads	100	points

OPTION

- Replace all Maultier trucks with Sd Kfz 251/7 (Pioneer) half-tracks for +15 points per half-track.

Pionier platoons assigned to support Tigers and panzers are trained to quickly remove obstacles that hinder their advance. Additionally, their assistance prevents enemy infantry from bothering your tanks.

The equipment and specialist training afforded to these troops more than makes up for their lesser numbers and with the ability to both attack and defend they make an excellent support platoon for your tanks.

LEUTNANT

LEUTNANT

Command Pioneer MG team

Schwimmwagen

HQ SECTION

UNTEROFFIZIER

Pioneer MG team

Opel Maultier

PIONEER SQUAD

UNTEROFFIZIER

Pioneer MG team

Opel Maultier

PIONEER SQUAD

UNTEROFFIZIER

Pioneer MG team

Opel Maultier

PIONEER SQUAD

PANZER PIONIER PLATOON

PANZER ANTI-AIRCRAFT GUN PLATOON

PLATOON

3 Sd Kfz 7/1 (Quad 2cm)	120 points
2 Sd Kfz 7/1 (Quad 2cm)	80 points
3 Armoured Sd Kfz 7/1 (Quad 2cm)	150 points
2 Armoured Sd Kfz 7/1 (Quad 2cm)	100 points
3 Flakpanzer 38(t)	120 points
2 Flakpanzer 38(t)	80 points
3 Möbelwagen (3.7cm)	165 points
2 Möbelwagen (3.7cm)	110 points
3 Wirbelwind (Quad 2cm)	165 points
2 Wirbelwind (Quad 2cm)	110 points

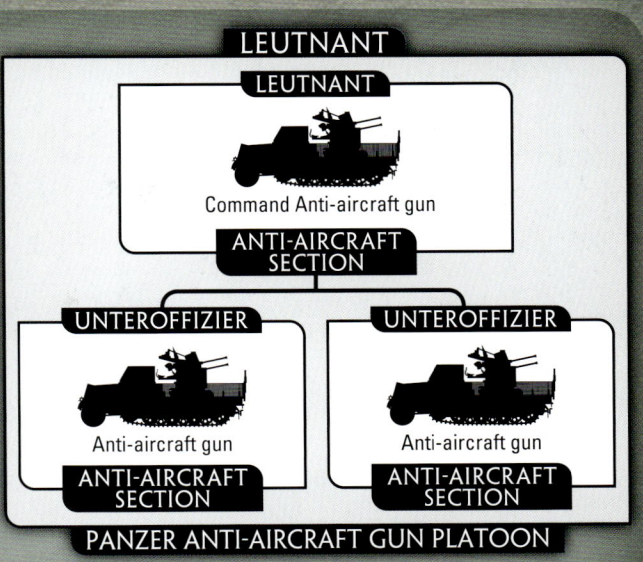

Keeping enemy fighters from harassing your panzers can be a full time job. Just make sure your anti-aircraft guns are far enough away from enemy armour not to become easy targets yet close enough to cover your panzers.

PANZER SCOUT PLATOON

PLATOON

HQ Section with:

3 Scout Squads	145 points
2 Scout Squads	110 points

OPTION

- Replace the Command Motorcycle MG team with a Command Panzerfaust Motorcycle SMG team for +10 points.

Panzer Scout Platoons use the Motorcycle Reconnaissance rules on page 187 and are Reconnaissance Platoons while mounted.

You may model your Motorcycle MG teams with Kübelwagen jeeps instead of motorcycles, they are based the same way as the Motorcycle MG teams and use the same rules.

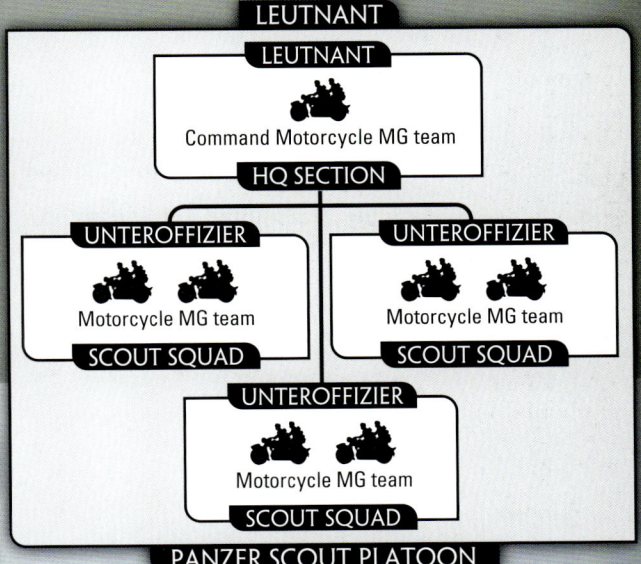

The Panzer Scout Platoon's job is to prevent an ambush of your Tigers from enemy tank-hunters. Protect your heavies and they will protect you from any armoured threat.

GEPANZERTE PANZERGRENADIERKOMPANIE
AMOURED INFANTRY COMPANY

(MECHANISED COMPANY)

HEADQUARTERS

Gepanzerte Panzer-
grenadierkompanie HQ — 77

You must field one platoon from each box shaded black and may field one platoon from each box shaded grey.

DIVISIONAL SUPPORT PLATOONS

COMBAT PLATOONS

INFANTRY
Gepanzerte Panzergrenadier Platoon — 77

INFANTRY
Gepanzerte Panzergrenadier Platoon — 77

INFANTRY
Gepanzerte Panzergrenadier Platoon — 77

INFANTRY
Gepanzerte Heavy Platoon — 78

REGIMENTAL SUPPORT PLATOONS

RECONNAISSANCE
Panzer Scout Platoon — 75

FLAME-THROWER
Armoured Flame-thrower Platoon — 84

ARTILLERY
Heavy Infantry Gun Platoon — 84
Self-propelled Infantry Gun Platoon — 85

WEAPONS PLATOONS

ARMOUR
Gepanzerte Cannon Platoon — 78

ARTILLERY
Gepanzerte Mortar Platoon — 79
Gepanzerte Heavy Mortar Platoon — 79

ARTILLERY
Gepanzerte Infantry Gun Platoon — 79

ANTI-TANK
Panzergrenadier Anti-tank Gun Platoon — 83

ANTI-AIRCRAFT
Panzergrenadier Anti-aircraft Gun Platoon — 84

INFANTRY
Gepanzerte Panzerpionier Platoon — 87

ALLIED PLATOONS

Luftwaffe Platoons in your force are Allies and follow the Allies rules in the rulebook.

ARMOUR

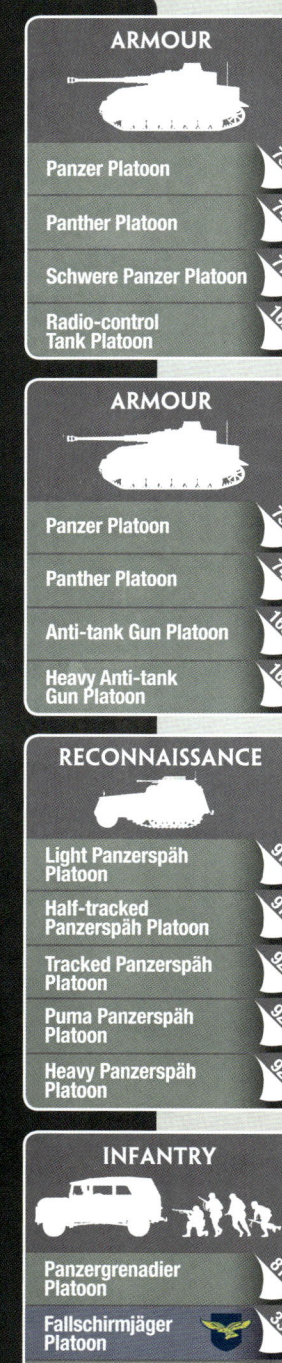

Panzer Platoon — 73
Panther Platoon — 74
Schwere Panzer Platoon — 71
Radio-control Tank Platoon — 162

ARMOUR

Panzer Platoon — 73
Panther Platoon — 74

ANTI-TANK
Anti-tank Gun Platoon — 165
Heavy Anti-tank Gun Platoon — 165

RECONNAISSANCE
Light Panzerspäh Platoon — 91
Half-tracked Panzerspäh Platoon — 91
Tracked Panzerspäh Platoon — 92
Puma Panzerspäh Platoon — 92
Heavy Panzerspäh Platoon — 92

INFANTRY
Panzergrenadier Platoon — 81
Fallschirmjäger Platoon — 33
Gepanzerte Aufklärungs Platoon — 93

INFANTRY

Gepanzerte Panzerpionier Platoon — 87

ARTILLERY
Motorised Artillery Battery — 167
Motorised Heavy Artillery Battery — 167
Armoured Artillery Battery — 168
Armoured Heavy Artillery Battery — 168
Armoured Train — 122

ARTILLERY
Rocket Launcher Battery — 169
Armoured Rocket Launcher Battery — 170
Motorised Artillery Battery — 167

ANTI-AIRCRAFT
Anti-aircraft Gun Platoon — 171

ANTI-AIRCRAFT
Heavy Anti-aircraft Gun Platoon — 171
Luftwaffe Heavy Anti-aircraft Gun Platoon — 173

AIRCRAFT
Air Support — 172

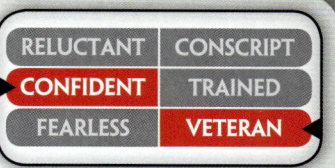
MOTIVATION AND SKILL

A Panzer division's gepanzerte, or armoured, Panzergrenadier battalion mounts its best men in armoured half-tracks to allow them to keep pace with tanks in an attack. A Gepanzerte Panzergrenadierkompanie is rated as **Confident Veteran.**

RELUCTANT	CONSCRIPT
CONFIDENT	TRAINED
FEARLESS	**VETERAN**

HEADQUARTERS

GEPANZERTE PANZERGRENADIERKOMPANIE HQ

HEADQUARTERS

Company HQ	65 points

OPTIONS

- Replace Command SMG teams with Command Panzerknacker SMG teams for +5 points per team or Command Panzerfaust SMG teams for +10 points per team.
- Add an Anti-tank Section for +40 points.
- Replace Panzerschreck team with a 8.8cm RW43 Püppchen rocket launcher at no cost.

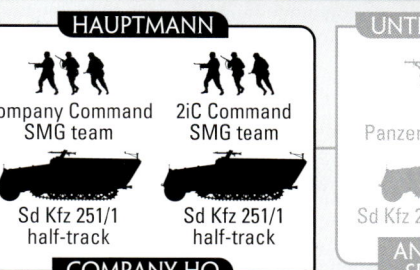

GEPANZERTE PANZERGRENADIERKOMPANIE HQ

The Company HQ of a Gepanzerte Panzergrenadier-kompanie may use the Mounted Assault special rule.

COMBAT PLATOONS

GEPANZERTE PANZERGRENADIER PLATOON

PLATOON

HQ Section with:

3 Panzergrenadier Squads	220 points
2 Panzergrenadier Squads	155 points

OPTIONS

- Replace the Command MG team with a Command Panzerknacker SMG team for +5 points or a Command Panzerfaust SMG team for +10 points.
- Replace Sd Kfz 251/1 half-track in HQ Section with a Sd Kfz 251/10 (3.7cm) half-track at no cost.

Gepanzerte Panzergrenadier Platoons may use the Mounted Assault special rule.

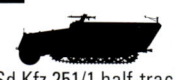

GEPANZERTE PANZERGRENADIER PLATOON

GEPANZERTE HEAVY PLATOON

PLATOON

HQ Section with:

2 Machine-gun Sections	165	points
1 Machine-gun Section	90	points
No Machine-gun Sections	15	points

OPTIONS

- Add a Gun Section for +80 points.
- Add a Mortar Section for +85 points.

A Gepanzerte Heavy Platoon must have a Gun or Mortar Section if it has no Machine-gun Sections.

Gepanzerte Heavy Platoons may make Combat Attachments to Gepanzerte Panzergrenadier Platoons.

Like their *Panzergrenadier* platoon, the *Schwere*, or heavy platoon, can fight from their half-tracks. This gives the company the firepower to assault most targets without slowing the pace of the advance.

LEUTNANT

LEUTNANT

Command SMG team

Sd Kfz 251/1 half-track

HQ SECTION

UNTEROFFIZIER

Sd Kfz 251/9 (7.5cm) half-track

Sd Kfz 251/9 (7.5cm) half-track

GUN SECTION

UNTEROFFIZIER

MG42 HMG

MG42 HMG

Sd Kfz 251/1 (HMG) half-track

MACHINE-GUN SECTION

UNTEROFFIZIER

MG42 HMG　　MG42 HMG

Sd Kfz 251/1 (HMG) half-track

MACHINE-GUN SECTION

UNTEROFFIZIER

Observer Rifle team　Kübelwagen　Sd Kfz 251/2 (8cm) half-track　Sd Kfz 251/2 (8cm) half-track

MORTAR SECTION

GEPANZERTE HEAVY PLATOON

WEAPONS PLATOONS

GEPANZERTE CANNON PLATOON

PLATOON

3 Gun Sections	240	points
2 Gun Sections	160	points
1 Gun Section	80	points

The cannon platoon gives the *Panzergrenadierkompanie* their own assault guns. This platoon will neutralise a machine-gun nest or a light anti-tank gun in a few shots. Do not expect them to fight tanks, as their light armour will lead them to a quick grave.

LEUTNANT

LEUTNANT

Command Sd Kfz 251/9 (7.5cm) half-track

Sd Kfz 251/9 (7.5cm) half-track

GUN SECTION

UNTEROFFIZIER

Sd Kfz 251/9 (7.5cm) half-track

Sd Kfz 251/9 (7.5cm) half-track

GUN SECTION

UNTEROFFIZIER

Sd Kfz 251/9 (7.5cm) half-track

Sd Kfz 251/9 (7.5cm) half-track

GUN SECTION

GEPANZERTE CANNON PLATOON

GEPANZERTE MORTAR PLATOON

PLATOON

HQ Section with:

3 Mortar Sections	240 points
2 Mortar Sections	175 points
1 Mortar Section	90 points

Mortars provide local artillery support to your advancing Panzergrenadiers. With three full sections of 8cm mortars you can expect to silence or blind even well hidden enemy machine-guns the instant they make themselves known.

The *Panzergrenadierdivisionen*, such as *Großdeutschland*, who were converted into *Panzerdivisionen* later were unique in having fully armoured mortar platoons.

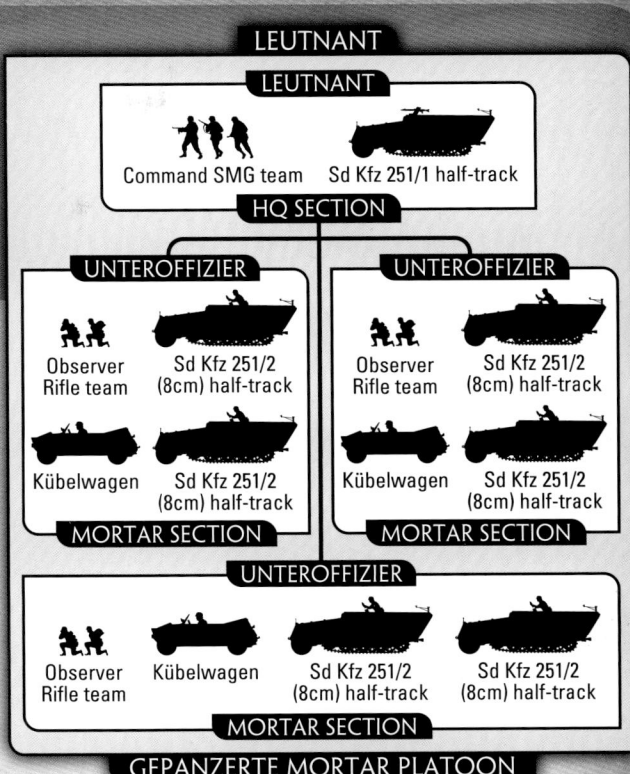

GEPANZERTE HEAVY MORTAR PLATOON

PLATOON

HQ Section with:

2 Mortar Sections	165 points
1 Mortar Section	85 points

OPTION

- Add Sd Kfz 251/1 half-tracks for +5 points per half-track.

Armed with 12cm sGW43 mortars, the armoured mortar platoon is able to blast away dug in troops and guns to give your troops deadly fire support.

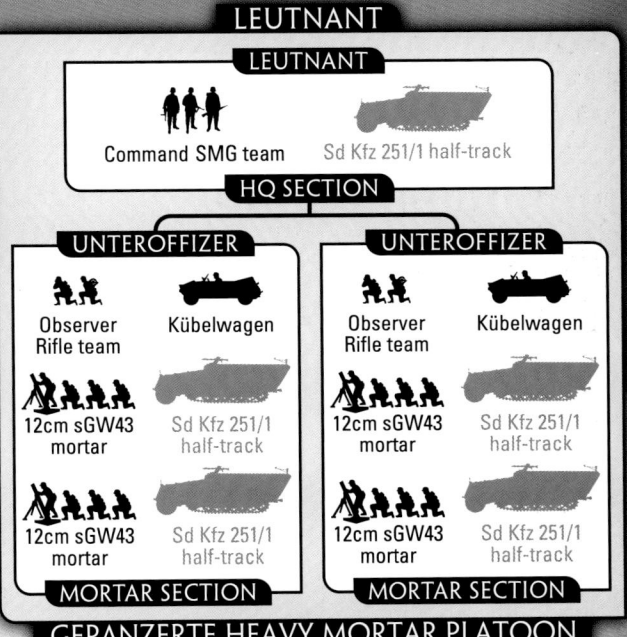

GEPANZERTE INFANTRY GUN PLATOON

PLATOON

HQ Section with:

2 7.5cm leIG18	70 points

OPTION

- Add Sd Kfz 251/1 half-tracks for +5 points per half-track.

Infantry gun crew in a *Gepanzerte Panzergrenadier* battalion have the same gun as those in normal *Panzergrenadier* battalions, but they are towed by Sd Kfz 251/1 half-tracks.

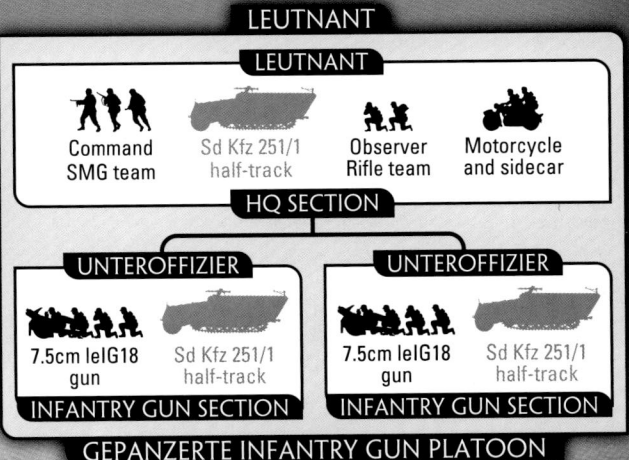

PANZERGRENADIERKOMPANIE

MOTORISED INFANTRY COMPANY

(INFANTRY COMPANY)

HEADQUARTERS

Panzergrenadier-kompanie HQ — 81

You must field one platoon from each box shaded black and may field one platoon from each box shaded grey.

COMBAT PLATOONS

INFANTRY

Panzergrenadier Platoon — 81

INFANTRY

Panzergrenadier Platoon — 81

INFANTRY

Panzergrenadier Platoon — 81

MACHINE-GUNS

Panzergrenadier Heavy Platoon — 82

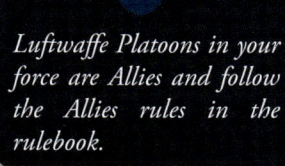

ALLIED PLATOONS

Luftwaffe Platoons in your force are Allies and follow the Allies rules in the rulebook.

WEAPONS PLATOONS

ARTILLERY

Panzergrenadier Mortar Platoon — 82

ARTILLERY

Panzergrenadier Infantry Gun Platoon — 83

ANTI-TANK

Panzergrenadier Anti-tank Gun Platoon — 83

ANTI-AIRCRAFT

Panzergrenadier Anti-aircraft Gun Platoon — 84

INFANTRY

Panzerpionier Platoon — 89

REGIMENTAL SUPPORT PLATOONS

RECONNAISSANCE

Panzer Scout Platoon — 75

FLAME-THROWER

Armoured Flame-thrower Platoon — 84

ARTILLERY

Heavy Infantry Gun Platoon — 84

Self-propelled Infantry Gun Platoon — 85

DIVISIONAL SUPPORT PLATOONS

ARMOUR

Panzer Platoon — 73

Panther Platoon — 74

Schwere Panzer Platoon — 71

Radio-control Tank Platoon — 162

ARMOUR

Sturmpanzer Platoon — 162

Tank-hunter Platoon — 163

Anti-tank Gun Platoon — 165

Heavy Anti-tank Gun Platoon — 165

RECONNAISSANCE

Light Panzerspäh Platoon — 91

Half-tracked Panzerspäh Platoon — 91

Tracked Panzerspäh Platoon — 92

Puma Panzerspäh Platoon — 92

Heavy Panzerspäh Platoon — 92

ARTILLERY

Motorised Artillery Battery — 167

Motorised Heavy Artillery Battery — 167

Armoured Artillery Battery — 168

Armoured Heavy Artillery Battery — 168

Armoured Train — 122

ARTILLERY

Rocket Launcher Battery — 169

Armoured Rocket Launcher Battery — 170

Motorised Artillery Battery — 167

INFANTRY

Gepanzerte Panzergrenadier Platoon — 77

Fallschirmjäger Platoon — 33

Aufklärungs Platoon — 93

Gepanzerte Aufklärungs Platoon — 93

INFANTRY

Panzerpionier Platoon — 89

Gepanzerte Panzerpionier Platoon — 87

ANTI-AIRCRAFT

Anti-aircraft Gun Platoon — 171

ANTI-AIRCRAFT

Heavy Anti-aircraft Gun Platoon — 171

Luftwaffe Heavy Anti-aircraft Gun Platoon — 173

AIRCRAFT

Air Support — 172

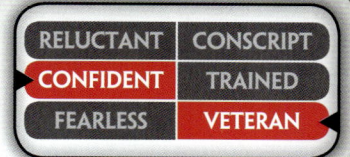

MOTIVATION AND SKILL

The Panzergrenadiers have been fighting and winning since the war began. These experienced soldiers keep pace with the armoured troops in their trucks. A Panzergrenadierkompanie is rated as **Confident Veteran.**

RELUCTANT	CONSCRIPT
CONFIDENT	TRAINED
FEARLESS	**VETERAN**

HEADQUARTERS

PANZERGRENDIERKOMPANIE HQ

HEADQUARTERS

Company HQ	45 points

OPTIONS

- Replace Command SMG teams with Command Panzerknacker SMG teams for +5 points per team or Command Panzerfaust SMG teams for +10 points per team.
- Add an Anti-tank Section for +20 points.
- Replace Panzerschreck team with a 8.8cm RW43 Püppchen rocket launcher at no cost.

Despite their name, most Panzergrenadier companies are motorised in trucks rather then mounted in armoured vehicles. None-the-less, they are still the core of the Panzer and Panzergrenadier divisions. Panzergrenadiers have a large amount of close fire support. Use this in attacks to knock out enemy machine-gun nests and infantry guns as your soldiers advance. Move quickly and decisively from cover to cover until you reach your assault positions. Then under covering fire from your machine-guns, storm the objective.

COMBAT PLATOONS

PANZERGRENADIER PLATOON

PLATOON

HQ Section with:

3 Panzergrenadier Squads	185 points
2 Panzergrenadier Squads	135 points

OPTIONS

- Replace the Command MG team with a Command Panzerknacker SMG team for +5 points or a Command Panzerfaust SMG team for +10 points.
- Remove Kfz 15 field car and replace all Kfz 70 trucks with 3-ton trucks at no cost.

Do not attempt to fight from your trucks. They should be used to move your troops up to the fighting zone. Dismount under cover and send them to the rear before assaulting on foot.

While the Panzer divisions have individual trucks for each Panzergrenadier section and a car for the platoon leader, Panzergrenadier divisions have fewer bigger trucks with the leader riding with the troops.

PANZERGRENADIER HEAVY PLATOON

PLATOON

HQ Section with:

2 Machine-gun Sections	140	points
1 Machine-gun Section	75	points
No Machine-gun Sections	10	points

OPTIONS

- Add a Mortar Section for +65 points.
- Add a second Kfz 70 truck per Machine-gun Section at no cost.

A Panzergrenadier Heavy Platoon must have a Mortar Section if it has no Machine-gun Sections.

Panzergrenadier Heavy Platoons may make Combat Attachments to Panzergrenadier Platoons.

Your heavy platoon must operate well forward in an attack. The mortars engage distant targets while the machine-guns hammer the target to keep the enemy pinned down as the Panzergrenadiers assault. Make sure you use all available cover to protect your vulnerable heavy weapons as they get into firing positions.

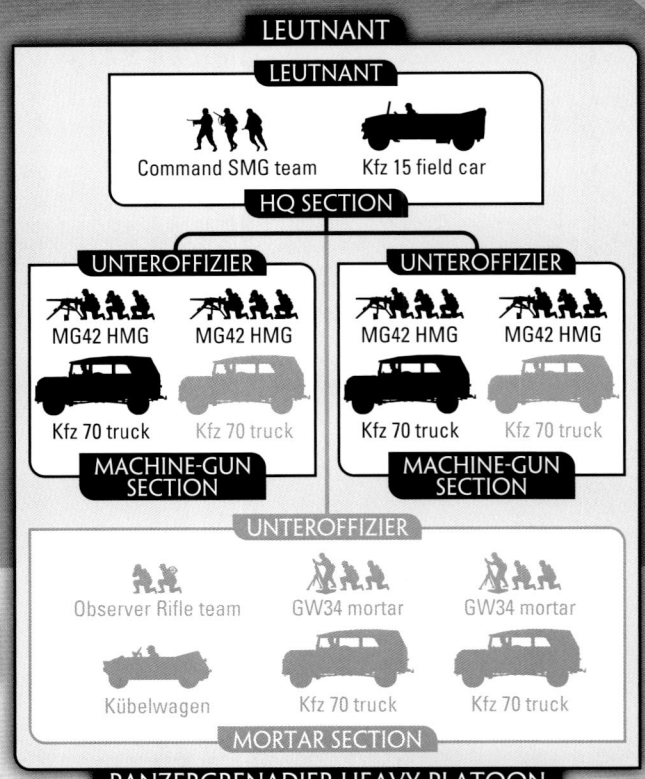

WEAPONS PLATOONS

PANZERGRENADIER MORTAR PLATOON

PLATOON

HQ Section with:

6 8cm GW34	180	points
4 8cm GW34	125	points
2 8cm GW34	65	points
4 12cm sGW43	165	points
2 12cm sGW43	85	points

OPTION

- Add Kfz 15 field car, Kübelwagen jeeps, and 3-ton trucks for +5 points for the platoon.

Mortar platoons are supposed to be equipped with 12cm heavy mortars. Unfortunately these are in very short supply and many units must make do with greater numbers of 8cm mortars instead. Use your mortars to engage enemy heavy weapons to protect your Panzergrenadiers as they advance.

The Panzergrenadier Mortar Platoon was usually found in the *Panzergreandierdivisionen*, while the Panzergreanadiers of the *Panzerdivisionen* usually had Heavy Platoons in support.

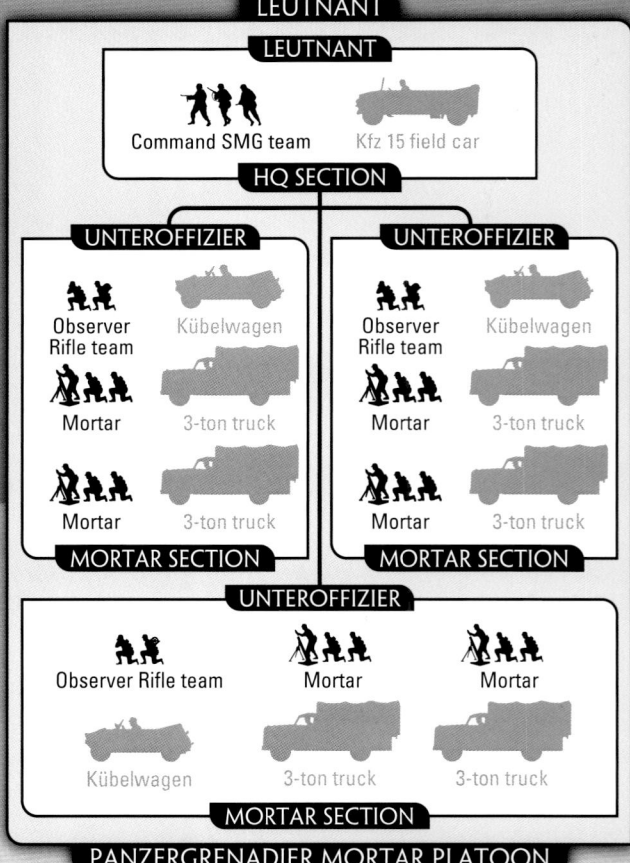

PANZERGRENADIER INFANTRY GUN PLATOON

PLATOON

HQ Section with:

2 7.5cm leIG18	70 points

OPTION

- Add Kfz 15 field car and Kfz 70 trucks +5 points for the platoon.

Though not overpowering, the 7.5cm leIG18 gun provides flexibility to your company. Adding a light infantry gun platoon provides an answer to a number of field problems. It can provide smoke and artillery support for advancing infantry while also protecting the front lines against assaults. It can dig out enemy machine guns and anti-tank guns as well as provide some anti-tank capability against assaulting tanks. A simple solution to some perplexing problems.

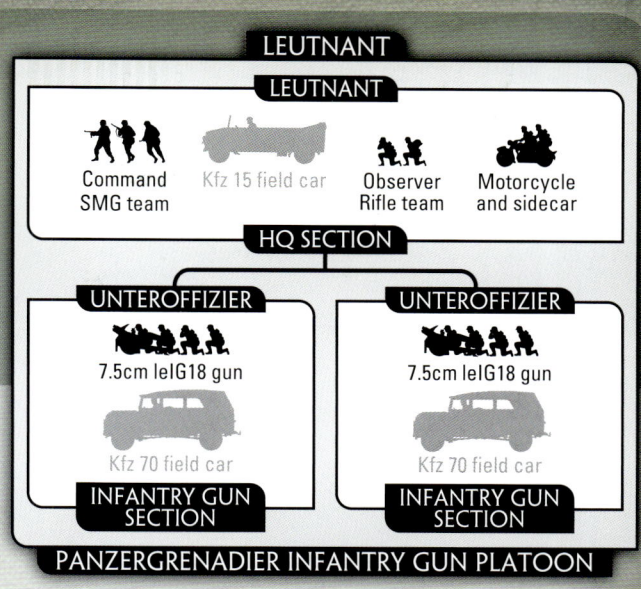

PANZERGRENADIER INFANTRY GUN PLATOON

PANZERGRENADIER ANTI-TANK GUN PLATOON

PLATOON

HQ Section with:

3 5cm PaK38	90 points
2 5cm PaK38	60 points
3 7.5cm PaK40	155 points
2 7.5cm PaK40	105 points

OPTION

- Add Kfz 15 field car and Kfz 70 trucks for +5 points for the platoon.

While the panzers are advancing you may need protection against wandering enemy armour. A light anti-tank platoon could be the best answer. Dug-in and well-concealed these guns can punish enemy tanks foolish enough to try and punch through your lines.

Each Panzergrenadier battalion has a light anti-tank gun platoon armed with either light 5cm PaK38 anti-tank guns or the more powerful 7.5cm PaK40 anti-tank guns.

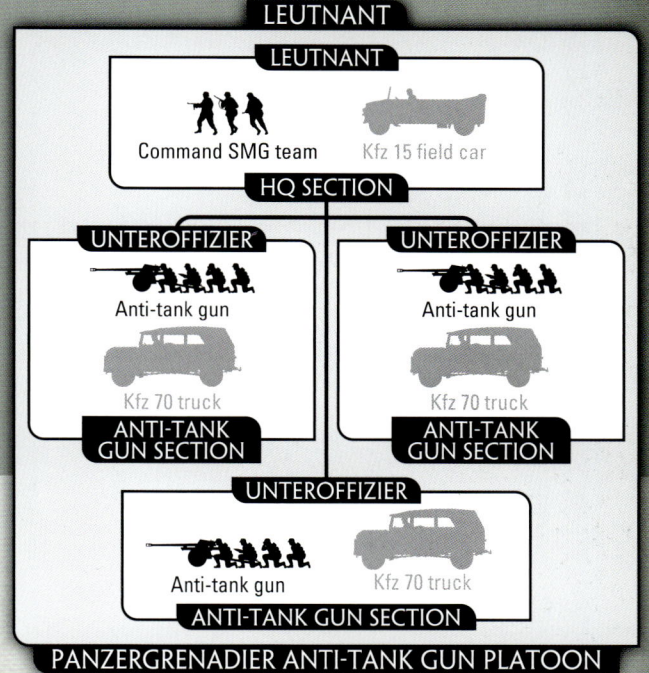

PANZERGRENADIER ANTI-TANK GUN PLATOON

83

PANZERGRENADIER ANTI-AIRCRAFT GUN PLATOON

PLATOON

3 Sd Kfz 10/5 (2cm)	85 points
2 Sd Kfz 10/5 (2cm)	55 points
3 Armoured Sd Kfz 10/5 (2cm)	115 points
2 Armoured Sd Kfz 10/5 (2cm)	75 points

Ground troops need to be self-sufficient in their anti-aircraft protection. These mobile anti-aircraft guns allow the Panzergrenadiers to manoeuvre even if an enemy Jabo attempts to interfere.

LEUTNANT

LEUTNANT

Command SdKfz 10/5 (2cm) half-track

HQ SECTION

UNTEROFFIZIER

SdKfz 10/5 (2cm) half-track

ANTI-AIRCRAFT SECTION

UNTEROFFIZIER

SdKfz 10/5 (2cm) half-track

ANTI-AIRCRAFT SECTION

PANZERGRENADIER ANTI-AIRCRAFT GUN PLATOON

REGIMENTAL SUPPORT PLATOONS

ARMOURED FLAME-THROWER PLATOON

PLATOON

3 Flame Sections	315 points
2 Flame Sections	210 points
1 Flame Section	105 points

Sd Kfz 251/16 (Flamm) half-tracks may not launch assaults, nor may they Counterattack if assaulted.

The Sd Kfz 251/16 mounts one flame-thrower on each side of the body. These can both fire at the same time, but must fire at the same enemy platoon. Each flame-thrower can fire at any target on its side of the half-track, from straight ahead to straight behind.

Flame-throwers are offensive weapons designed to take the initiative away from the enemy and provide deadly support to attacking Panzergrenadiers. The Sd Kfz 251/16 *Flammpanzerwagen* armoured flame-thrower half-track is a terrifying weapon by itself. Six of them can open a wide hole in the enemy's defence allowing the assaulting infantry the luxury of mopping up. Make use of them as a shock weapon to force the enemy into inactivity just as your Panzergrenadiers slam into their lines.

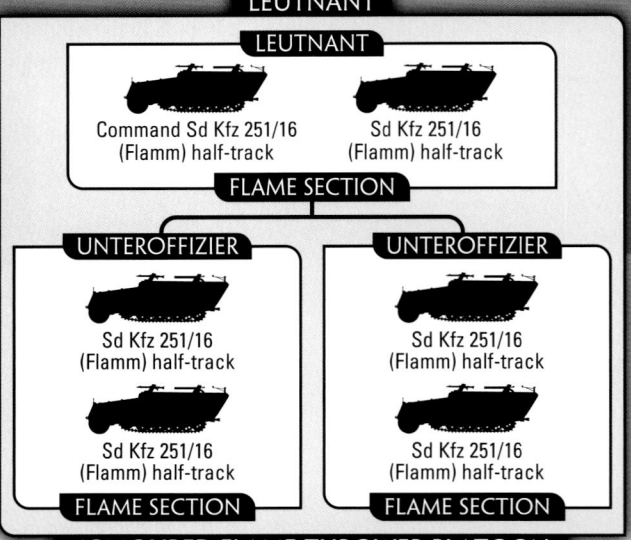

LEUTNANT

LEUTNANT

Command Sd Kfz 251/16 (Flamm) half-track

Sd Kfz 251/16 (Flamm) half-track

FLAME SECTION

UNTEROFFIZIER

Sd Kfz 251/16 (Flamm) half-track

Sd Kfz 251/16 (Flamm) half-track

FLAME SECTION

UNTEROFFIZIER

Sd Kfz 251/16 (Flamm) half-track

Sd Kfz 251/16 (Flamm) half-track

FLAME SECTION

ARMOURED FLAME-THROWER PLATOON

HEAVY INFANTRY GUN PLATOON

PLATOON

HQ Section with:

2 15cm sIG33	145 points

OPTION

- Add Kfz 15 field car and Sd Kfz 11 half-tracks to the platoon for +5 points for the platoon.

Heavy infantry guns can and will destroy any enemy nest of resistance. Even bunkers are not immune. Use them to reinforce the centre of gravity of your attacks to quickly eliminate enemy weapons that are holding up the Panzergrenadiers. Do not be afraid of using them over open sights to finish off particularly stubborn resistance.

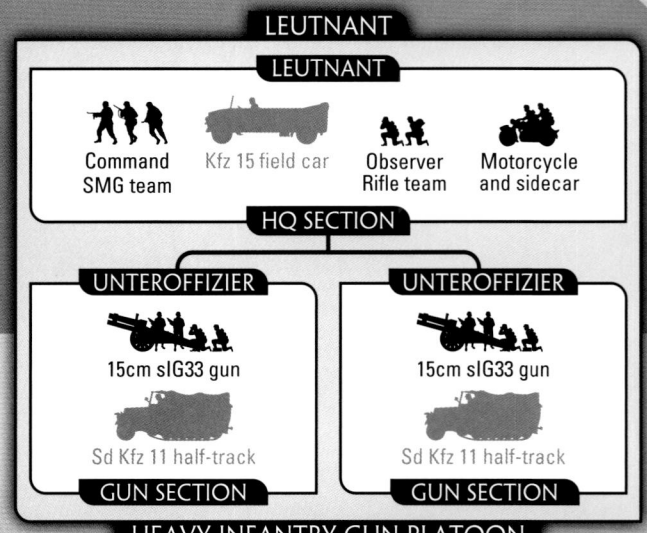

LEUTNANT

LEUTNANT

Command SMG team · Kfz 15 field car · Observer Rifle team · Motorcycle and sidecar

HQ SECTION

UNTEROFFIZIER

15cm sIG33 gun

Sd Kfz 11 half-track

GUN SECTION

UNTEROFFIZIER

15cm sIG33 gun

Sd Kfz 11 half-track

GUN SECTION

HEAVY INFANTRY GUN PLATOON

SELF-PROPELLED INFANTRY GUN PLATOON
PLATOON

HQ Section with:

2 Grille H	175 points
2 Grille K	170 points

Self-propelled infantry guns are far more flexible than the towed models. They can still operate as artillery, but when needed can drive forward, using the protection of their armour, to bring the enemy under direct fire for faster results.

LEUTNANT

LEUTNANT

Command SMG team | Sd Kfz 251/1 half-track | Observer Rifle team | Motorcycle and sidecar

HQ SECTION

UNTEROFFIZIER | UNTEROFFIZIER

Self-propelled infantry gun | Self-propelled infantry gun

GUN SECTION | GUN SECTION

SELF-PROPELLED INFANTRY GUN PLATOON

GEPANZERTE PANZERPIONIERKOMPANIE
ARMOURED ENGINEER COMPANY

(MECHANISED COMPANY)

HEADQUARTERS

Gepanzerte Panzer-pionierkompanie HQ   87

You must field one platoon from each box shaded black and may field one platoon from each box shaded grey.

COMBAT PLATOONS

INFANTRY

Gepanzerte Panzerpionier Platoon   87

INFANTRY

Gepanzerte Panzerpionier Platoon   87

INFANTRY

Gepanzerte Panzerpionier Platoon   87

WEAPONS PLATOONS

INFANTRY

Panzerpionier Platoon   89

ALLIED PLATOONS

Luftwaffe Platoons in your force are Allies and follow the Allies rules in the rulebook.

The combat engineers of the *Gepanzerte Panzerpionier-kompanie* are heavily-armed assault specialists. They have more firepower per man than any other infantry force.

German Iron Cross

DIVISIONAL SUPPORT PLATOONS

ARMOUR

Panzer Platoon   73
Panther Platoon   74
Schwere Panzer Platoon   71
Radio-control Tank Platoon   162

ARMOUR

Sturmpanzer Platoon   162
Tank-hunter Platoon   163
Anti-tank Gun Platoon   165
Heavy Anti-tank Gun Platoon   165

RECONNAISSANCE

Light Panzerspäh Platoon   91
Half-tracked Panzerspäh Platoon   91
Tracked Panzerspäh Platoon   92
Puma Panzerspäh Platoon   92
Heavy Panzerspäh Platoon   92

INFANTRY

Gepanzerte Panzergrenadier Platoon   77
Panzergrenadier Platoon   81
Aufklärungs Platoon   93
Gepanzerte Aufklärungs Platoon   93
Fallschirmjäger Platoon   33

ARTILLERY

Motorised Artillery Battery   167
Motorised Heavy Artillery Battery   167
Armoured Artillery Battery   168
Armoured Heavy Artillery Battery   168

ARTILLERY

Rocket Launcher Battery   169
Armoured Rocket Launcher Battery   170
Motorised Artillery Battery   167

ANTI-AIRCRAFT

Anti-aircraft Gun Platoon   171

ANTI-AIRCRAFT

Heavy Anti-aircraft Gun Platoon   171
Luftwaffe Heavy Anti-aircraft Gun Platoon   173

AIRCRAFT

Air Support   172

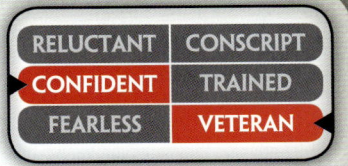
MOTIVATION AND SKILL

A Panzer division's gepanzerte, or armoured, pioneers mount their best men in armoured half-tracks to allow them to keep pace with tanks in an attack. A Gepanzerte Panzerpionier-kompanie is rated as **Confident Veteran.**

RELUCTANT	CONSCRIPT
CONFIDENT	TRAINED
FEARLESS	**VETERAN**

HEADQUARTERS

GEPANZERTE PANZERPIONIERKOMPANIE HQ

HEADQUARTERS

Company HQ	65 points

OPTIONS

- Replace Command SMG teams with Command Panzerknacker SMG teams for +5 points per team or Command Panzerfaust SMG teams for +10 points per team.
- Add a Mortar Section for +85 points.
- Add an Anti-tank Section for +35 points.

The Company HQ of a Gepanzerte Panzerpionier-kompanie and Gepanzerte Panzerpionier Platoons may use the Mounted Assault special rule.

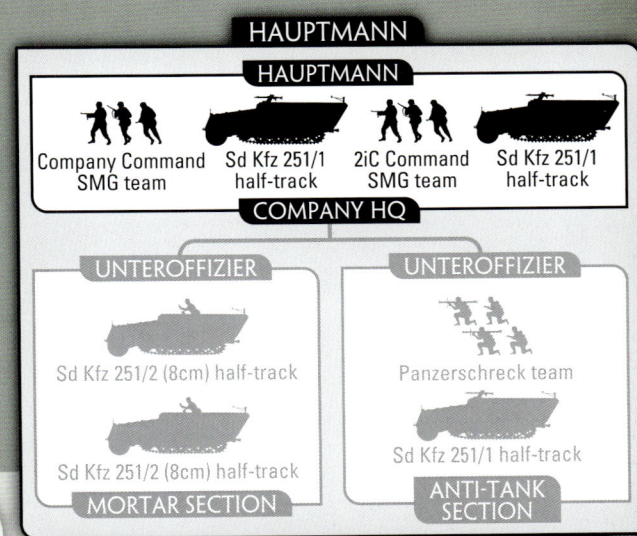

HAUPTMANN

Company Command SMG team — Sd Kfz 251/1 half-track — 2iC Command SMG team — Sd Kfz 251/1 half-track

COMPANY HQ

UNTEROFFIZIER — Sd Kfz 251/2 (8cm) half-track — Sd Kfz 251/2 (8cm) half-track — **MORTAR SECTION**

UNTEROFFIZIER — Panzerschreck team — Sd Kfz 251/1 half-track — **ANTI-TANK SECTION**

GEPANZERTE PANZERPIONIERKOMPANIE HQ

COMBAT PLATOONS

GEPANZERTE PANZERPIONIER PLATOON

PLATOON

HQ Section with:

3 Pionier Squads	265 points
2 Pionier Squads	190 points

OPTIONS

- Replace the Command Pioneer MG team with a Command Panzerknacker Pioneer SMG team for +5 points or a Command Panzerfaust Pioneer SMG team for +10 points.
- Equip one Pioneer MG team with a Goliath demolition carrier in addition to its normal weapons for +30 points.
- Add an additional Sd Kfz 251/7 half-track to each squad for +10 points per half-track.
- Replace Sd Kfz 251/1 half-track in HQ Section with a Sd Kfz 251/11 (2.8cm) half-track for +5 points.
- Add Pioneer Supply 3-ton truck for +25 points or Pioneer Supply Maultier for +30 points.
- Replace any or all Sd Kfz 251/7 half-tracks with Sd Kfz 251/1 (Stuka) half-tracks for +35 points per half-track.

Only one Gepanzerte Panzerpionier Platoon in your Company may be equipped with Sd Kfz 251/1 (Stuka) half-tracks.

LEUTNANT

Command Pioneer MG team — Sd Kfz 251/1 half-track — Pioneer Supply 3-ton truck

HQ SECTION

UNTEROFFIZIER — Pioneer MG team — Sd Kfz 251/7 half-track — Pioneer MG team — Sd Kfz 251/7 half-track — **PIONEER SQUAD**

UNTEROFFIZIER — Pioneer MG team — Sd Kfz 251/7 half-track — Pioneer MG team — Sd Kfz 251/7 half-track — **PIONEER SQUAD**

UNTEROFFIZIER — Pioneer MG team — Pioneer MG team — Sd Kfz 251/7 half-track — Sd Kfz 251/7 half-track — **PIONEER SQUAD**

GEPANZERTE PIONIER PLATOON

You may replace up to one Pioneer MG team per Pioneer Squad with a Flame-thrower team at the start of the game before deployment.

PANZERPIONIERKOMPANIE
MOTORISED ENGINEER COMPANY

(INFANTRY COMPANY)

HEADQUARTERS

Panzerpionierkompanie HQ 89

You must field one platoon from each box shaded black and may field one platoon from each box shaded grey.

COMBAT PLATOONS

INFANTRY

Panzerpionier Platoon 89

INFANTRY

Panzerpionier Platoon 89

INFANTRY

Panzerpionier Platoon 89

WEAPONS PLATOONS

INFANTRY

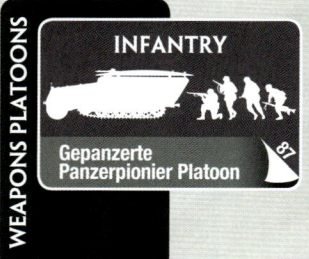

Gepanzerte Panzerpionier Platoon 87

ALLIED PLATOONS

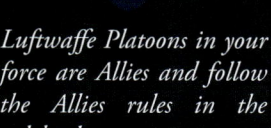

Luftwaffe Platoons in your force are Allies and follow the Allies rules in the rulebook.

DIVISIONAL SUPPORT PLATOONS

ARMOUR

Panzer Platoon 73
Panther Platoon 74
Schwere Panzer Platoon 71
Radio-control Tank Platoon 162

ARMOUR

Sturmpanzer Platoon 162
Tank-hunter Platoon 163
Anti-tank Gun Platoon 165
Heavy Anti-tank Gun Platoon 165

RECONNAISSANCE

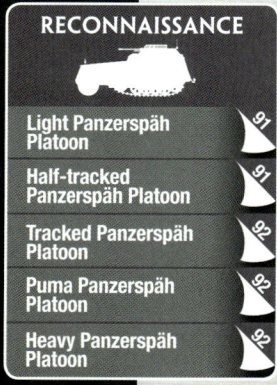

Light Panzerspäh Platoon 91
Half-tracked Panzerspäh Platoon 91
Tracked Panzerspäh Platoon 92
Puma Panzerspäh Platoon 92
Heavy Panzerspäh Platoon 92

INFANTRY

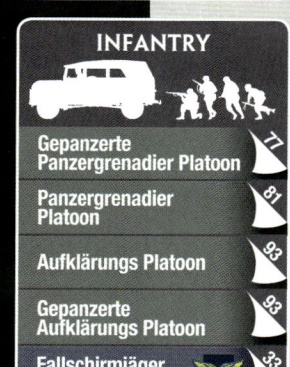

Gepanzerte Panzergrenadier Platoon 77
Panzergrenadier Platoon 81
Aufklärungs Platoon 83
Gepanzerte Aufklärungs Platoon 83
Fallschirmjäger Platoon 33

ARTILLERY

Motorised Artillery Battery 167
Motorised Heavy Artillery Battery 167
Armoured Artillery Battery 168
Armoured Heavy Artillery Battery 168

ARTILLERY

Rocket Launcher Battery 169
Armoured Rocket Launcher Battery 170
Motorised Artillery Battery 167

ANTI-AIRCRAFT

Anti-aircraft Gun Platoon 171

ANTI-AIRCRAFT

Heavy Anti-aircraft Gun Platoon 171
Luftwaffe Heavy Anti-aircraft Gun Platoon 173

AIRCRAFT

Air Support 172

German Infantry Assault Badge.

PANZERPIONIERKOMPANIE

MOTIVATION AND SKILL

The bulk of a Panzerdivision's pioneers ride in trucks and takes care of enemy obstacles, minefields, and fortifications before the Panzerdivision attacks. A Panzerpionierkompanie is rated as **Confident Veteran.**

RELUCTANT	CONSCRIPT
CONFIDENT	TRAINED
FEARLESS	**VETERAN**

HEADQUARTERS

PANZERPIONIERKOMPANIE HQ

HEADQUARTERS

Company HQ	45 points

OPTIONS

- Replace either or both Command SMG teams with Command Panzerknacker SMG teams for +5 points per team or Command Panzerfaust SMG teams for +10 points per team.
- Add Anti-tank Section for +25 points.
- Add Mortar Section for +55 points.
- Add Machine-gun Section for +65 points.

Note: The Infantry teams of the Company HQ are not Pioneer teams, they are too busy commanding for pioneer work.

The *Panzerpionier* troops are called forward when enemy fortified positions need to be cleared.

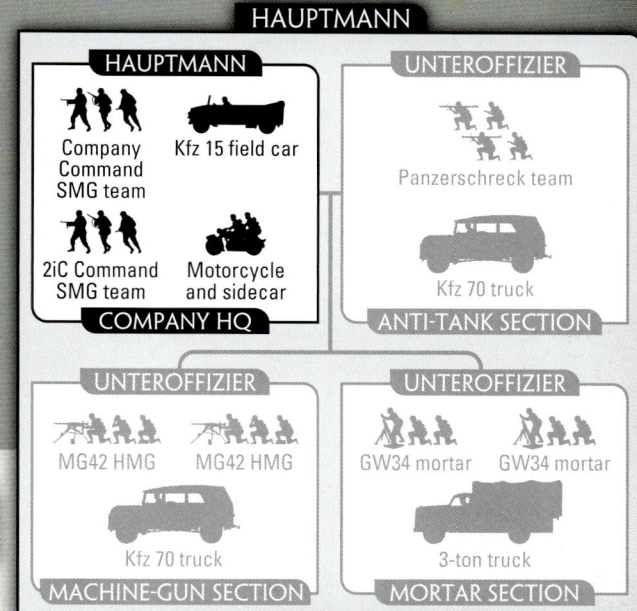

PANZERPIONIERKOMPANIE HQ

COMBAT PLATOONS

PANZERPIONIER PLATOON

PLATOON

HQ Section with:

3 Pionier Squads	205 points
2 Pionier Squads	145 points

OPTIONS

- Replace the Command Pioneer Rifle/MG team with a Command Panzerknacker Pioneer SMG team for +5 points or a Command Panzerfaust Pioneer SMG team for +10 points.
- Equip one Pioneer Rifle/MG team with a Goliath demolition carrier in addition to its normal weapons for +30 points.
- Add Pioneer Supply 3-ton truck for +25 points or Pioneer Supply Maultier for +30 points.

You may replace up to one Pioneer Rifle/MG team per Pioneer Squad with a Flame-thrower team at the start of the game before deployment.

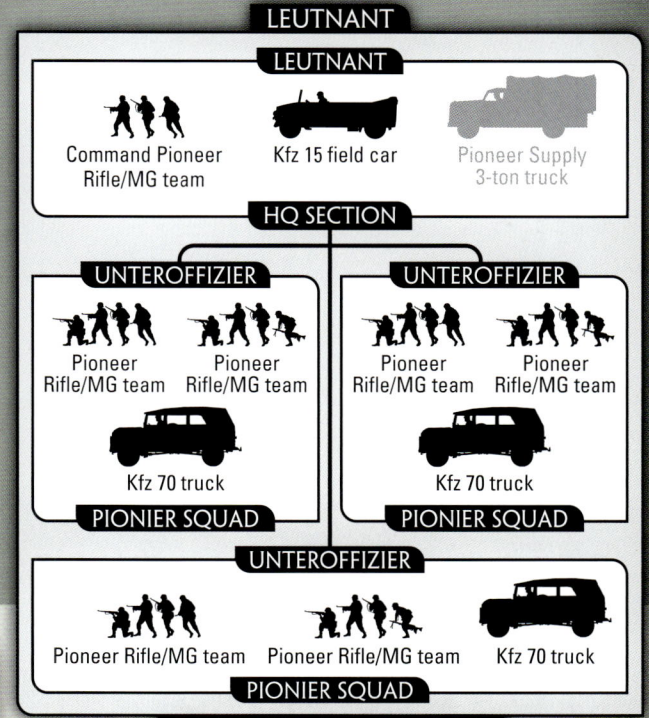

PANZERPIONIER PLATOON

The *Panzerpionier* platoon is invaluable in assaults on fortifications, being able to clear mines and breach obstacles with ease. In defence, they are equally good at creating barriers and obstacles around strategic positions. Well armed with rifles, machine-guns, mines and demolition carriers, these pioneers are able to take on anything that the Allies can throw against them.

GERMAN 89

PANZERSPÄHKOMPANIE
ARMOURED CAR COMPANY

(MECHANISED COMPANY)

HEADQUARTERS

Panzerspähkompanie HQ — 91

You must field one platoon from each box shaded black and may field one platoon from each box shaded grey.

COMBAT PLATOONS

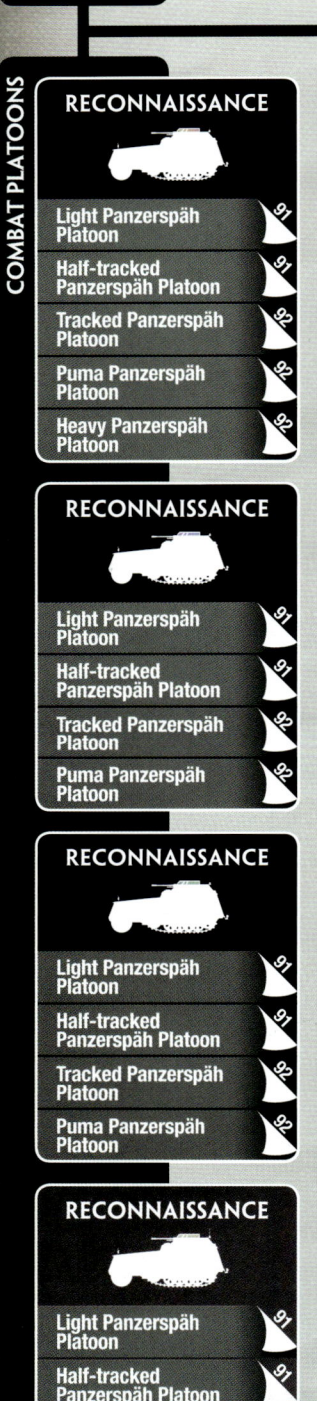

RECONNAISSANCE

Light Panzerspäh Platoon — 91
Half-tracked Panzerspäh Platoon — 91
Tracked Panzerspäh Platoon — 92
Puma Panzerspäh Platoon — 93
Heavy Panzerspäh Platoon — 92

RECONNAISSANCE

Light Panzerspäh Platoon — 91
Half-tracked Panzerspäh Platoon — 91
Tracked Panzerspäh Platoon — 92
Puma Panzerspäh Platoon — 92

RECONNAISSANCE

Light Panzerspäh Platoon — 91
Half-tracked Panzerspäh Platoon — 91
Tracked Panzerspäh Platoon — 92
Puma Panzerspäh Platoon — 92

RECONNAISSANCE

Light Panzerspäh Platoon — 91
Half-tracked Panzerspäh Platoon — 91
Tracked Panzerspäh Platoon — 92
Puma Panzerspäh Platoon — 92

WEAPONS PLATOONS

ARMOUR

7.5cm Armoured Car Platoon — 93

INFANTRY

Gepanzerte Aufklärungs Platoon — 93
Aufklärungs Platoon — 93

INFANTRY

Gepanzerte Aufklärungs Platoon — 93
Aufklärungs Platoon — 93

ANTI-TANK

Panzergrenadier Anti-tank Gun Platoon — 83

ARTILLERY

Gepanzerte Cannon Platoon — 78
Gepanzerte Infantry Gun Platoon — 79

INFANTRY

Panzerpionier Platoon — 89

DIVISIONAL SUPPORT PLATOONS

ARMOUR

Panzer Platoon — 73
Panther Platoon — 74
Schwere Panzer Platoon — 71
Radio-control Tank Platoon — 162

ANTI-TANK

Tank-hunter Platoon — 163
Anti-tank Gun Platoon — 165
Heavy Anti-tank Gun Platoon — 165

INFANTRY

Gepanzerte Panzergrenadier Platoon — 77
Panzergrenadier Platoon — 81

INFANTRY

Gepanzerte Panzerpionier Platoon — 87

ALLIED PLATOONS

Luftwaffe Platoons in your force are Allies and follow the Allies rules in the rulebook.

ARTILLERY

Motorised Artillery Battery — 167
Motorised Heavy Artillery Battery — 167
Armoured Artillery Battery — 168
Armoured Heavy Artillery Battery — 168

ARTILLERY

Rocket Launcher Battery — 169
Armoured Rocket Launcher Battery — 170
Motorised Artillery Battery — 167

ANTI-AIRCRAFT

Anti-aircraft Gun Platoon — 171

ANTI-AIRCRAFT

Heavy Anti-aircraft Gun Platoon — 171
Luftwaffe Heavy Anti-aircraft Gun Platoon — 173

AIRCRAFT

Air Support — 172

MOTIVATION AND SKILL

Armoured cars do not generally engage the enemy directly except when confronted by a weak enemy or at the point of attack. They attack rear-area troops and generally cause havoc while finding enemy troop concentrations. The armoured car crews of a Panzerspähkompanie know they are the first to meet the enemy and are rated **Confident Veteran.**

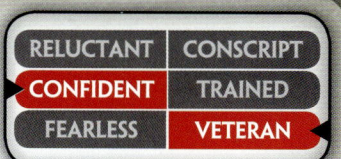

RELUCTANT	CONSCRIPT
CONFIDENT	TRAINED
FEARLESS	**VETERAN**

HEADQUARTERS

PANZERSPÄHKOMPANIE HQ

HEADQUARTERS

1 Sd Kfz 250 (Recce)	35 points
1 Sd Kfz 223 (radio)	30 points
1 Sd Kfz 231 (8-rad)	40 points
1 Sd Kfz 234/2 Puma	50 points
1 Panzer II L Luchs	45 points

HAUPTMANN

HAUPTMANN

Company Command Armoured Car

COMPANY HQ

PANZERSPÄHKOMPANIE HQ

Armoured cars are the true reconnaissance troops of an *Aufklärungsabteilung* (pronounced ouf-klear-ungs-ab-tile-ung, reconnaissance detachment or battalion). While the rest of the battalion is beating a path through the enemy the armoured cars exploit the gaps in the lines to get behind the enemy and scout out his positions.

You must field at least one Combat Platoon equipped at least in part with the same vehicle as the Company HQ.

The Company Command vehicle of a Panzerspähkompanie Company HQ is a Reconnaissance team.

COMBAT PLATOONS

LIGHT PANZERSPÄH PLATOON

PLATOON

2 Panzerspäh Patrols	220 points
1 Panzerspäh Patrol	110 points

- Replace any or all Sd Kfz 222 (2cm) with Sd Kfz 221 (2.8cm) at no cost.

Light Panzerspäh Platoons are Reconnaissance Platoons.

Panzerspäh Patrols of a Light Panzerspäh Platoon operate as separate platoons, each with their own command team.

LEUTNANT

LEUTNANT

Command Sd Kfz 223 (radio) — Sd Kfz 222 (2cm) — Sd Kfz 222 (2cm)

PANZERSPÄH PATROL

FELDWEBEL

Command Sd Kfz 223 (radio) — Sd Kfz 222 (2cm) — Sd Kfz 222 (2cm)

PANZERSPÄH PATROL

LIGHT PANZERSPÄH PLATOON

HALF-TRACKED PANZERSPÄH PLATOON

PLATOON

2 Panzerspäh Patrols	230 points
1 Panzerspäh Patrol	115 points

Half-tracked Panzerspäh Platoons are Reconnaissance Platoons.

Panzerspäh Patrols of a Half-tracked Panzerspäh Platoon operate as separate platoons, each with their own command team.

LEUTNANT

LEUTNANT

Command Sd Kfz 250 (Recce) — Sd Kfz 250/9 (2cm) — Sd Kfz 250/9 (2cm)

PANZERSPÄH PATROL

FELDWEBEL

Command Sd Kfz 250 (Recce) — Sd Kfz 250/9 (2cm) — Sd Kfz 250/9 (2cm)

PANZERSPÄH PATROL

HALF-TRACKED PANZERSPÄH PLATOON

TRACKED PANZERSPÄH PLATOON

PLATOON

2 Panzerspäh Patrols	290 points
1 Panzerspäh Patrol	145 points

Tracked Panzerspäh Platoons are Reconnaissance Platoons.

Panzerspäh Patrols of a Tracked Panzerspäh Platoon operate as separate platoons, each with their own command team.

The Luchs (Lynx) is an excellent fully-tracked reconnaissance vehicle. It can go anywhere and is very fast. The Panzer II L Luchs saw service on both fronts during 1944 to 45. Amongst its users were *4. Panzerdivision* on the Eastern Front.

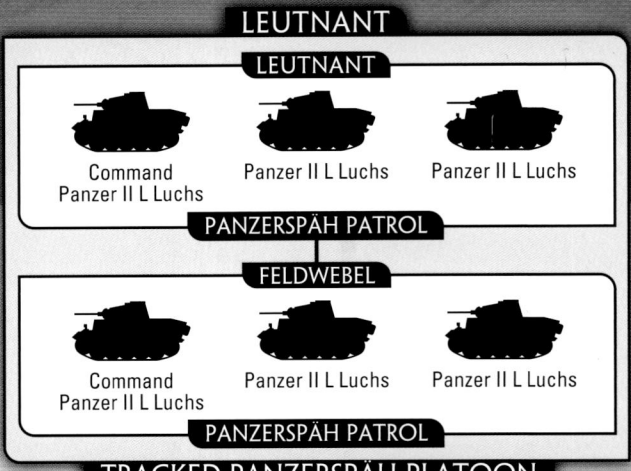

TRACKED PANZERSPÄH PLATOON

PUMA PANZERSPÄH PLATOON

PLATOON

With Sd Kfz 234/2 Puma:

2 Panzerspäh Patrols	300 points
1 Panzerspäh Patrol	150 points

Puma Panzerspäh Platoons are Reconnaissance Platoons.

Panzerspäh Patrols of a Puma Panzerspäh Platoon operate as separate platoons, each with their own command team.

The new Puma are heavy armoured cars. They have great mobility and traction due to their eight wheels. The *Puma* is armed with a 5cm KwK39 gun that can defeat any enemy light armour it may encounter. On the eastern front *20. Panzerdivision* received 16 and the *7. Panzerdivision* only 6 new Puma armoured cars.

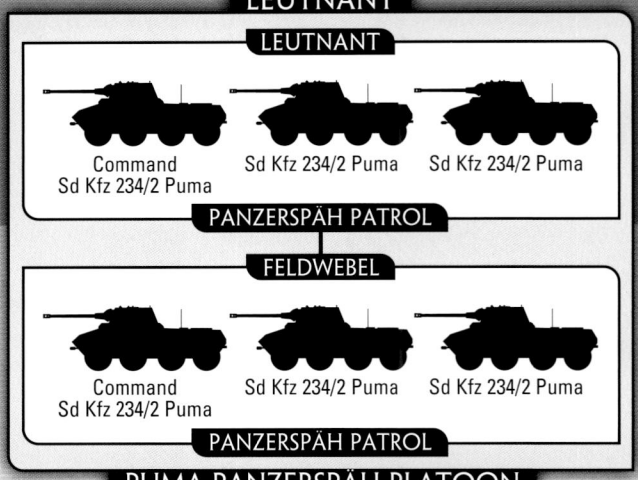

PUMA PANZERSPÄH PLATOON

If your Panzerspähkompanie contains any Puma Panzerspäh Platoons you may not take any Tracked Panzerspäh Platoons.

HEAVY PANZERSPÄH PLATOON

PLATOON

With Sd Kfz 231 (8-rad):

3 Panzerspäh Patrols	240 points
2 Panzerspäh Patrols	160 points

Heavy Panzerspäh Platoons are Reconnaissance Platoons.

Panzerspäh Patrols of a Heavy Panzerspäh Platoon operate as separate platoons, each with their own command team.

The old and trusted *8-rad* heavy armoured cars have great mobility and traction due to their eight wheels.

HEAVY PANZERSPÄH PLATOON

WEAPONS PLATOONS

7.5CM ARMOURED CAR PLATOON

PLATOON

3 Sd Kfz 233 (7.5cm)	145 points
2 Sd Kfz 233 (7.5cm)	95 points

*7.5cm Armoured Car Platoons are **not** reconnaissance platoons.*

The battalion's 7.5cm armoured car platoon gives the armoured cars the direct firepower they need to knock out light anti-tank guns and other enemy hazards blocking their way forward.

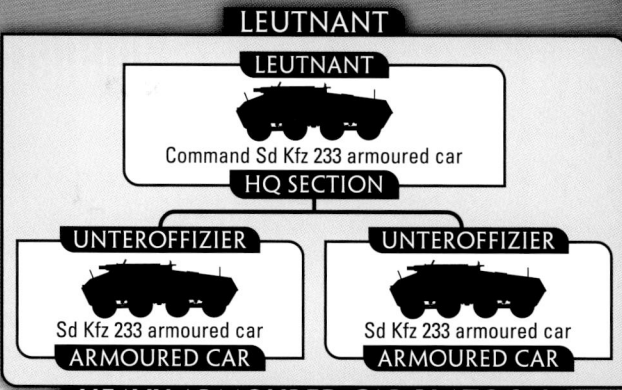

AUFKLÄRUNGS PLATOON

PLATOON

HQ Section with:

3 Aufklärungs Squads	255 points
2 Aufklärungs Squads	180 points

OPTIONS

- Replace the Command Motorcycle MG team with a Command Panzerfaust Motorcycle SMG team for +10 points.

- Replace all Motorcycle teams with the equivalent Schwimmwagen teams for +5 points for the platoon.

Aufklärungs Platoons use the Motorcycle Reconnaissance rules on page 187 and are Reconnaissance Platoons while mounted.

You may model your Motorcycle MG teams with Kübelwagen jeeps instead of motorcycles, they are based the same way as the Motorcycle MG teams and use the same rules.

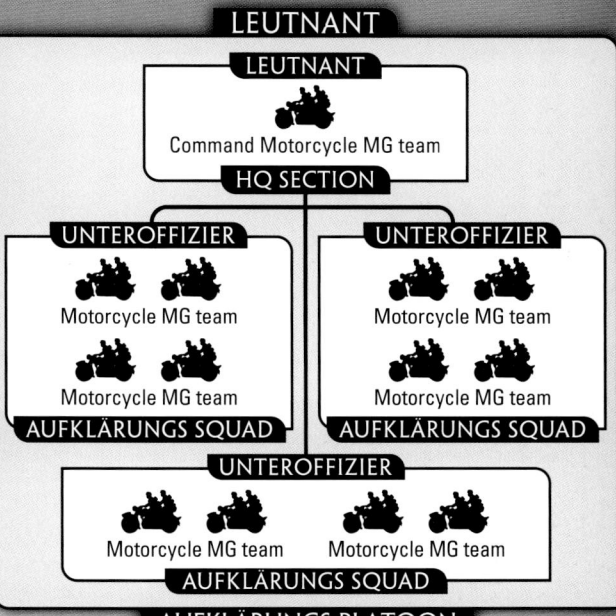

GEPANZERTE AUFKLÄRUNGS PLATOON

PLATOON

HQ Section with:

3 Aufklärungs Squads	220 points
2 Aufklärungs Squads	155 points

OPTIONS

- Replace the Command MG team with Command Panzerknacker SMG team for +5 points or Command Panzerfaust SMG team for +10 points.

- Add an additional Sd Kfz 250 half-track to each squad for +10 points per half-track.

- Replace Sd Kfz 250 half-track in HQ Section with Sd Kfz 250/10 (3.7cm) or Sd Kfz 250/11 (2.8cm) half-track at no cost.

Gepanzerte Aufklärungs Platoons may use the Mounted Assault special rule.

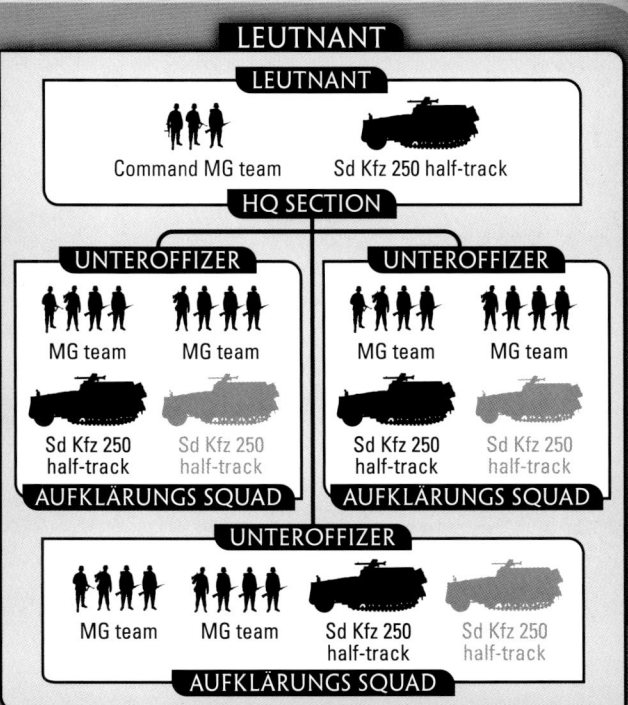

SPERRVERBAND
BLOCKING FORCE

(INFANTRY COMPANY)

HEADQUARTERS

Sperrverband HQ   95

You must field one platoon from each box shaded black and may field one platoon from each box shaded grey.

DIVISIONAL SUPPORT PLATOONS

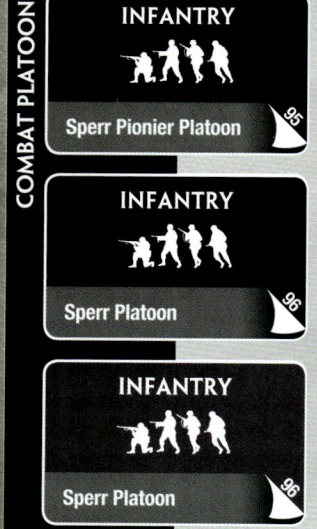

COMBAT PLATOONS

INFANTRY

Sperr Pionier Platoon   95

INFANTRY

Sperr Platoon   96

INFANTRY

Sperr Platoon   96

WEAPONS PLATOONS

ARTILLERY

Sperr Mortar Platoon   97

ANTI-TANK

Sperr Anti-tank Gun Platoon   96

ARTILLERY

Sperr Infantry Gun Platoon   97

ARMOUR

Panzer Platoon   73

Sperr Assault Gun Platoon   97

Veteran Tank-hunter Platoon   164

ARMOUR

Panzer Platoon   73

Veteran Tank-hunter Platoon   164

Panzergrenadier Platoon   81

Panzerpionier Platoon   89

INFANTRY

Gepanzerte Panzergrenadier Platoon   77

Panzergrenadier Platoon   81

Gepanzerte Panzerpionier Platoon   87

Panzerpionier Platoon   89

ARTILLERY

Motorised Artillery Battery   167

Motorised Heavy Artillery Battery   167

Armoured Artillery Battery   168

Armoured Heavy Artillery Battery   168

Self-Propelled Infantry Gun Platoon   85

ANTI-AIRCRAFT

Panzer Anti-aircraft Gun Platoon   75

Luftwaffe Heavy Anti-aircraft Gun Platoon   173

Luftwaffe Light Anti-aircraft Gun Platoon   173

ANTI-AIRCRAFT

Luftwaffe Heavy Anti-aircraft Gun Platoon   173

Luftwaffe Light Anti-aircraft Gun Platoon   173

ANTI-AIRCRAFT

Luftwaffe Light Anti-aircraft Gun Platoon   173

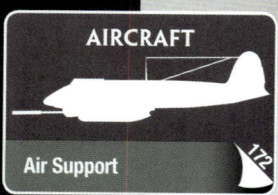

AIRCRAFT

Air Support   172

TIME'S UP

A *Sperrverband* has the most experienced infantry remaining in the division. It is equipped with the greatest amount of firepower they can muster. To buy time for the rest of the division to retreat, each member of the *Sperrverband* realizes their time is up and are resigned to their Fate.

The Sperrverband HQ, and all Sperr and Sperr Pionier Platoons from a Sperrverband are considered Fearless when taking Motivation Tests to launch an Assault or Counterattack a platoon that contains Armoured Tank teams.

ALLIED PLATOONS

Luftwaffe Platoons in your force are Allies and follow the Allies rules in the rulebook.

MOTIVATION AND SKILL

The Sperrverband are rearguard forces organized around the remaining battalion veterans with the grim task of holding while the rest of the division escapes. A Sperrverband is rated as **Reluctant Veteran**.

RELUCTANT	CONSCRIPT
CONFIDENT	TRAINED
FEARLESS	VETERAN

HEADQUARTERS

SPERRVERBAND HQ

HEADQUARTERS

Company HQ	40 points

OPTIONS

- Replace Command SMG teams with Command Panzerfaust SMG teams for +10 points per team.
- Add up to one sniper team for +50 points.
- Add Anti-tank Section with up to two Panzerschreck teams for +25 points per team.

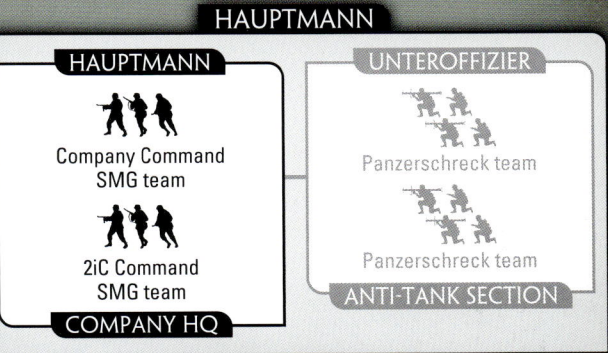

The Company HQ, and all Sperr Platoons and Sperr Pionier Platoons from a Sperrverband are considered Fearless when taking Motivation Tests to launch an assault against Armoured Tank teams or to conduct a Counterattack.

In an attempt to halt the rapid advance of Soviet mechanised forces, German infantry divisions would select the best forces available to form *Sperrverband* (or blocking company).

COMBAT PLATOONS

SPERR PIONIER PLATOON

PLATOON

HQ Section with:

3 Pioneer Squads	190 points
2 Pioneer Squads	135 points

OPTIONS

- Replace Command Pioneer Rifle/MG team with a Command Pioneer Panzerfaust SMG team for +10 points.
- Add Pioneer Supply Maultier half-track for +30 points.
- Add Schwimmwagen amphibious jeep and Maultier half-tracks for +5 points for the platoon.

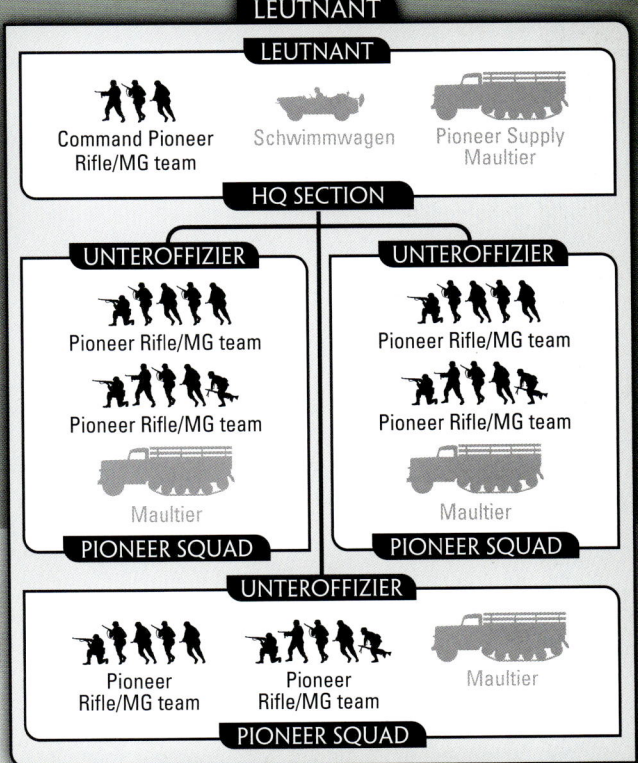

You may replace up to one Pioneer Rifle/MG per Pioneer Squad with a Flame-thrower team at the start of the game before deployment.

PANZERKNACKER BADGE

Experienced pioneers have already earned their *Panzerknacker Abzeichen* (Tank-killer Badge) and are prepared for assaulting Soviet armour by gathering extra anti-tank mines.

All Pioneer teams in a Sperr Pionier Platoon have Tank Assault 5.

The pioneer platoon of a *Sperrverband* constituted the elite of the divisional pioneers. These experienced combat engineers outfitted themselves with everything their extensive combat knowledge and dwindling supply support would allow.

SPERR PLATOON

PLATOON

HQ Section with:

2 Sperr Squads	100 points

OPTIONS

- Replace Command Rifle/MG team with Command Panzerfaust SMG team for +10 points.
- Replace all Rifle/MG teams with MG teams for +30 points for the platoon.
- Add up to two MG42 HMG teams for +30 points per team.

The infantry of the *Sperrverband* provide the firepower to deal with assaulting Soviet infantry. They too were the finest the German infantry battalions had left and were experienced and savvy enough to upgrade their weapons if the opportunity presented itself.

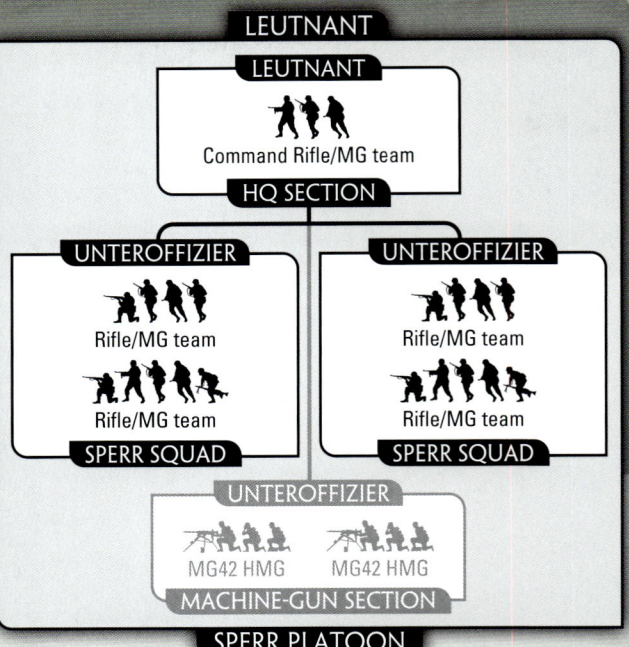

WEAPONS PLATOONS

SPERR ANTI-TANK GUN PLATOON

PLATOON

HQ Section with:

2 Anti-tank Gun Sections	185 points
1 Anti-tank Gun Section	95 points

OPTION

- Add 3-ton trucks and Kfz 15 field car for +5 points for the platoon.

Any and all available anti-tank support was welcome in a *Sperrverband*. The 7.5cm PaK40 will break the back of any tank assault if proper tactics and surprise can be effected.

Place your anti-tank assets in good cover behind your front lines, then spring your attack when the tanks close to overrun your position. Your concentrated fire will stop the assault in its tracks.

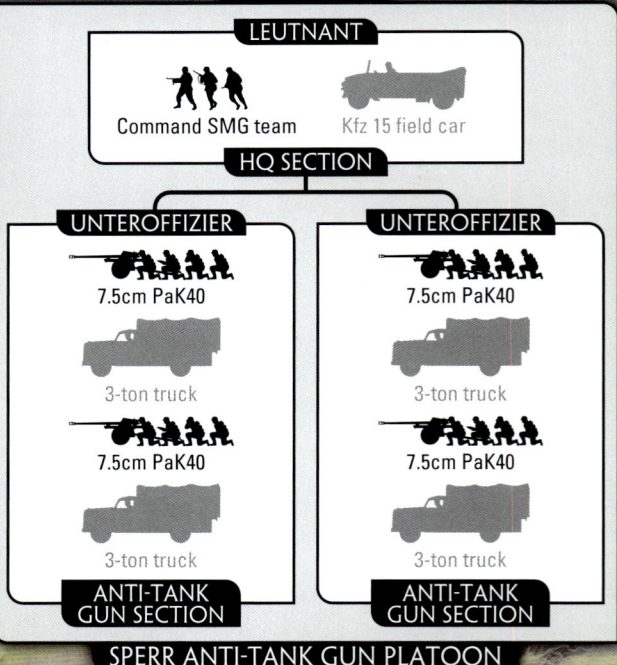

SPERR MORTAR PLATOON

PLATOON

HQ Section with:

2 Mortar Sections	110 points
1 Mortar Section	55 points

OPTION

- Replace all 8cm GW34 mortars with 12cm sGW43 mortars carried by a 3-ton truck or RSO tractor for +15 points per Mortar Section

Providing critical bombardments and smoke in support of the *Sperrverband* platoons will prove the difference in holding off the enemy's assault, allowing the *Sperrverband* to fight on.

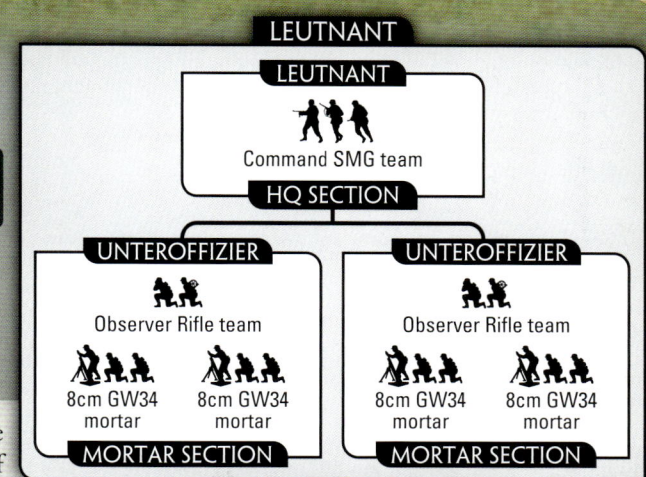

SPERR INFANTRY GUN PLATOON

PLATOON

HQ Section with:

2 7.5cm leIG18	60 points
2 15cm sIG33	125 points

OPTION

- Add RSO tractors for +5 points for the platoon.

Infantry guns bolster the last-ditch defence of the *Sperrverband*. Providing the smoke needed to launch a final assault or pinning enemy infantry before they attack places these guns in a critical support role.

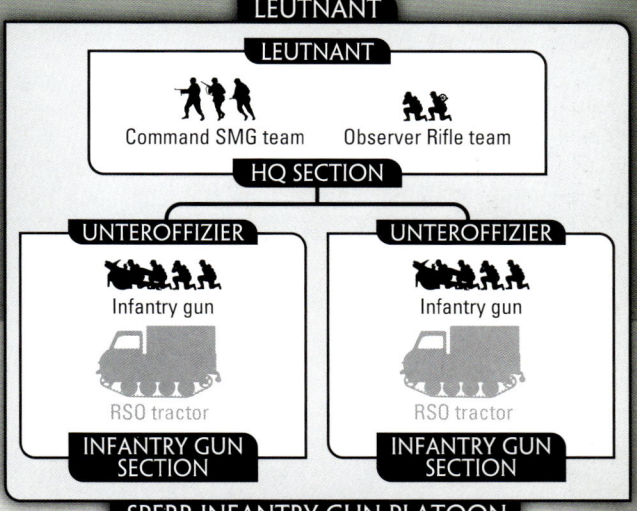

DIVISIONAL SUPPORT PLATOONS

SPERR ASSAULT GUN PLATOON

PLATOON

3 StuG G or StuG IV	255 points
2 StuG G or StuG IV	170 points

OPTION

- Replace one or all StuG G assault guns with StuH42 assault guns at no cost.

You have the honour of supporting the Sperrverband in their defence of the Fatherland. We are all counting on you. Do not fail them.
— *General der Infanterie H. Jordan, Kommandant 9. Armee, July 1944*

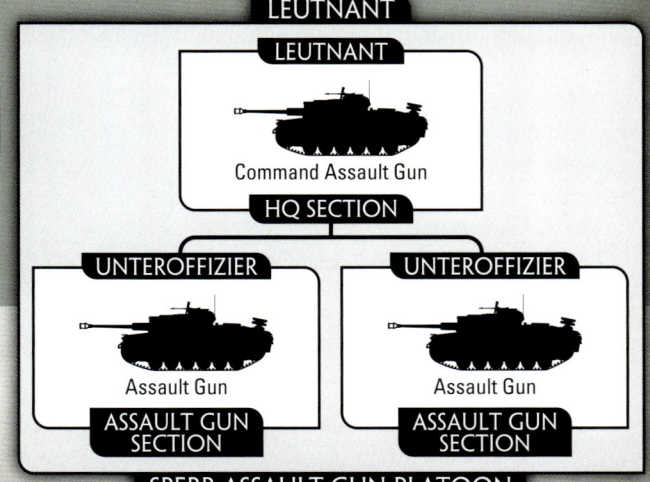

ERSATZ PIONIERKOMPANIE
RESERVE ENGINEER COMPANY

(INFANTRY COMPANY)

HEADQUARTERS

Ersatz Pionierkompanie HQ *99*

You must field one platoon from each box shaded black and may field one platoon from each box shaded grey.

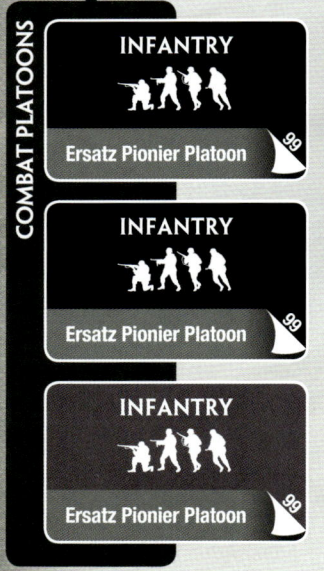

COMBAT PLATOONS

INFANTRY
Ersatz Pionier Platoon *99*

INFANTRY
Ersatz Pionier Platoon *99*

INFANTRY
Ersatz Pionier Platoon *99*

ALLIED PLATOONS

Luftwaffe Platoons in your force are Allies and follow the Allies rules in the rulebook.

DIVISIONAL SUPPORT PLATOONS

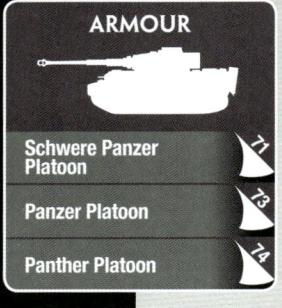

ARMOUR

Schwere Panzer Platoon *71*

Panzer Platoon *73*

Panther Platoon *74*

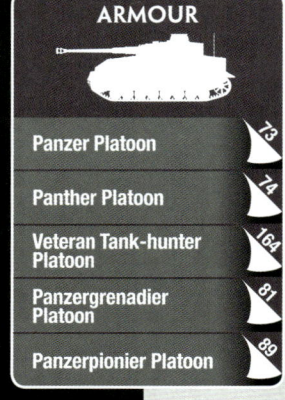

ARMOUR

Panzer Platoon *73*

Panther Platoon *74*

Veteran Tank-hunter Platoon *164*

Panzergrenadier Platoon *81*

Panzerpionier Platoon *89*

ARMOUR

Veteran Tank-hunter Platoon *164*

Tank-hunter Platoon *163*

ARTILLERY

Motorised Artillery Battery *167*

Motorised Heavy Artillery Battery *167*

Armoured Artillery Battery *168*

Armoured Heavy Artillery Battery *168*

Self-Propelled Infantry Gun Platoon *85*

INFANTRY

Gepanzerte Panzergrenadier Platoon *77*

Panzergrenadier Platoon *81*

Gepanzerte Panzerpionier Platoon *87*

Panzerpionier Platoon *89*

ANTI-AIRCRAFT

Panzer Anti-aircraft Gun Platoon *75*

Luftwaffe Heavy Anti-aircraft Gun Platoon *173*

Luftwaffe Light Anti-aircraft Gun Platoon *173*

ANTI-AIRCRAFT

Luftwaffe Heavy Anti-aircraft Gun Platoon *173*

Luftwaffe Light Anti-aircraft Gun Platoon *173*

ANTI-AIRCRAFT

Luftwaffe Light Anti-aircraft Gun Platoon *173*

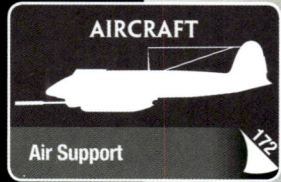

AIRCRAFT

Air Support *172*

ERSATZ PIONIERKOMPANIE

If you meet the 5. Panzerdivision do not engage, go around.
— Soviet order, July 1944

To build a defence against the inevitable, begin with a solid foundation. Use the asset that can take the first attack and return destruction with zeal. If your target is tanks begin with pioneers. Add additional anti-tank support, throw in some armour, anti-air, and a mobile reserve. Set up an ambush and you have a plan for success.
— General der Panzertruppen von Saucken

These thoughts guided von Saucken's defence against advancing Soviet mechanised and tank formations. Using reserve engineers, he held bridgeheads and forced Soviet forward detachments to find new routes to the west.

Forming a solid defensive position is the job of the pioneer. He used the bridge and road builders, the experts at removing obstacles, to create obstacles. He supplied them with extra materials, enhancing their ingenuity for invention.

He added retreating *Luftwaffe* Flak battalions and support from the *5. Panzerdivision* and the *505. Schwere Panzerabteilung* to increase the ability of pioneers to abruptly halt even the most energetic Soviet attack.

In addition he used his *Panzers* as a mobile counterattacking force to provide the firepower to not only stop the offensive but destroy it as well.

MOTIVATION AND SKILL

An Ersatz Pionierkompanie could build hasty fortifications but lacked combat experience. Nevertheless they knew they must stop the Soviet spearheads and fought well. An Ersatz Pionierkompanie is rated as **Confident Trained.**

RELUCTANT	CONSCRIPT
CONFIDENT	**TRAINED**
FEARLESS	VETERAN

HEADQUARTERS

ERSATZ PIONIERKOMPANIE HQ

HEADQUARTERS

Company HQ	40 points

OPTIONS

- Replace Command SMG teams with Command Panzerfaust SMG teams for +10 points per team.
- Add up to two Pioneer Supply Maultier half-tracks for +30 points per half-track.
- Add a Machine-gun Section with up to two MG42 HMG teams for +25 points per team.

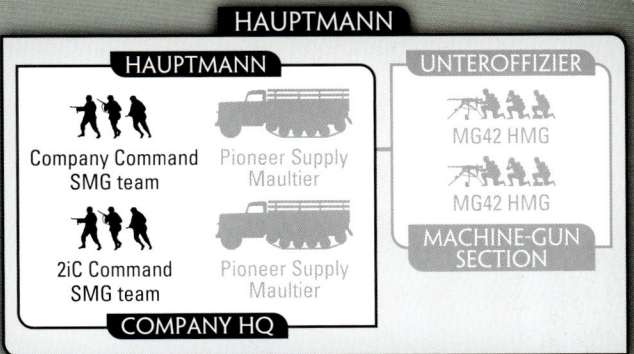

When deploying fortifications using the Pioneer Supply Vehicles rule (see the main rulebook), the Company HQ of an Ersatz Pionierkompanie may, instead of placing Minefields or Barbed Wire, use both of their Pioneer Supply Maultier half-tracks to place one Anti-tank Obstacle.

The *Ersatz Pionierkompanie* or substitute company (pronounced er-zats pi-o-neer kom-pan-ee) possessed a full complement of manpower. However they were limited in the amount of firepower they could bring to bear, and their experience was behind the lines building roads and bridges.

COMBAT PLATOONS

ERSATZ PIONIER PLATOON

PLATOON

HQ Section with:

3 Pioneer Squads	230 points
2 Pioneer Squads	160 points

OPTIONS

- Replace Command Pioneer Rifle team with a Command Panzerfaust Pioneer SMG team for +10 points.
- Equip one Pioneer team with a Goliath demolition carrier in addition to its normal equipment for +30 points.
- Add Pioneer Supply Maultier half-track for +30 points.

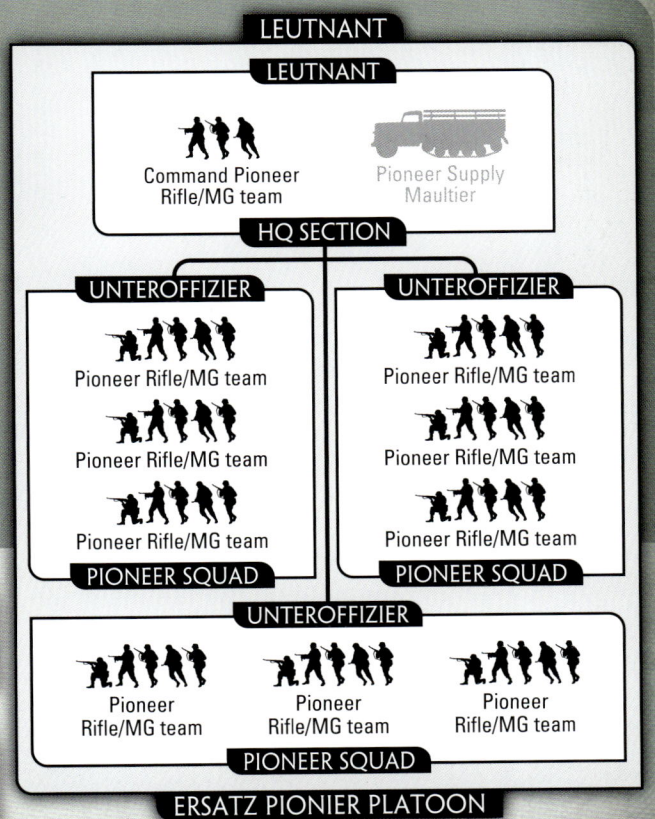

You may replace up to one Pioneer Rifle/MG per Pioneer Squad with a Flame-thrower team at the start of the game before deployment.

PANZERKNACKER BADGE

All Pioneer teams in a Ersatz Pionier Platoon have Tank Assault 5.

IV SS-Panzerkorps advances to meet the Soviet spearhead...

Sturmbanführer Biermeyer blazes into and through the Red armour.

...blunting their efforts to break through the defence.

SS-Panzerpionier complete the counterattack, denying Ivan his bridgehead.

BATTLE FOR POLAND

IV SS-PANZERKORPS

IV SS-Panzerkorps was formed in August 1943 as a headquarters for reforming SS divisions. It also enabled SS divisions to fight together instead of being attached piecemeal to *Heer* (Army) higher formations.

In response to the Soviet successes in Operation Bagration, *IV SS-Panzerkorps* took command of *3. SS-Panzerdivision Totenkopf* and *5. SS-Panzerdivision Wiking* under the leadership of *SS-Obergruppenführer* Herbert-Otto Gille. Until that time, Gille had commanded the *Wiking SS-Division*.

Its first task was to assist *XXXIX Panzerkorps* under *General der Panzertruppen* von Saucken in halting the lead Soviet spearhead east of Warsaw, Poland. These two *Panzerkorps* were successful in virtually destroying the Soviet 2nd Tank Corps at Wolomin and Radzymin 20 kilometres east Warsaw.

Elements of *IV SS-Panzerkorps*, primarily SS security troops from the *Totenkopf SS-Division* were dispatched and took part in the putting down Warsaw Uprising. *IV SS-Panzerkorps* remained in the defence of Warsaw until December 1944.

It was then shifted south to join *6. Armee* in Operation *Konrad*, an attempt to free the Hungarian capital, Budapest, from encirclement. *IV SS-Panzerkorps* continued fighting the Soviet advance through Hungary and Austria into 1945.

After escaping encirclement near Vienna, Austria, the remnants of the *Panzerkorps* surrendered to the Americans on 9 May 1945

HEER FIRE BRIGADES

In the belief that the *Panzerdivisionen* could turn the tide against the Red Army as it stormed westward, the Germans used mobile *Panzerkorps* designed to counterattack the spearheads of advancing Soviet armour. The concept grew out of *General der Panzertruppen* von Saucken's and *Feldmarschall* Model's efforts to create the semblance of a front line using *Sperrverbände* and *Kampfgruppen* fortified by a *Panzerdivision*. These *Panzerkorps*, because of the ever dwindling manpower of Germany, became the last efforts of the dying Nazis' regime to defend the fatherland. These divisions were combined into *Panzerkorps* and readily used as fire brigades in the German attempts to stem the tide of Allied incursions towards the *Reich*.

Four of these armoured divisions, *4. Panzerdivision*, *5. Panzerdivision*, *16. Panzerdivision* and *19. Panzerdivision*, became the backbone of the *Heer* units committed in defending Poland against the Soviet onslaught from the East.

Across the entire line of the Vistula River, these four armoured divisions fought to the death with advancing Soviet armour exacting a heavy toll on the Soviet Army in tanks and manpower and doing their part to finally bring Operation Bagration to a halt at the end in October 1944. The Germans could finally catch their breath and rearm their battered forces.

THE WAFFEN-SS

As well as those featured in this book, other SS formations fought on the Eastern Front in 1944 to 1945.

1. SS-PANZERDIVISION LEIBSTANDARTE SS ADOLF HITLER

In early 1944 the *1. SS-Panzerdivision* fought to hold the line at Zhitomir in the Ukraine. Then as part of *III Panzerkorps* relieved the Korsun Pocket in February. Afterwards most of the division was withdrawn to Belgium, but a *Kampfgruppe* stayed on the front to be encircled in the Kamenets-Poldolsky Pocket in March. The *Kampfgruppe* fought to escape the Soviet encirclement to finally link up with *II SS-Panzerkorps* on 6 April 1944. They then joined the rest of the division in Belgium. In March 1945 the division returned to the east for Operation *Frühlingerwachen* (Spring Awakening), Hitler's last desperate attempt to push the Soviets back in Hungary.

2. SS-PANZERDIVISION DAS REICH

In early 1944 only a portion of *2. SS-Panzerdivision* was fighting in the east as *Kampfgruppe Das Reich*. It was encircled along with other German units in the Kamenets-Poldolsky

Pocket in March. The *II SS-Panzerkorps* relief assault freed the *Kampfgruppe*, and the remnants were sent to France to rest and refit. However, a small number of *Das Reich* units remained in the east until April 1944.

9. SS-PANZERDIVISION HOHENSTAUFEN

Along with *10. SS-Panzerdivision Frundsberg*, the division fought in *II SS-Panzerkorps* during the relief of the Kamenets-Poldolsky Pocket in March 1944. The two divisions provided the spearhead for the operation and were first to link up with the pocket's trapped forces. After suffering heavy casualties during the operation the *Hohenstaufen* division was pulled out of the line to refit in April 1944. It performed fire brigade duties, stamping out Soviet attacks, until it was moved to France in June. They returned to the east to take part in Operation *Frühlingerwachen* (Spring Awakening).

10. SS-PANZERDIVISON FRUNDSBERG

The division was formed from conscripts and first saw action at Tarnopol in April 1944. It took part in the rescue of German troops cut off in the Kamenets-Poldolsky Pocket as part of *II SS-Panzerkorps*. In June it was transferred to the west.

VISTULA BRIGDEHEADS

NAREW BRIDGEHEAD

On 18 July the First Byelorussian Front struck northeast of Warsaw where the bend in the Vistula met the Narew River. A Soviet bridgehead here threatened to cut off German forces in East Prussia. *XXXIX. Panzerkorps* and *IV. SS-Panzerkorps* launched a counterattack on 1 August to stop the Soviets. The battle raged for nearly ten days and when the smoke cleared the Soviet 3rd Tank Corps was encircled and severely mauled. The Soviet 8th Guards Tank Corps suffered similar losses in an effort to break through to their comrades.

WARKA BRIDGEHEAD

Though halted in their dash to the Narew River, the Soviet 8th Guards Army began an assault across the Vistula at Warka just south of Warsaw. The Germans moved the *Herman Göring Panzerdivision* and *19. Panzerdivision* to try and eliminate the bridgehead. Additional Soviet forces were pushed across the river to expand the bridgehead. Fierce fighting continued through August and September, and though the Soviets took heavy casualties they managed to maintain their toehold.

PULAWY BRIDGEHEAD

The Pulawy Bridgehead, nearly 100 kilometres south of Warsaw, was secured by the Soviets on 28 August. Here the Soviets found the weakened remains of the *26. Infanteriedivision* supported by the *19. Panzerdivision*. Using their assault engineers to force the river, the Soviets gained another toehold.

SANDOMIERZ BRIDGEHEAD

Further south along the Vistula River another bridgehead developed. The Soviet First Ukrainian Front had been hammering its way through Army Group North Ukraine since mid-July. By 29 July, they had crossed the Vistula near the town of Sandomierz where the Germans counterattacked. *4. Panzerarmee* with ten German divisions, three Hungarian divisions, six assault gun battalions and *501. Schwere Panzerabteilung* with the new Tiger II heavy tanks rushed to eliminate the Sandomierz bridgehead. After bitter fighting, the Red Army held, but both sides were totally exhausted.

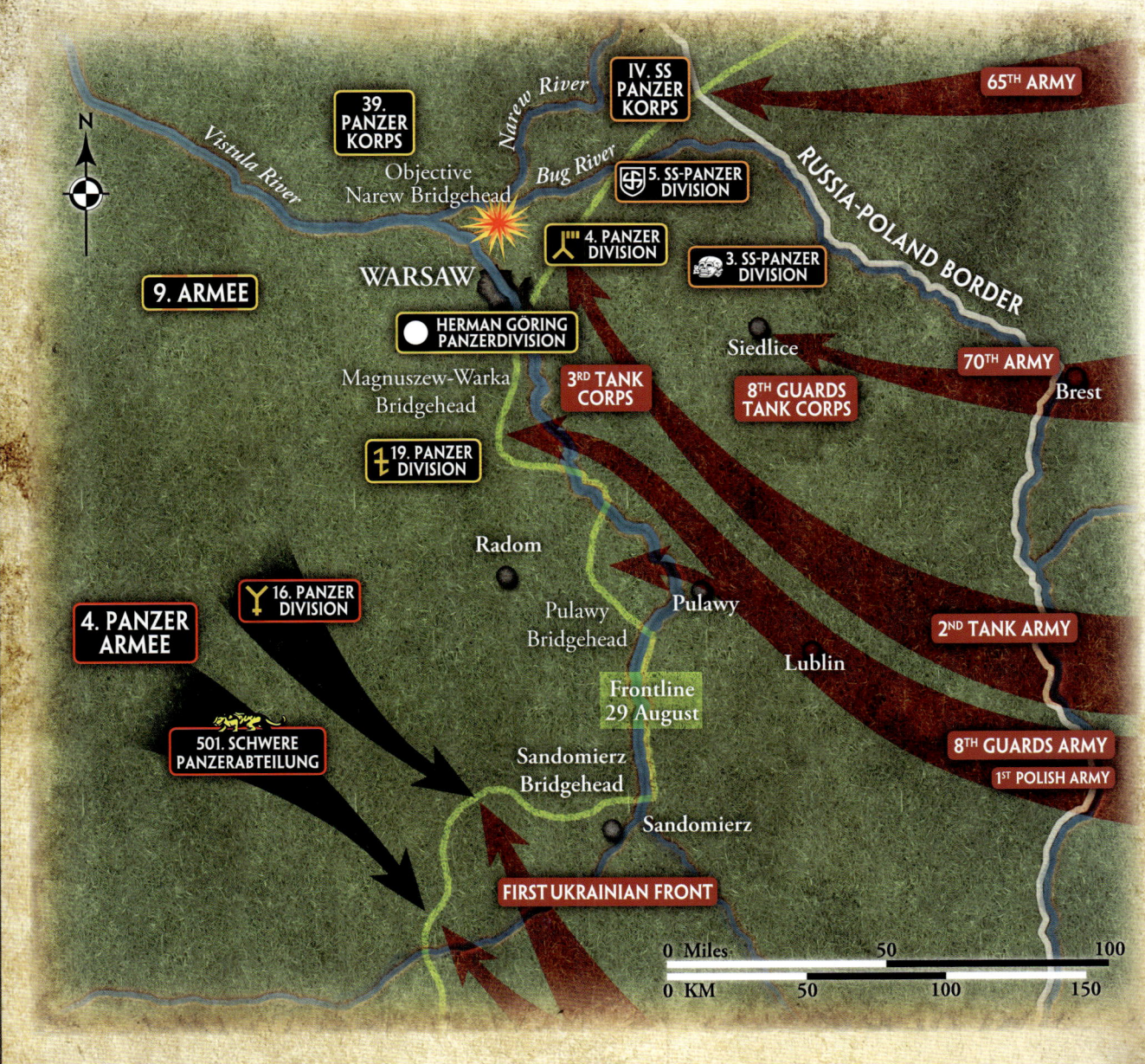

The infamous *3. SS-Panzerdivision Totenkopf* began its history in an inauspicious manner. Initially performing poorly during the campaign in France it was relegated to a support role for the Soviet invasion in 1941. Its reputation as a group of thugs and prison guards brought them nothing but contempt from their SS peers and German Army officers.

By the end of the war the *3. SS-Panzerdivision Totenkopf* had sustained more casualties, nearly 57,000, than any other German division. From nearly being annihilated at Demjansk in 1941, it fought through most of the bloodiest battles on the Eastern Front. Its reputation aside, the soldiers believed themselves to be the elite of the German SS. As such, they gave and received no quarter from their foes.

IV. SS-PANZERKORPS

Initially developed to build new SS divisions in France, *IV. SS Panzerkorps* became the battle headquarters for the *Totenkopf* and *Wiking SS-Panzerdivisionen* in Poland and Hungary during 1944 to 1945.

3. SS-PANZERDIVISION TOTENKOPF OPERATIONAL HISTORY

1940 - Assisted *7. Panzerdivision* in France, suffered heavy losses. Replacements supplied via regular SS recruitment.

1941 - Reserve unit to Army Group North for Operation Barbarrosa, the invasion of the Soviet Union.

Feb-April, 1942 - Surrounded at Demyansk, suffered 13,000 casualties. Surviving 2700 men pulled out in October and sent to France for refit. Eleven Iron Crosses awarded.

November, 1942 - Participated in Vichy France takeover. Received Panzer-regiment and designated *3. SS-Panzergrenadierdivision Totenkopf.*

February, 1943 - Transferred to Kharkov on the Eastern Front. *3. SS-Panzerregiment* receives company of Tiger I heavy tanks.

July, 1943 - *Totenkopf* covers advance on the left flank towards Kursk. Fought in the largest tank battle of the war at Prokhorovka. Division lost half its armour.

July-August, 1943 - Eliminated Soviet bridgehead near Stepanovka. Lost 1500 men and reduced to 20 tanks.

August, 1943 - Along with *Das Reich*, defends Kharkov. Halts offensive and inflicts heavy casualties destroying 800 tanks. However, Soviets outflank defenders, forcing them to abandon city on 23 August.

September, 1943 - Thrown into action against the Kremenchug bridgehead on the Dniepr River.

October-November, 1943 - Reformed as *3. SS-Panzerdivision Totenkopf* Engaged in defensive actions at Krivoi-Rog west of the Dnepr. Destroyed 500 tanks and suffered 20-25% casualties.

January - February, 1944 - Relieved encircled forces at Cherkassy with *1. Panzerdivision*.

April, 1944 - Issued Panther tanks. Fought off heavy Soviet attacks towards Târgul Frumos.

May, 1944 - Reorganized. Received 6000 replacements, 2700 men from the *16. SS-Panzergrenadierdivision Reichsfuhrer*. Strength reaches nearly 20,000.

July-August, 1944 - Held line near Siedlice for 11 days against 9:1 odds. Teamed with *5. SS-Panzerdivision Wiking* to assist in destruction of Soviet 3rd Tank Corps near Wolomin. Held off Soviet attempts to relieve Warsaw.

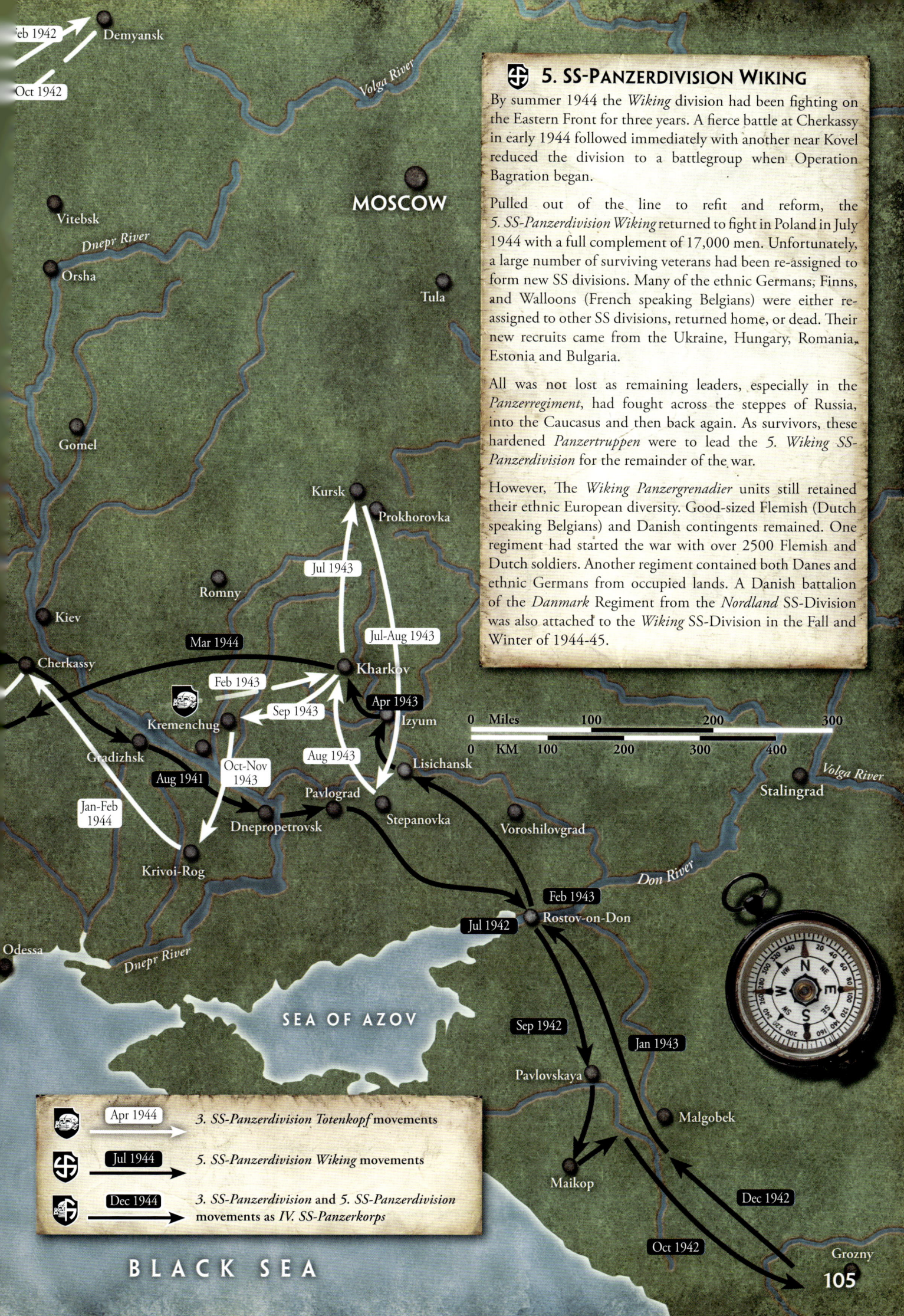

5. SS-PANZERDIVISION WIKING

By summer 1944 the *Wiking* division had been fighting on the Eastern Front for three years. A fierce battle at Cherkassy in early 1944 followed immediately with another near Kovel reduced the division to a battlegroup when Operation Bagration began.

Pulled out of the line to refit and reform, the *5. SS-Panzerdivision Wiking* returned to fight in Poland in July 1944 with a full complement of 17,000 men. Unfortunately, a large number of surviving veterans had been re-assigned to form new SS divisions. Many of the ethnic Germans, Finns, and Walloons (French speaking Belgians) were either re-assigned to other SS divisions, returned home, or dead. Their new recruits came from the Ukraine, Hungary, Romania, Estonia and Bulgaria.

All was not lost as remaining leaders, especially in the *Panzerregiment*, had fought across the steppes of Russia, into the Caucasus and then back again. As survivors, these hardened *Panzertruppen* were to lead the 5. *Wiking SS-Panzerdivision* for the remainder of the war.

However, The *Wiking Panzergrenadier* units still retained their ethnic European diversity. Good-sized Flemish (Dutch speaking Belgians) and Danish contingents remained. One regiment had started the war with over 2500 Flemish and Dutch soldiers. Another regiment contained both Danes and ethnic Germans from occupied lands. A Danish battalion of the *Danmark* Regiment from the *Nordland* SS-Division was also attached to the *Wiking* SS-Division in the Fall and Winter of 1944-45.

Map labels

Volga River
MOSCOW
Demyansk
Feb 1942
Oct 1942
Vitebsk
Dnepr River
Orsha
Tula
Gomel
Kursk
Prokhorovka
Jul 1943
Romny
Jul-Aug 1943
Kiev
Cherkassy
Mar 1944
Feb 1943
Kharkov
Sep 1943
Apr 1943
Kremenchug
Izyum
Gradizhsk
Aug 1943
Lisichansk
Aug 1941
Oct-Nov 1943
Pavlograd
Volga River
Stalingrad
Jan-Feb 1944
Dnepropetrovsk
Stepanovka
Krivoi-Rog
Voroshilovgrad
Feb 1943
Don River
Odessa
Jul 1942
Rostov-on-Don
Dnepr River
SEA OF AZOV
Sep 1942
Jan 1943
Pavlovskaya
Malgobek
Maikop
Dec 1942
Oct 1942
Grozny

Miles 0   100   200   300
KM 100 200 300 400

BLACK SEA

Legend

	Apr 1944	3. SS-Panzerdivision Totenkopf movements
	Jul 1944	5. SS-Panzerdivision Wiking movements
	Dec 1944	3. SS-Panzerdivision and 5. SS-Panzerdivision movements as IV. SS-Panzerkorps

STURMBANNFÜHRER
FRITZ BIERMEYER

A native of Augsburg, Fritz Biermeyer joined the SS in 1933. He was wounded in France fighting with the *Totenkopf SS-Division* as a infantry platoon commander. After recovering he was transferred to the *Das Reich SS-Division* as a tank company commander. When 3. *SS-Panzerregiment* was formed in 1942, he returned to *Totenkopf*.

Biermeyer won his Knight's Cross for leading his company against Soviet tanks near the city of Krasno Konstantinovka in November 1943. The battle began with him supporting infantry repelling an attack. With the enemy repelled, he advanced ahead of the infantry and attacked 38 T-34 tanks and about 800 infantry with his *Panzerkompanie*. While he personally accounted for six destroyed T-34 tanks, his company's total reached 31 through his aggressive leadership.

His aggressive, attacking reputation continued when he later took on another Soviet tank battalion with his *Panzerkompanie* supported by an armoured pioneer company and two assault guns. By the end of the day his company had destroyed eleven M4 Sherman, a T-34 tank, a KV-85 heavy tank and numerous guns.

Finally, during the defence of Warsaw in August 1944, he led his company in five separate attacks to stop the Bolsheviks. His 5. *Kompanie* reported seven T-34 tanks, eight KV-85 and one IS-2 heavy tanks, and four 4.5cm guns destroyed. His tactical skill and unwavering cold-bloodedness won the day, though he had only a few panzers left at the end of it.

Biermeyer was killed near Modlin on 10 November 1944 and was posthumously awarded the Knight's Cross with Oak Leaves.

CHARACTERISTICS

Sturmbannführer Fritz Biermeyer is a Warrior Tank team who leads a Panzer Platoon in a *Totenkopf SS-Panzerkampfgruppe* or *SS-Panzergrenadierkampfgrppe*. You must purchase him and his platoon together. Biermeyer and his platoon are rated **Fearless Veteran**.

BIERMEYER'S 3 SS-PANZER PLATOON

Biermeyer's Panzer Platoon was always found in the thick of battle and suffered the greatest losses within the division.

Biermeyer and his platoon cannot be joined by Independent teams. The platoon is organized as a standard Panzer Platoon as on page 109 with Biermeyer as the Platoon Command Tank for the following point costs.

BIERMEYER'S PANZER IV PLATOON

PLATOON

5 Panzer IV H	325 points
4 Panzer IV H	260 points
3 Panzer IV H	195 points

RECKLESS

Sturmbannführer Biermeyer drove his subordinates on to victory regardless of the danger or tactical situation.

*Sturmbannführer Biermeyer and his platoon are hit on a 3+ as if they were **Trained**.*

NEW RECRUITS

Sturmbannführer Biermeyer's losses required his platoons to be refit with newly trained *Panzertruppen*.

*If Biermeyer is Destroyed then the platoon becomes **Confident Trained** for the remainder of the game.*

TOWARDS THE ENEMY

When Biermeyer spotted his enemy he immediately ordered his panzers to close and attack.

In the Assault Step, a platoon led by Biermeyer must assault if they can. If they cannot assault they must attempt to make a Stormtroopers Move as far as possible towards the closest enemy team or closest Objective.

PANZER KANONEN

Thirteen members of the *5. SS-Panzerregiment* of the *5. SS-Panzerdivision Wiking* were awarded the Knight's Cross. Though individual kill numbers are not available, *Obersturmführer* Senghas led them with over 30 armoured kills. The division's first *Panzerabteilung* (tank battalion) was formed in early 1942 when *Wiking* was upgraded to a *SS-Panzergrenadierdivision*. The cadre of these *Panzer Kanonen* (Panzer Aces) began then and stayed with the division throughout the war. Nearly all of these *Panzer Kanonen* survived the war, despite the regiment suffering severe casualties in fighting the Soviets across Russia, the Ukraine, Poland, and Hungary.

Panzer Aces of 5. *SS-Panzerdivision Wiking*: from left to right, **Top row:** *Oberscharführer* Helmut Bauer, *Obersturmführer* Fritz Darges, *Hauptscharführer* Sepp Draxenberger, *Hauptsturmführer* Hans Flugel, *Untersturmführer* Alfred Grossrock, *Obersturmführer* Willi Hein, *Hauptsturmführer* Karl-Heinz Lichte. **Bottom row:** *Sturmbannführer* Johannes Mühlenkamp, *Obersturmführer* Karl Nicolussi-Leck, *Obersturmführer* Karl Picus, *Obersharführer* Hugo Ruf, *Untersturmführer* Kurt Schumacher, *Obersturmführer* Paul Senghas.

CHARACTERISTICS

You may replace any or all Platoon Command teams from any SS-Panzer Platoons in the Wiking SS-Division with a Panzer Kanone for +65 points per platoon. A Panzer Kanone is a Warrior with the following abilities:

PANZER KANONE

At the start of the game roll a die for each Panzer Kanone to determine the medal he has been awarded. That Panzer Kanone possesses the skills shown for that medal in the table below.

 ### KNIGHT'S CROSS

 A Panzer Kanone with a Knight's Cross may:
- Have Tank teams in their platoon re-roll failed Skill Tests.

 ### KNIGHT'S CROSS WITH OAK LEAVES

 A Panzer Kanone with a Knight's Cross with Oak Leaves may:
- Move and Shoot with full ROF, and
- Have Tank teams in their platoon re-roll failed Skill Tests.

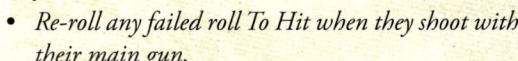 ### KNIGHT'S CROSS WITH SWORDS

 A Panzer Kanone with a Knight's Cross with Swords may:
- Re-roll any failed roll To Hit when they shoot with their main gun,
- Move and Shoot with full ROF, and
- Have Tank teams in their platoon re-roll failed Skill Tests.

 ### KNIGHT'S CROSS WITH DIAMONDS

 A Panzer Kanone with a Knight's Cross with Diamonds may:
- Re-roll any failed roll To Hit when they shoot with their main gun,
- Move and Shoot with full ROF,
- Only be destroyed on an enemy roll of 5+ rather than a roll of 4+ when using the Warrior Tank Team Casualties rule in the rulebook, and
- Have Tank teams in their platoon re-roll failed Skill Tests.

SS-PANZERKAMPFGRUPPE
SS ARMOURED BATTLEGROUP

(TANK COMPANY)

HEADQUARTERS

SS-Panzerkampfgruppe HQ — 109

You must field one platoon from each box shaded black and may field one platoon from each box shaded grey.

Your Company HQ, Combat and Support platoons must be all either from 3. SS-Panzerdivision 'Totenkopf' (Marked ☠) or 5. SS-Panzerdivision 'Wiking' (Marked ⛨), unless otherwise noted.

COMBAT PLATOONS

ARMOUR

SS-Panzer Platoon — 109

Biermeyer's 3. SS-Panzer Platoon — 106

ARMOUR

SS-Panzer Platoon — 109

INFANTRY

Gepanzerte SS-Panzergrenadier Platoon — 111

SS-Panzergrenadier Platoon — 112

ALLIED PLATOONS

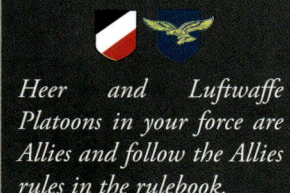

Heer and Luftwaffe Platoons in your force are Allies and follow the Allies rules in the rulebook.

ARMOUR

Armoured SS-Cannon Platoon — 116

WEAPONS PLATOONS

ANTI-AIRCRAFT

Armoured SS-Anti-aircraft Gun Platoon — 112

MACHINE-GUNS

SS-Heavy Platoon — 113

Armoured SS-Heavy Platoon — 113

ARTILLERY

SS-Mortar Platoon — 114

Armoured SS-Mortar Platoon — 114

ARTILLERY

SS-Self-propelled Infantry Gun Platoon — 115

ANTI-TANK

SS-Anti-tank Gun Platoon — 115

RECONNAISSANCE

SS-Scout Platoon — 116

FLAME-THROWER

SS-Armoured Flame-thrower Platoon — 116

DIVISIONAL SUPPORT PLATOONS

ARMOUR

Heavy SS-Tank Platoon — 174

ARMOUR

SS-Panzer Platoon — 109

SS-Tank-hunter Platoon — 176

INFANTRY

Gepanzerte SS-Panzergrenadier Platoon — 111

SS-Panzergrenadier Platoon — 112

SS-Aufklarüngs Platoon — 175

SS-Panzergrenadier Pioneer Platoon — 177

Gepanzerte SS-Panzerpionier Platoon — 177

RECONNAISSANCE

SS-Scout platoon — 116

SS-Panzerspäh Platoon — 176

Heavy SS-Panzerspäh Platoon — 176

ARTILLERY

SS-Kampfgruppe Artillery Battery — 117

SS-Kampfgruppe Heavy Artillery Battery — 117

Armoured SS-Artillery Battery — 179

Armoured Heavy SS-Artillery Battery — 179

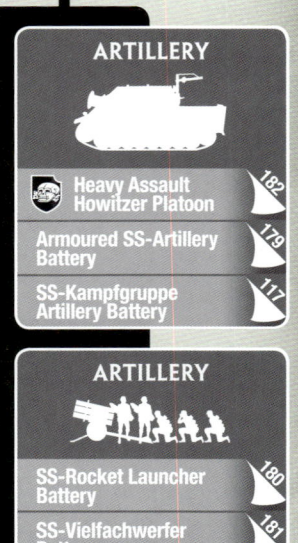

ARTILLERY

Heavy Assault Howitzer Platoon — 182

Armoured SS-Artillery Battery — 179

SS-Kampfgruppe Artillery Battery — 117

ARTILLERY

SS-Rocket Launcher Battery — 180

SS-Vielfachwerfer Battery — 181

ANTI-TANK

Heavy SS-Anti-aircraft Gun Platoon — 180

Heavy Anti-tank Gun Platoon — 165

ANTI-AIRCRAFT

Luftwaffe Heavy Anti-aircraft Gun Platoon — 173

Luftwaffe Light Anti-aircraft Gun Platoon — 173

AIRCRAFT

Air Support — 172

MOTIVATION AND SKILL

 3. SS-Panzerdivision Totenkopf *retained its core of combat veterans. A Totenkopf SS-Panzerkampfgruppe is rated* **Fearless Veteran.**

A Totenkopf SS-Panzerkampfgruppe uses the Totenkopf Waffen-SS special rules on page 188 as well as all the normal German special rules on pages 183 to 187.

 5. SS-Panzerdivision Wiking *lost many of its veterans in combat and to outfit new SS-Divisions. A Wiking SS-Panzerkampfgruppe is rated* **Fearless Trained.**

A Wiking SS-Panzerkampfgruppe uses the Wiking Waffen-SS special rules on page 188 as well as all the normal German special rules on pages 183 to 187.

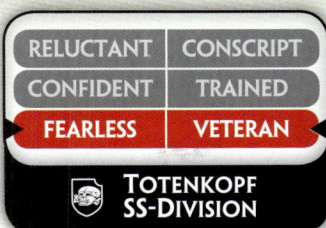
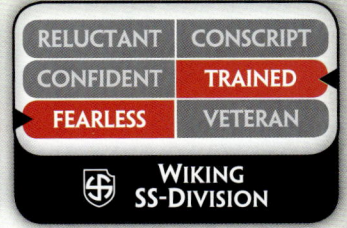

RELUCTANT	CONSCRIPT
CONFIDENT	TRAINED
FEARLESS	**VETERAN**

TOTENKOPF SS-DIVISION

RELUCTANT	CONSCRIPT
CONFIDENT	**TRAINED**
FEARLESS	VETERAN

WIKING SS-DIVISION

HEADQUARTERS

SS-PANZERKAMPFGRUPPE HQ

SS-PANZER HQ

2 Panzer IV H	200 points	155 points
2 Panther A or G	425 points	325 points

OPTION

• Add a Famo or Bergepanther recovery vehicle for +15 points.

 Remember to roll for your Panzer Ace Skill for your Totenkopf Company Command tank before each game.

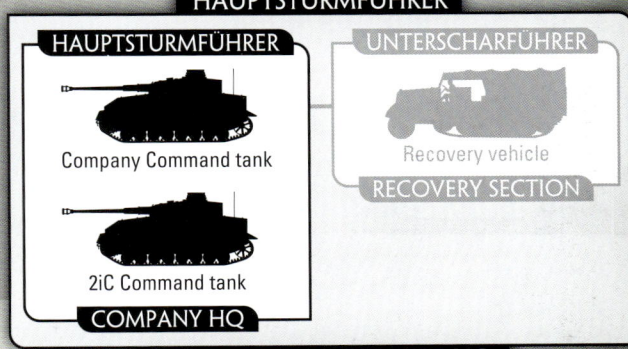

HAUPTSTURMFÜHRER

HAUPTSTURMFÜHRER — Company Command tank

UNTERSCHARFÜHRER — Recovery vehicle — RECOVERY SECTION

2iC Command tank — COMPANY HQ

SS-PANZERKAMPFGRUPPE HQ

COMBAT PLATOONS

SS-PANZER PLATOON

PLATOON

5 Panzer IV H	500 points	390 points
4 Panzer IV H	400 points	310 points
3 Panzer IV H	300 points	230 points
5 Panther A or G	1060 points	815 points
4 Panther A or G	850 points	655 points
3 Panther A or G	640 points	490 points
4 StuG G or StuG IV	435 points	335 points
3 StuG G or StuG IV	325 points	250 points

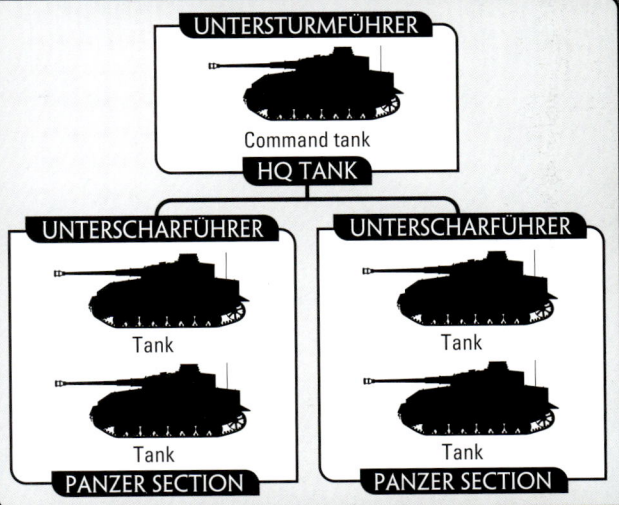

UNTERSTURMFÜHRER

UNTERSTURMFÜHRER — Command tank — HQ TANK

UNTERSCHARFÜHRER — Tank / Tank — PANZER SECTION

UNTERSCHARFÜHRER — Tank / Tank — PANZER SECTION

SS-PANZER PLATOON

SS-PANZERGRENADIERKAMPFGRUPPE
SS MECHANISED INFANTRY BATTLEGROUP

(MECHANISED COMPANY)

HEADQUARTERS

HEADQUARTERS
SS-Panzergrenadier-kampfgruppe HQ — 111

You must field one platoon from each box shaded black and may field one platoon from each box shaded grey.

Your Company HQ, Combat and Support platoons must be all either from 3. SS-Panzerdivision 'Totenkopf' (Marked ⊡) or 5. SS-Panzerdivision 'Wiking' (Marked ⊞), unless otherwise noted.

DIVISIONAL SUPPORT PLATOONS

COMBAT PLATOONS

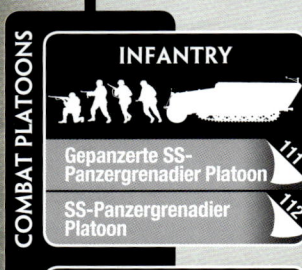

INFANTRY
- Gepanzerte SS-Panzergrenadier Platoon — 111
- SS-Panzergrenadier Platoon — 112

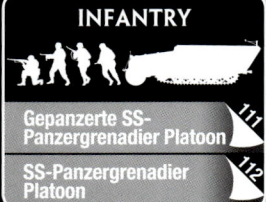

INFANTRY
- Gepanzerte SS-Panzergrenadier Platoon — 111
- SS-Panzergrenadier Platoon — 112

ARMOUR
- SS-Panzer Platoon — 109
- Biermeyer's 3. SS-Panzer Platoon — 106

ALLIED PLATOONS

Heer and Luftwaffe Platoons in your force are Allies and follow the Allies rules in the rulebook.

ARMOUR
- Armoured SS-Cannon Platoon — 116

WEAPONS PLATOONS

ANTI-AIRCRAFT
- Armoured SS-Anti-aircraft Gun Platoon — 112

MACHINE-GUNS
- SS-Heavy Platoon — 113
- Armoured SS-Heavy Platoon — 113

ARTILLERY
- SS-Mortar Platoon — 114
- Armoured SS-Mortar Platoon — 114

ARTILLERY
- SS-Self-propelled Infantry Gun Platoon — 115

ANTI-TANK
- SS-Anti-tank Gun Platoon — 115

RECONNAISSANCE
- SS-Scout Platoon — 116

FLAME-THROWER
- Armoured SS-Flame-thrower Platoon — 116

ARMOUR
- Heavy SS-Tank Platoon — 174

ARMOUR
- SS-Panzer Platoon — 109
- SS-Tank-hunter Platoon — 175

INFANTRY
- Gepanzerte SS-Panzergrenadier Platoon — 111
- SS-Panzergrenadier Platoon — 112
- SS-Aufklarüngs Platoon — 175
- SS-Panzergrenadier Pioneer Platoon — 177
- Gepanzerte SS-Panzerpionier Platoon — 177

RECONNAISSANCE
- SS-Scout platoon — 116
- SS-Panzerspäh Platoon — 176
- Heavy SS-Panzerspäh Platoon — 176

ARTILLERY
- SS-Kampfgruppe Artillery Battery — 117
- SS-Kampfgruppe Heavy Artillery Battery — 117
- Armoured SS-Artillery Battery — 179
- Armoured Heavy SS-Artillery Battery — 179

ARTILLERY
- Heavy Assault Howitzer Platoon — 182
- Armoured SS-Artillery Battery — 179
- SS-Kampfgruppe Artillery Battery — 117

ARTILLERY
- SS-Rocket Launcher Battery — 180
- SS-Vielfachwerfer Battery — 181

ANTI-TANK
- Heavy SS-Anti-aircraft Gun Platoon — 180
- Heavy Anti-tank Gun Platoon — 165

ANTI-AIRCRAFT
- Luftwaffe Heavy Anti-aircraft Gun Platoon — 173
- Luftwaffe Light Anti-aircraft Gun Platoon — 173

AIRCRAFT
- Air Support — 172

MOTIVATION AND SKILL

 3. SS-Panzerdivision Totenkopf *retained its combat veterans. A Totenkopf SS-Panzergrenadierkampfgruppe is rated* **Fearless Veteran.**

A Totenkopf SS-Panzergrenadierkampfgruppe uses the Totenkopf Waffen-SS special rules on page 188 as well as all the normal German special rules on pages 183 to 187.

 5. SS-Panzerdivision Wiking *lost many of its veterans in combat and to outfit new SS-Divisions. A Wiking SS-Panzergrenadierkampfgruppe is rated* **Fearless Trained.**

A Wiking SS-Panzergrenadierkampfgruppe uses the Wiking Waffen-SS special rules on page 188 as well as all the normal German special rules on pages 183 to 187.

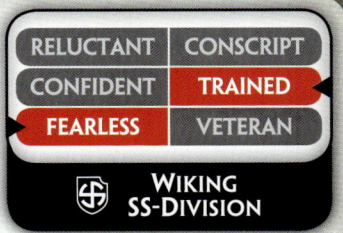

RELUCTANT	CONSCRIPT
CONFIDENT	TRAINED
FEARLESS	**VETERAN**

TOTENKOPF SS-DIVISION

RELUCTANT	CONSCRIPT
CONFIDENT	**TRAINED**
FEARLESS	VETERAN

WIKING SS-DIVISION

HEADQUARTERS

SS-PANZERKAMPFGRUPPE HQ

SS-PANZERGRENADIER HQ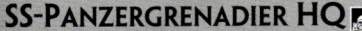

Company HQ	75 points	60 points

OPTION

- Replace either or both Command SMG teams with Command Panzerfaust SMG teams for +10 points per team.
- Add an Anti-tank Section with:

2 Panzerschreck teams	+75 points	+65 points
1 Panzerschreck team	+40 points	+35 points

A SS-Panzergrenadierkampfgruppe HQ may use the Mounted Assault special rule.

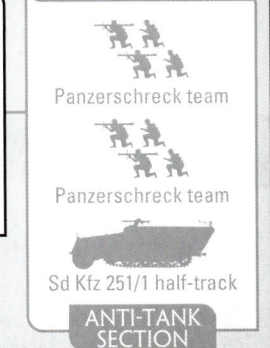

HAUPTSTURMFÜHRER

HAUPTSTURMFÜHRER

Company Command SMG team — Sd Kfz 251/1 half-track
2iC Command SMG team — Sd Kfz 251/1 half-track

COMPANY HQ

UNTERSCHARFÜHRER

Panzerschreck team
Panzerschreck team
Sd Kfz 251/1 half-track

ANTI-TANK SECTION

SS-PANZERGRENADIERKAMPGRUPPE HQ

COMBAT PLATOONS

GEPANZERTE SS-PANZERGRENADIER PLATOON

PLATOON

HQ Section with:		
3 Panzergrenadier Squads	260 points	200 points
2 Panzergrenadier Squads	190 points	145 points

OPTIONS

- Replace the Command MG team with Command Panzerfaust SMG team for +10 points.
- Replace up to one MG team per squad with a Panzerfaust MG team for +10 points per squad.

Gepanzerte SS-Panzergrenadier Platoons may use the Mounted Assault special rule.

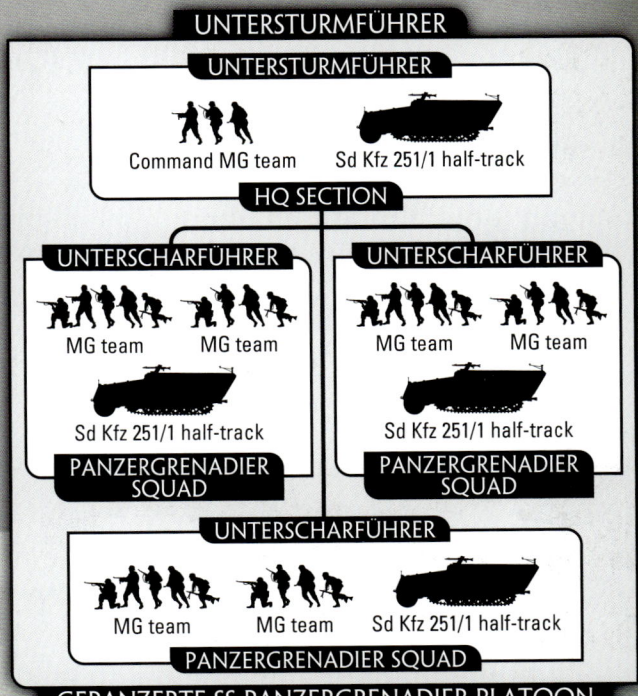

UNTERSTURMFÜHRER

UNTERSTURMFÜHRER

Command MG team — Sd Kfz 251/1 half-track

HQ SECTION

UNTERSCHARFÜHRER

MG team — MG team
Sd Kfz 251/1 half-track

PANZERGRENADIER SQUAD

UNTERSCHARFÜHRER

MG team — MG team
Sd Kfz 251/1 half-track

PANZERGRENADIER SQUAD

UNTERSCHARFÜHRER

MG team — MG team — Sd Kfz 251/1 half-track

PANZERGRENADIER SQUAD

GEPANZERTE SS-PANZERGRENADIER PLATOON

SS-Panzergrenadier Platoon

Platoon

HQ Section with:

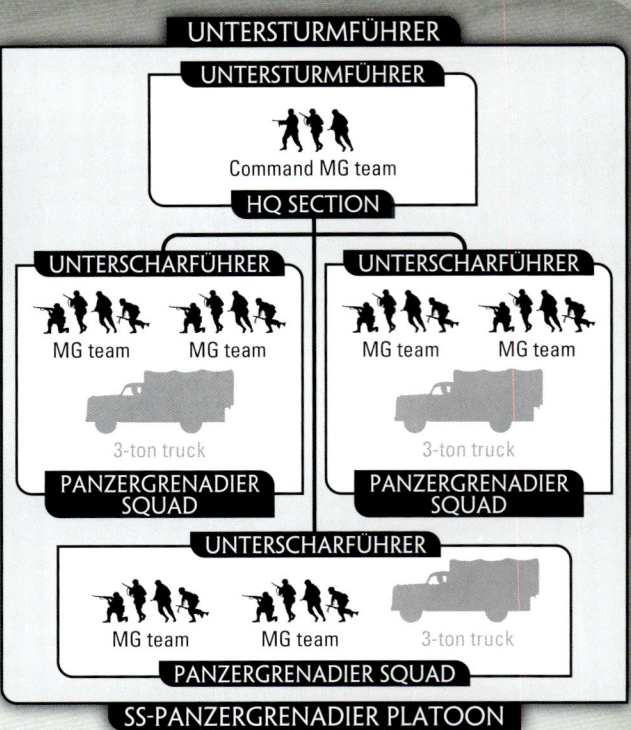

3 Panzergrenadier Squads	220 points	165 points
2 Panzergrenadier Squads	155 points	115 points

Options

- Replace the Command MG team with a Command Panzerfaust SMG team for +10 points.
- Replace up to one MG team per squad with a Panzerfaust MG team for +10 points per squad.
- Add 3-ton trucks for +5 points for the platoon.

Panzergrenadiers are the perfect complement to Panzers. The SS divisions found that mixing both together provided the best flexibility in fighting the Soviets hordes.

Whether the mission was offence, defence, or counterattacking Soviet armoured thrusts, having panzergrenadiers in your force allows you to hold objectives, protect your panzers, and advance against the enemy.

WEAPONS PLATOONS

Armoured SS-Anti-aircraft Gun Platoon

Platoon

3 Armoured Sd Kfz 7/1 (Quad 2cm)	185 points	140 points
2 Armoured Sd Kfz 7/1 (Quad 2cm)	125 points	95 points
3 Armoured Sd Kfz 10/5 (2cm)	135 points	100 points
2 Armoured Sd Kfz 10/5 (2cm)	90 points	70 points

SS-HEAVY PLATOON

PLATOON

HQ Section with:	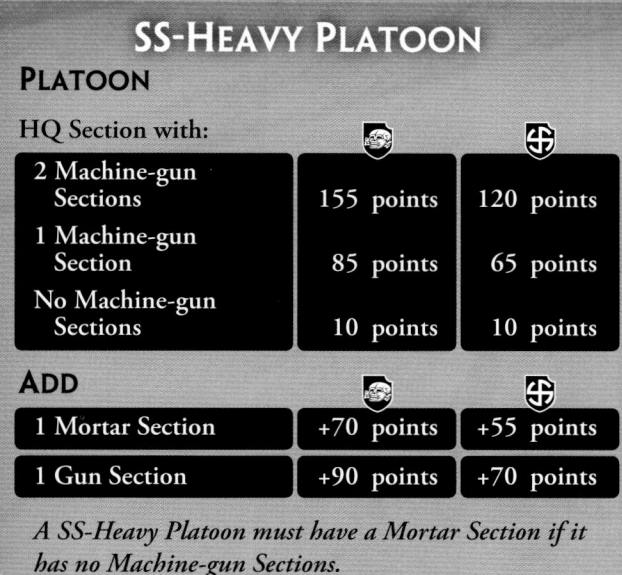	
2 Machine-gun Sections	155 points	120 points
1 Machine-gun Section	85 points	65 points
No Machine-gun Sections	10 points	10 points

ADD

1 Mortar Section	+70 points	+55 points
1 Gun Section	+90 points	+70 points

A SS-Heavy Platoon must have a Mortar Section if it has no Machine-gun Sections.

SS-Heavy Platoons may make Combat Attachments to SS-Panzergrenadier Platoons.

In one platoon you can consolidate all the heavy weapons support you may need to complement your SS-Panzergrenadiers whether they be on offence, defence or in counterattack. This one platoon can provide the necessary firepower to halt most any infantry assaults against you.

The heavy machine-guns can provide covering fire for your advancing infantry, while the mortars can keep the enemy's heads down.

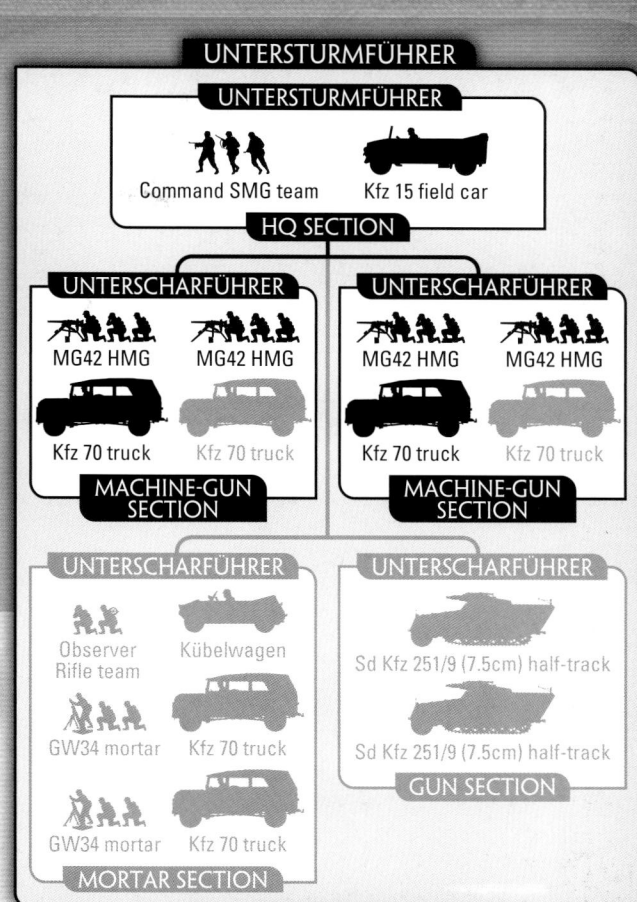

ARMOURED SS-HEAVY PLATOON

PLATOON

HQ Section with:		
2 Machine-gun Sections	200 points	130 points
1 Machine-gun Section	110 points	75 points
No Machine-gun Sections	20 points	15 points

ADD

1 Mortar Section	+100 points	+75 points
1 Gun Section	+90 points	+70 points

An Armoured SS-Heavy Platoon must have a Mortar Section if it has no Machine-gun Sections.

Armoured SS-Heavy Platoons may make Combat Attachments to Gepanzerte SS-Panzergrenadier Platoons.

If you assault or counterattack a heavy weapons platoon can provide the edge required to ensure your assaulting forces reach the enemy lines with minimal casualties.

A cannon platoon can silence enemy anti-tank guns and infantry guns that threaten your own armour or half-tracks.

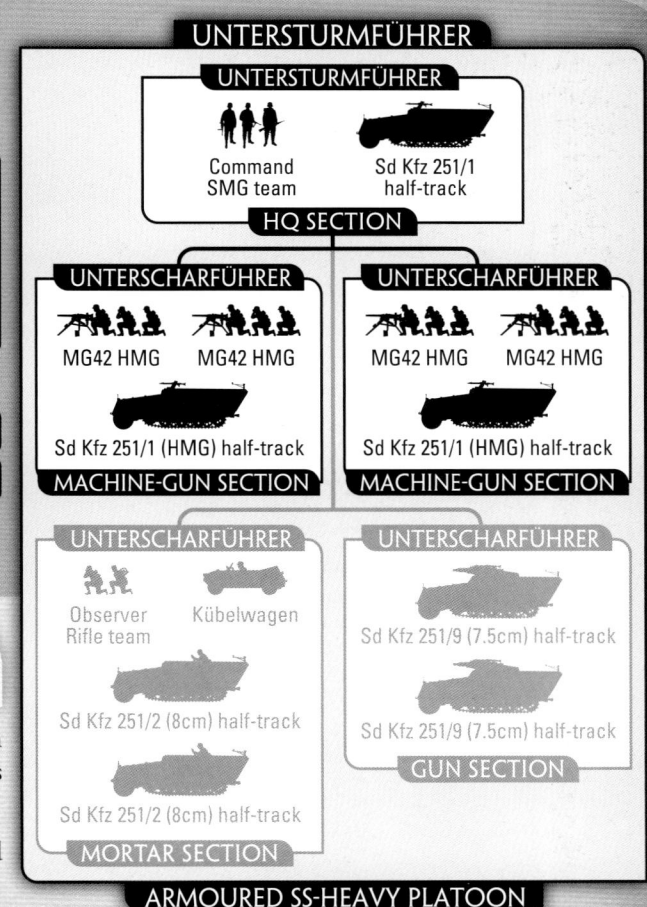

SS-Mortar Platoon

PLATOON

HQ Section with:

3 Mortar Sections	215 points	165 points
2 Mortar Sections	155 points	120 points
1 Mortar Section	85 points	60 points

The 8cm GW34 mortar can pin down enemy attacks as well as deliver covering smoke. The SS mortar crew move their tubes up in trucks to keep pace with the panzergrenadiers. Positioned just behind the infantry, they are instantly ready to give fire support.

Whether they are breaking up enemy attacks, supporting a *SS-Panzergrenadier* counterattack, or dropping smoke on the enemy to conceal their company's movement, these weapons give the infantry artillery support at short notice.

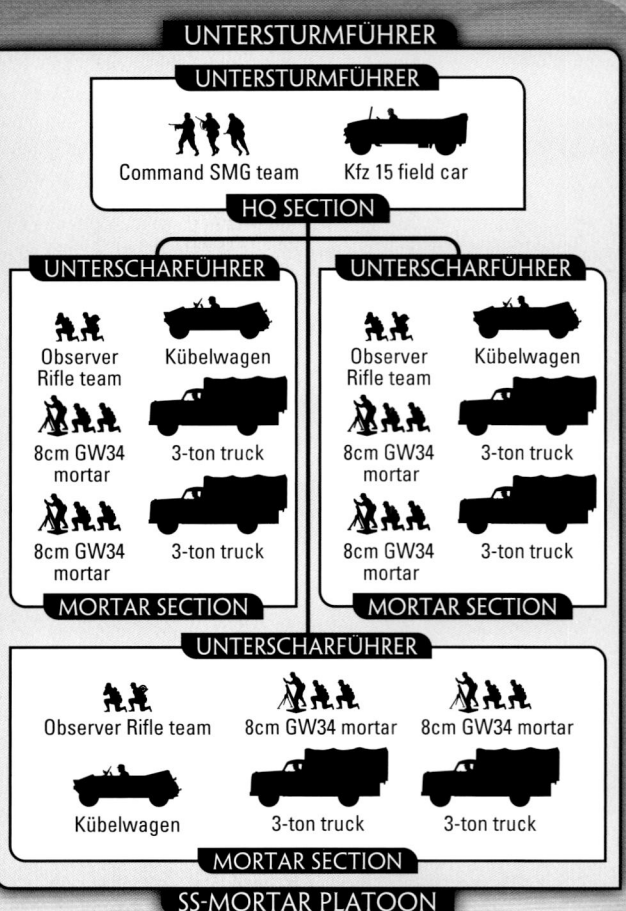

ARMOURED SS-Mortar Platoon

PLATOON

HQ Section with:

3 Mortar Sections	285 points	220 points
2 Mortar Sections	205 points	155 points
1 Mortar Section	110 points	85 points

Mobile mortars with armoured protection can provide the needed artillery support for your mechanised company. With their overabundance of spotters, this platoon is most assured of finding and hitting any enemy unit in range regardless of terrain.

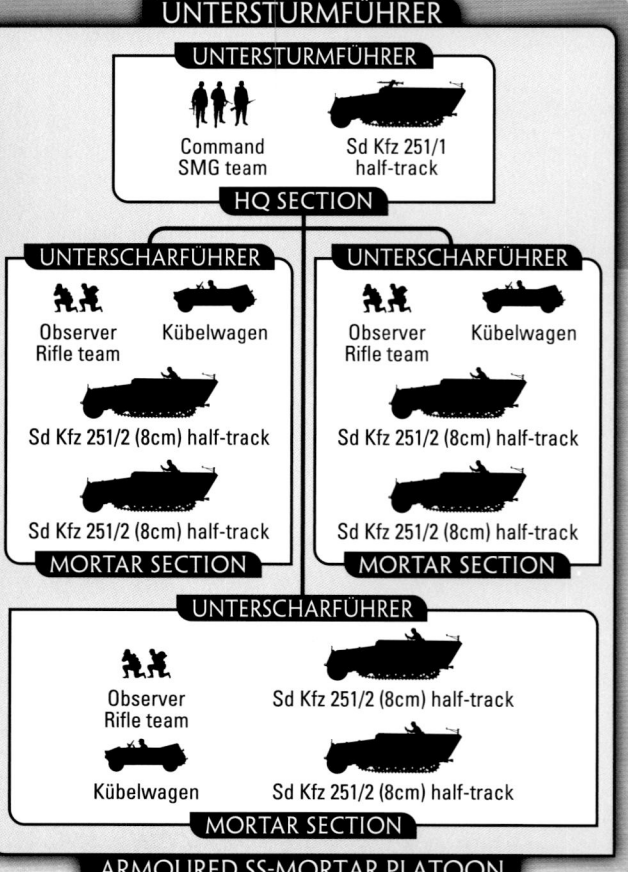

SS-SELF-PROPELLED INFANTRY GUN PLATOON

PLATOON

HQ Section with:		
2 Gun Sections	195 points	145 points
1 Gun Section	115 points	85 points

The self-propelled infantry gun platoon has a flexibility surpassing most other combat units. The ability to provide artillery support is but one aspect where this unit excels.

If your target is enemy infantry these guns can provide both indirect and direct fire support. It can deal with dug-in enemy infantry, harassing anti-tank and machine guns, as well as bringing a heavy shell against an enemy tank at close-quarters.

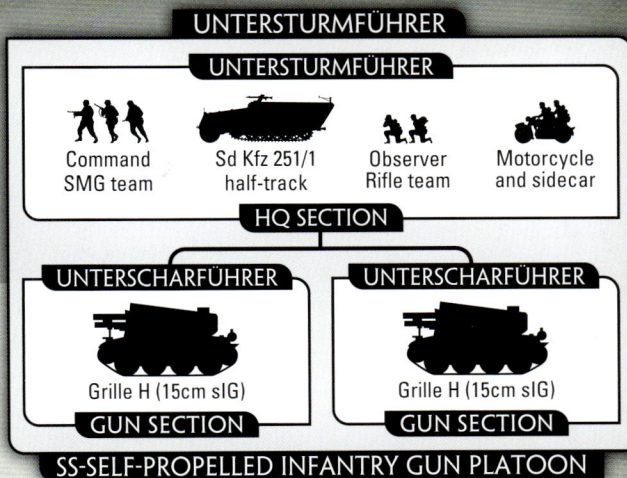

SS-ANTI-TANK GUN PLATOON

PLATOON

HQ Section with:		
3 7.5cm PaK40	185 points	140 points
2 7.5cm PaK40	125 points	95 points
1 7.5cm PaK40	65 points	50 points
3 7.6cm PaK 36(r)	140 points	105 points
2 7.6cm PaK 36(r)	95 points	70 points
1 7.6cm PaK 36(r)	50 points	35 points

OPTION

- Replace Kubelwagen and all 3-ton trucks with Sd Kfz 251/1 half-tracks for +10 points per section.

SS-Anti-tank Gun Platoons may make Combat Attachments to SS-Panzergrenadier and Gepanzerte SS-Panzergrenadier Combat Platoons.

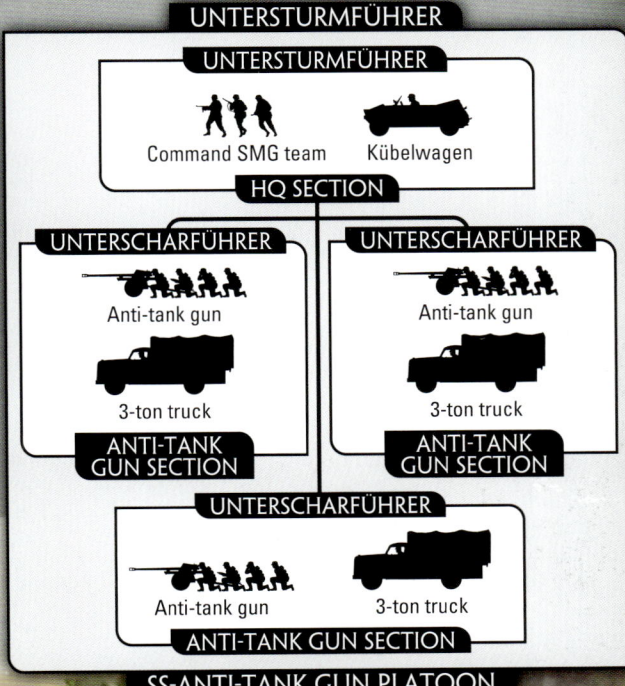

SS-Scout Platoon

Platoon

HQ Section with:

3 Scout Squads	175 points	135 points
2 Scout Squads	130 points	100 points

Option

- Replace all Motorcycle teams with the equivalent Schwimmwagen teams for +5 points for the platoon.

SS-Scout Platoons use the Motorcycle Reconnaissance rules on page 187 and are Reconnaissance Platoons while mounted.

You may replace all Motorcycle MG teams with Panzerfaust MG teams at the start of the game before deployment. If you do this, the SS-Scout Platoon is no longer a Motorcycle Reconnaissance Platoon. Instead, the platoon may make Combat Attachments to Combat Platoons.

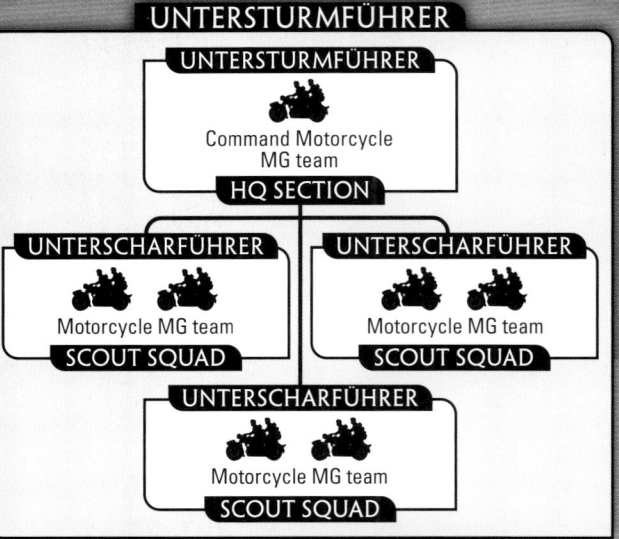

UNTERSTURMFÜHRER

UNTERSTURMFÜHRER

Command Motorcycle MG team

HQ SECTION

UNTERSCHARFÜHRER

Motorcycle MG team

SCOUT SQUAD

UNTERSCHARFÜHRER

Motorcycle MG team

SCOUT SQUAD

UNTERSCHARFÜHRER

Motorcycle MG team

SCOUT SQUAD

SS-PANZER SCOUT PLATOON

You may model your Motorcycle MG teams with Kübelwagen jeeps instead of motorcycles, they are based the same way as the Motorcycle MG teams and use the same rules.

Amoured SS-Flame-thrower Platoon

Platoon

3 Flame Sections	350 points	270 points
2 Flame Sections	235 points	180 points
1 Flame Section	120 points	90 points

Sd Kfz 251/16 (Flamm) half-tracks may not launch assaults, nor may they Counterattack if assaulted.

If you are fortunate enough to have armoured flame-thrower support then your chances of success against dug-in enemy infantry will be increased tremendously. The ability to rapidly burn a hole through a defensive line guarantees a quick advance into the enemy's rear areas.

UNTERSTURMFÜHRER

UNTERSTURMFÜHRER

Command Sd Kfz 251/16 (Flamm) half-track

Sd Kfz 251/16 (Flamm) half-track

FLAME SECTION

UNTERSCHARFÜHRER

Sd Kfz 251/16 (Flamm) half-track

Sd Kfz 251/16 (Flamm) half-track

FLAME SECTION

UNTERSCHARFÜHRER

Sd Kfz 251/16 (Flamm) half-track

Sd Kfz 251/16 (Flamm) half-track

FLAME SECTION

ARMOURED SS-FLAME-THROWER PLATOON

Armoured SS-Cannon Platoon

Platoon

3 Gun Sections	270 points	210 points
2 Gun Sections	180 points	140 points
1 Gun Section	90 points	70 points

Armoured direct fire support provides the needed firepower to dig out enemy anti-tank and machine-gun positions without too much difficulty. This allows your panzergrenadiers the opportunity to overrun the enemy's position and make the terrain safe for your advancing Panzers.

UNTERSTURMFÜHRER

UNTERSTURMFÜHRER

Command Sd Kfz 251/9 (7.5cm) half-track

Sd Kfz 251/9 (7.5cm) half-track

GUN SECTION

UNTERSCHARFÜHRER

Sd Kfz 251/9 (7.5cm) half-track

Sd Kfz 251/9 (7.5cm) half-track

GUN SECTION

UNTERSCHARFÜHRER

Sd Kfz 251/9 (7.5cm) half-track

Sd Kfz 251/9 (7.5cm) half-track

GUN SECTION

ARMOURED SS-CANNON PLATOON

DIVISIONAL SUPPORT PLATOONS

SS-KAMPFGRUPPE ARTILLERY BATTERY

PLATOON

HQ Section with:

6 10.5cm leFH18	330 points	255 points
4 10.5cm leFH18	240 points	185 points
3 10.5cm leFH18	180 points	140 points
2 10.5cm leFH18	130 points	100 points

OPTION

- Add Kfz 15 field car, Kfz 68 radio truck, and 3-ton trucks for +5 points for the platoon.

You must purchase all of the guns from one Gun Section before adding any extra teams from the second Gun Section.

IV. SS-PANZERKORPS ARTILLERIE

The *IV. SS-Panzerkorps* consolidated the divisional artillery of the *Totenkopf* and *Wiking* SS Panzer divisions at the corps level. This allowed them to provide artillery support in greater numbers in support of either division.

The Totenkopf and Wiking SS-Panzerdivision may be supported by SS-Kampfgruppe, SS-Kampfgruppe Heavy or SS-Armoured Artillery Batteries of either SS-Panzerdivision.

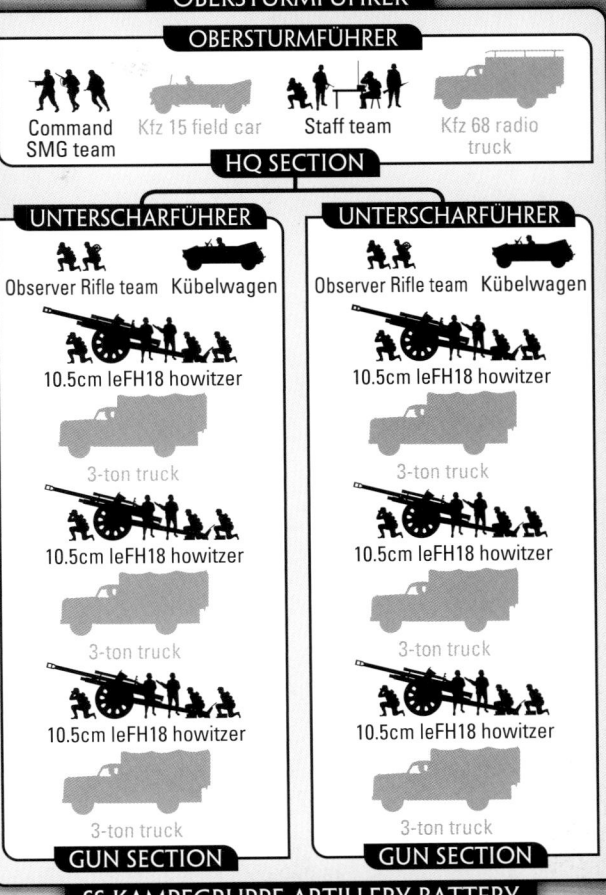

SS-KAMPFGRUPPE HEAVY ARTILLERY BATTERY

PLATOON

HQ Section with:

6 15cm sFH18	500 points	385 points
4 15cm sFH18	360 points	275 points
3 15cm sFH18	265 points	205 points
2 15cm sFH18	185 points	150 points

OPTION

- Add Kfz 15 field car, Kfz 68 radio truck, and Sd Kfz 7 (8t) half-tracks for +5 points for the platoon.

You must purchase all of the guns from one Gun Section before adding any extra teams from the second Gun Section.

SS-Kampfgruppe Heavy Artillery Batteries may not be placed from Ambush within 16"/40cm of enemy teams.

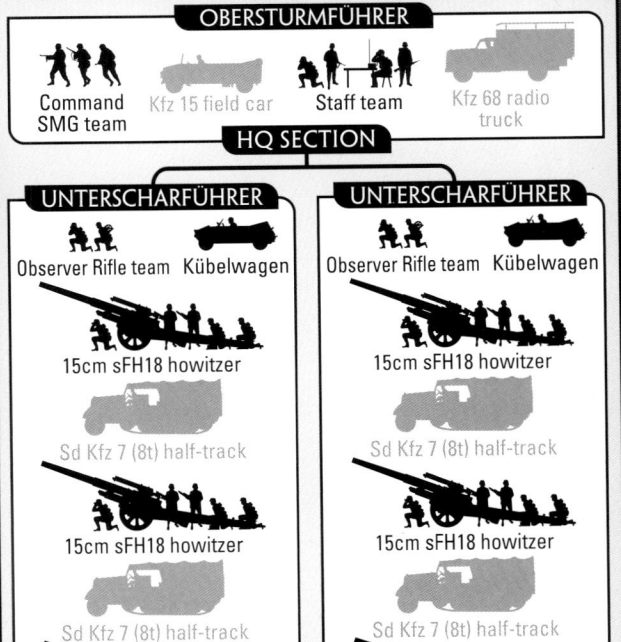

SICHERUNGSKOMPANIE
SECURITY COMPANY

(INFANTRY COMPANY)

HEADQUARTERS

Sicherungskompanie HQ — 119

You must field one platoon from each box shaded black and may field one platoon from each box shaded grey.

COMBAT PLATOONS

ARMOUR
Sicherungs Platoon — 119

ARMOUR
Sicherungs Platoon — 119

ARMOUR
Sicherungs Platoon — 119

ALLIED PLATOONS

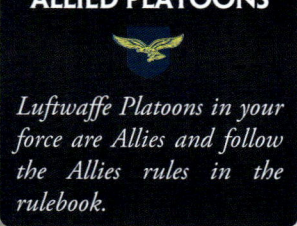

Luftwaffe Platoons in your force are Allies and follow the Allies rules in the rulebook.

WEAPONS PLATOONS

RECONNAISSANCE
Sicherungs Armoured Car Patrol — 120

ARTILLERY
Sicherungs Infantry Gun Platoon — 120

ANTI-TANK
Sicherungs Anti-tank Gun Platoon — 120

INFANTRY
Feldgendarmerie Platoon — 121

DIVISIONAL SUPPORT PLATOONS

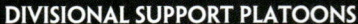

ARMOUR
Panzer Platoon — 73
Panther Platoon — 74
Schwere Panzer Platoon — 71
Veteran Tank-hunter Platoon — 164
Captured Tank Platoon — 121

ARMOUR
Tank-hunter Platoon — 163
Veteran Tank-hunter Platoon — 164
StuG Platoon — 59

INFANTRY
Panzergrenadier Platoon — 81
Panzerpionier Platoon — 89

INFANTRY
Panzergrenadier Platoon — 81
Panzerpionier Platoon — 89

ARTILLERY
Motorised Artillery Battery — 167
Motorised Heavy Artillery Battery — 167
Armoured Artillery Battery — 168
Armoured Heavy Artillery Battery — 168
Self-propelled Infantry Gun Platoon — 85
Heavy Assault Howitzer Platoon — 182
Armoured Train — 122

ANTI-AIRCRAFT
Panzer Anti-aircraft Gun Platoon — 75
Luftwaffe Heavy Anti-aircraft Gun Platoon — 173
Luftwaffe Light Anti-aircraft Gun Platoon — 173

ANTI-AIRCRAFT
Luftwaffe Heavy Anti-aircraft Gun Platoon — 173
Luftwaffe Light Anti-aircraft Gun Platoon — 173

ANTI-AIRCRAFT
Luftwaffe Light Anti-aircraft Gun Platoon — 173

AIRCRAFT
Air Support — 172

SICHERUNGSKOMPANIE

When the German *4. Armee* (Fourth Army) retreated from Orsha and Mogilev they headed towards Minsk. Behind them were the *Sicherungs* (zish er-roong) or Security Forces that had been ordered to hold and defend key transportation hubs from partisan forces behind the lines.

Unfortunately for the *Sicherungs*, they rapidly became the new front line. Though hastily reinforced with retreating *Luftwaffe* Flak battalions, they had to rely on the appearance of the *5. Panzerdivision* and the *505. Schwere Panzerabteilung* to have any hope of stopping the advancing Soviet Army. Armed with inadequate weapons for defending against front line Soviet troops, the *Sicherungskompanie* was nevertheless expected to hold until a new defensive line could be established with the retreating German infantry divisions.

MOTIVATION AND SKILL

A Sicherungskompanie was expected to police the rear areas and fight lightly-armed partisans, however they were thrown into the front lines in an attempt to halt the Soviet offensive. A Sicherungskompanie is rated as **Reluctant Trained.**

RELUCTANT	CONSCRIPT
CONFIDENT	TRAINED
FEARLESS	VETERAN

HEADQUARTERS

SICHERUNGSKOMPANIE HQ

HEADQUARTERS

Company HQ	25 points

OPTIONS

- Add a machine-gun section for +40 points.
- Add a mortar section for +35 points.

Although it is an Infantry Company, a Sicherungskompanie may not field Sniper teams.

The *Sicherungskompanie* was primarily used to hunt down partisans and rear area troops retreating from the front lines. However, as the front lines in Byelorussia crumbled German security forces were thrust into the front lines in a desperate effort to stem the Red tide. Security forces often found themselves confronting mechanised forces with little organic heavy support.

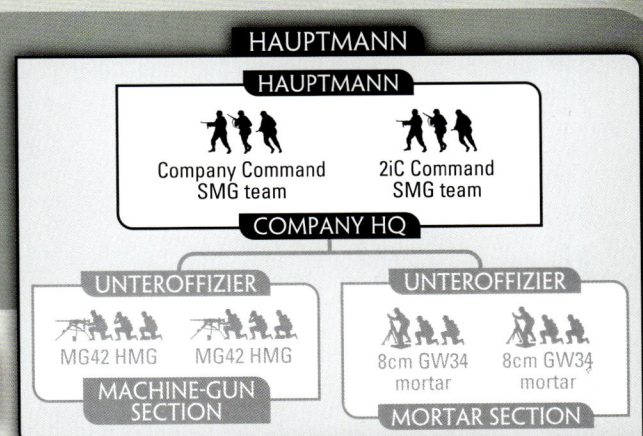

A *Sicherungskompanie* possessed some light support but any real hope of success would require immediate support from whatever counterattacking mechanised forces available. A *Sicherungskompanie* was also responsible for guarding the rail and roadways behind the lines but swiftly advancing Soviet spearheads sometimes forced German security forces to build a hasty defence.

COMBAT PLATOONS

SICHERUNGS PLATOON

PLATOON

HQ Section with:

3 Sicherungs Squads	70 points

OPTION

- Add 3-ton trucks for +5 points for the platoon.

Security battalions were formed by integrating military police forces, convalescing infantryman, and older recruits not fit for frontline duty. Though trained in local police work with some combat capability against local partisan forces they were totally unprepared to face the brunt of the Red Army.

SECURITY TROOPS

The *Sicherungs* and other German security forces bore the brunt of the Polish insurrection. They had to deal with the partisan uprising until relieved by better equipped reinforcements.

Sicherungs Platoons may only be placed in Reserve if no other types of platoon are available to fill the required number of platoons to be held in Reserve.

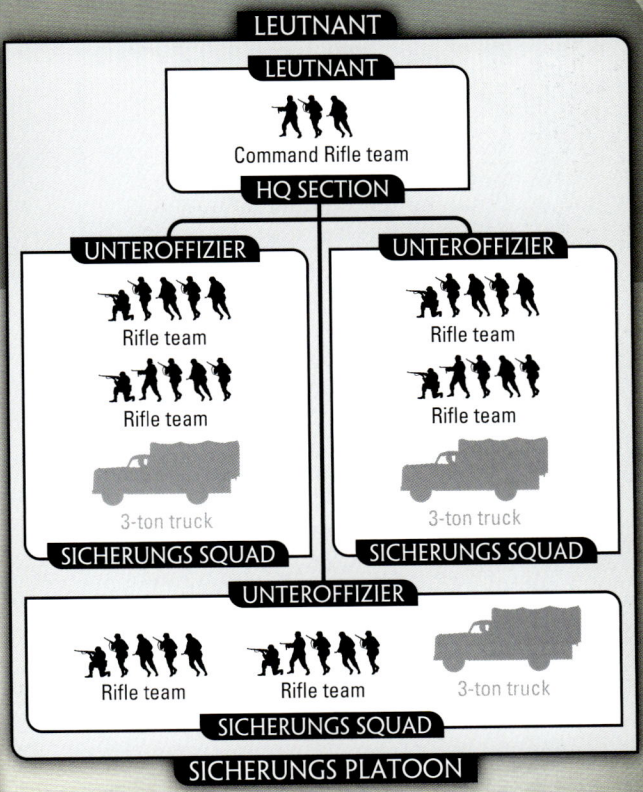

WEAPONS PLATOONS

SICHERUNGS ARMOURED CAR PATROL

PLATOON

3 Sd Kfz 221 (MG)	60 points
3 Panhard 178 (f)	65 points
1 BA 10M	25 points

Armoured Car Patrols are Reconnaissance Platoons.

Supporting your *Sicherungskompanie* with an armoured car patrol provides a reconnaissance capability as well as a light mobile counterattack force against partisans, light enemy reconnaissance, or mounted infantry.

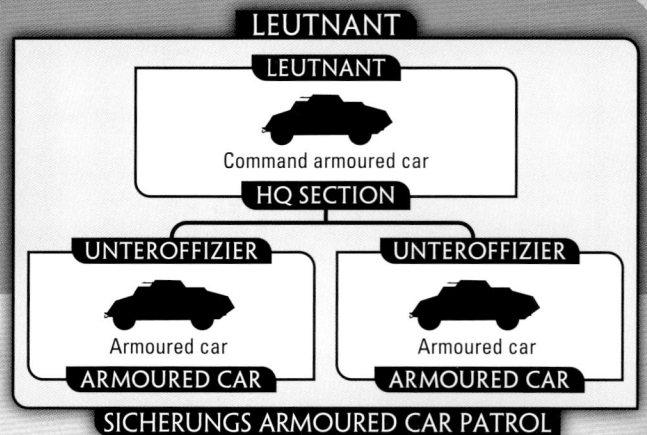

SICHERUNGS INFANTRY GUN PLATOON

PLATOON

HQ Section with:

2 7.5cm leIG18	45 points

OPTION

• Add horse-drawn limbers for +5 points for the platoon.

Attaching an infantry gun platoon to your security forces will give you additional direct and indirect fire support. They can bolster your defence while enhancing your ability to hit the enemy before they can close in on your *Sicherungs* platoons. Don't forget their smoke capability to hide your movement until you can get your counterattack forces into position.

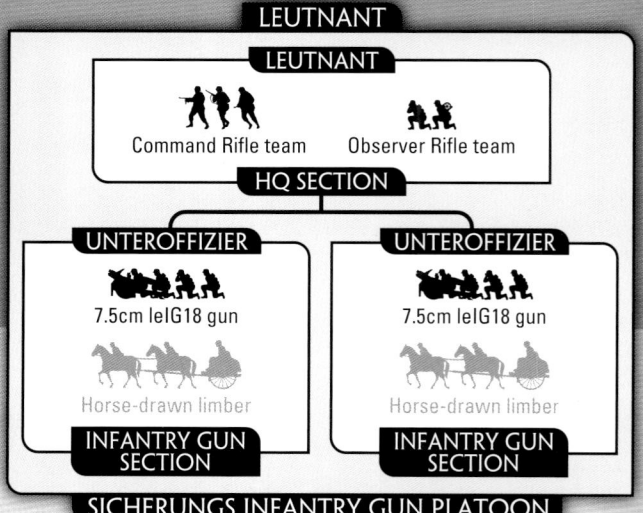

SICHERUNGS ANTI-TANK GUN PLATOON

PLATOON

HQ Section with:

3 3.7cm PaK 36	50 points

OPTIONS

• Add Kfz 15 field car and Kfz 70 trucks +5 points for the platoon.
• All 3.7cm PaK36 guns are equipped with Stielgranate ammunition at no cost.

As the main threat against security forces came from light armoured vehicles the older 3.7cm PaK36 was deemed sufficient for the *Sicherungskompanie*. While this remained true for advancing enemy mechanized transports it proved insufficient in stopping enemy armour columns in 1944.

You can use the *Stielgranate* ammunition in a last ditch effort to stop enemy tanks but they must be at close range and it works best if your units are dug-in and in cover.

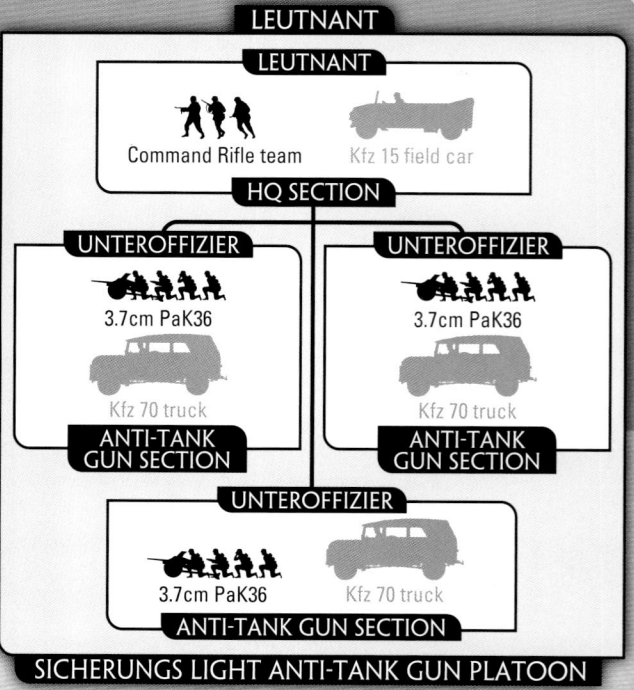

FELDGENDARMERIE PLATOON

PLATOON

2 Feldgendarmerie Sections	40 points
1 Feldgendarmerie Section	20 points

LEUTNANT

LEUTNANT

Feldgendarm team — Feldgendarm team

FELDGENDARMERIE SECTION

LEUTNANT

Feldgendarm team — Feldgendarm team

FELDGENDARMERIE SECTION

FELDGENDARMERIE PLATOON

FELDGENDARMERIE

Feldgendarmerie (Field Police) platoons must make Combat Attachments to Sicherungs Platoons, Gepanzerte Panzergrenadier Platoons, Panzergrenadier Platoons, Gepanzerte Panzerpionier Platoons, Panzerpionier Platoons, Sicherungs Infantry Gun Platoons, Sicherungs Anti-tank Gun Platoons, Motorised Artillery Batteries or Heavy Motorised Artillery Batteries with all their sections.

If a German platoon containing a Feldgendarmerie team that is In Command fails its Motivation Test to rally from being Pinned Down, to Counterattack in an assault, or to pass a Platoon Morale Check (but not to launch an assault against tank or any other Motivation Test), you may immediately Destroy any other Infantry or Gun team from the platoon within Command Distance of the Feldgendarmerie team and re-roll the Motivation Test.

If you roll a 1 for the new Motivation Test, however, the soldiers are pushed past the point of breaking and revolt against the Feldgendarmerie! In this case the Feldgendarmerie team is Destroyed as well.

If a Company Commander joins a platoon with a Feldgendarmerie team attached, the Company Commander overrules the Feldgendarmerie capability. Thus the unit may only use the Company Commander's ability to reroll motivation.

As with any unit, a German platoon Pinned Down by Defensive Fire fails to assault its target. It must wait until its next turn to rally as normal.

With losses mounting and reinforcements coming from non-line units, the ability of German units to hold the line became harder and harder. In a questionable attempt to keep front line units from retreating, the *Feldgendarmerie* were ordered to prevent deserters, traitors, and cowards from forsaking their duties to the Fatherland.

To accomplish this directive the *Feldgendarmerie* would position themselves behind the front lines and attempt to stop German units from retreating. They had strict orders to shoot any retreating German personnel.

This tactic was not always effective. The presence of German officers within the retreating formations would preclude the *Feldgendarmerie* from carrying out their orders.

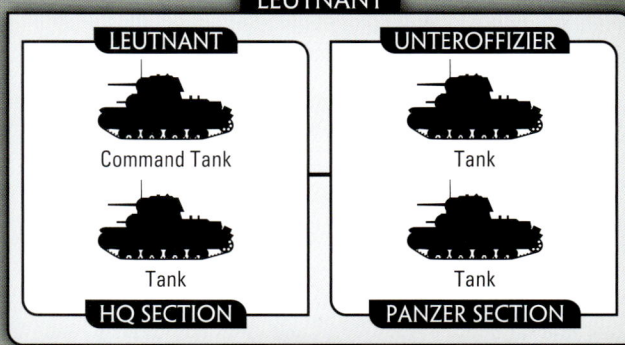

DIVISIONAL SUPPORT PLATOONS

CAPTURED TANK PLATOON

PLATOON

4 Panzer M14/41	100 points
3 Panzer M14/41	75 points
2 Panzer M14/41	50 points
4 T-34 obr 1942 (captured)	195 points
3 T-34 obr 1942 (captured)	145 points
2 T-34 obr 1942 (captured)	95 points

LEUTNANT

LEUTNANT

Command Tank

Tank

HQ SECTION

UNTEROFFIZIER

Tank

Tank

PANZER SECTION

CAPTURED TANK PLATOON

OPTION

- Upgrade T-34 obr 1942 tanks to have cupolas for +5 points for the platoon.

The *Sicherungs* forces had some limited armour of its own. While these tanks helped boost German morale, they were relatively ineffective in combat and often proved more of a liability than a help.

A Captured Tank Platoon is rated **Reluctant Trained**.

RELUCTANT	TRAINED

ARMOURED TRAIN

MOTIVATION AND SKILL

The troops and crews of the German armoured trains had had much experience fighting partisans, defending the rail network, and supporting the frontline troops by the beginning of 1944. An Armoured Train can be either a Captured Armoured Train or a BP44 Armoured Train, both are rated **Confident Veteran.**

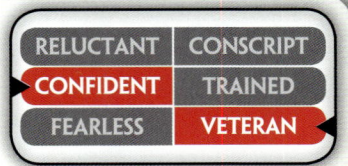

RELUCTANT	CONSCRIPT
CONFIDENT	TRAINED
FEARLESS	VETERAN

CAPTURED ARMOURED TRAIN

ARMOURED TRAIN

Locomotive, Assault Car and Panzergrenadier Platoon with:

2 Heavy Artillery Cars	330 points
1 Heavy Artillery Car and 1 Light Artillery Car	310 points
2 Light Artillery Cars	300 points
1 Heavy Artillery Car	260 points
1 Light Artillery Car	250 points

Locomotive with:

2 Heavy Artillery Cars	140 points
1 Heavy Artillery Car and 1 Light Artillery Car	120 points
2 Light Artillery Cars	110 points
1 Heavy Artillery Car	70 points
1 Light Artillery Car	60 points

OPTIONS

- Add Panzer Platoon for +75 points.
- Add up to two Tank-hunter Cars for +35 points each.

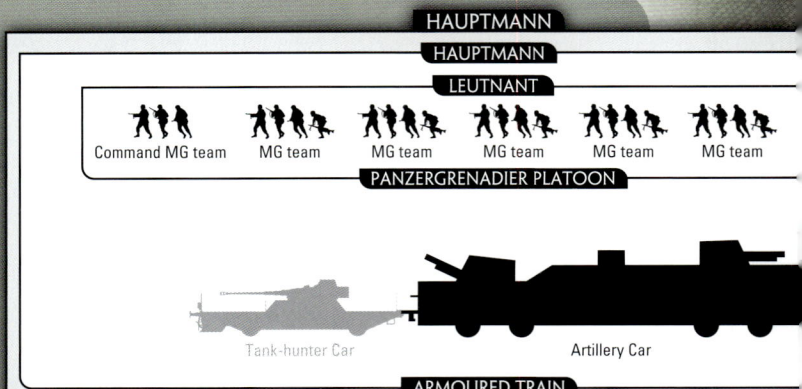

Command MG team · MG team · MG team · MG team · MG team · MG team

HAUPTMANN
HAUPTMANN
LEUTNANT
PANZERGRENADIER PLATOON

Tank-hunter Car · Artillery Car

ARMOURED TRAIN
CAPTURED ARMOURED TRAIN

- Add Pioneer Section to Panzergrenadier Platoon for +50 points.
- Add Mortar Section to Panzergrenadier Platoon for +50 points.
- Replace one 7.5cm FK02/26(p) gun in a Light Artillery Car with a 2cm FlaK38 (V) gun at no cost.

Armoured Trains may not be deployed in Ambush.

Each Artillery Car fires as a Battery (see page 127).

BP44 ARMOURED TRAIN

ARMOURED TRAIN

Locomotive, Infantry Car and Panzergrenadier Platoon with:

2 Artillery Cars and 2 Anti-aircraft Cars	485 points
1 Artillery Car and 1 Anti-aircraft Car	335 points

Locomotive with:

2 Artillery Cars and 2 Anti-aircraft Cars	295 points
1 Artillery Car and 1 Anti-aircraft Car	145 points

OPTIONS

- Add Panzer Platoon for +75 points.
- Add up to two Tank-hunter Cars for +35 points each.
- Add Staff Car for +15 points.
- Add Pioneer Section to Panzergrenadier Platoon for +50 points.
- Add Mortar Section to Panzergrenadier Platoon for +50 points.
- Replace 2cm FlaK38 (V) gun in Anti-aircraft Car with 3.7cm FlaK43 gun for +5 points per car.

HAUPTMANN
HAUPTMANN
LEUTNANT

Command MG team · MG team · MG team · MG team · MG team · MG team

PANZERGRENADIER PLATOON

Tank-hunter Car · Anti-aircraft Car · Infantry Car

ARMOURED TRAIN
BP44 ARMOURED TRAIN

You may replace up to one Pioneer Rifle team in the Pioneer Section with a Flame-thrower team at the start of the game before deployment.

Armoured Trains may not be deployed in Ambush.

Each pair of an Anti-aircraft Car and an Artillery Car fires as a Battery (see page 127).

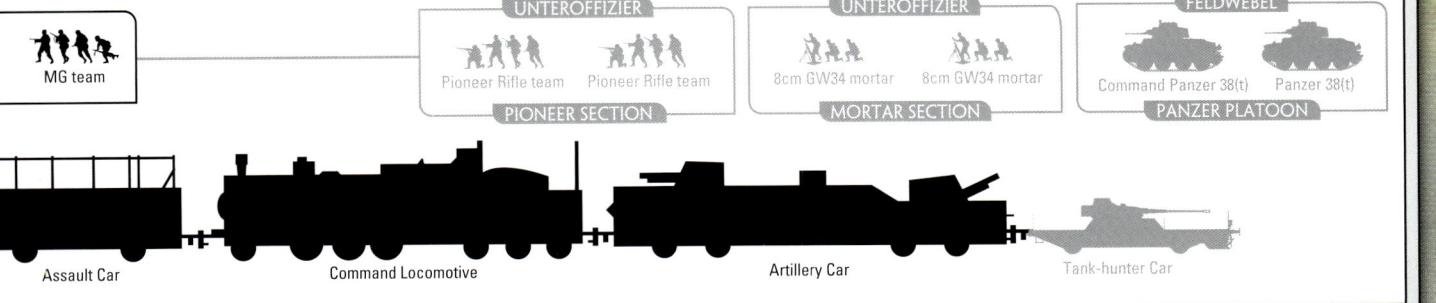

MG team

UNTEROFFIZIER
Pioneer Rifle team — Pioneer Rifle team
PIONEER SECTION

UNTEROFFIZIER
8cm GW34 mortar — 8cm GW34 mortar
MORTAR SECTION

FELDWEBEL
Command Panzer 38(t) — Panzer 38(t)
PANZER PLATOON

Assault Car — Command Locomotive — Artillery Car — Tank-hunter Car

With the outbreak of war in 1939 the Germans decided to set-up several armoured trains. By 1941 the Germans were also using two captured Polish and a number of Czech armoured trains, in addition to earlier German trains.

With captured Polish and Russian trains as templates the Germans went about creating a standard model of armoured train in 1941. This was known as the BP42, and with later additions, the BP44. These new trains borrowed from the captured designs' increased artillery capability over the existing German trains. They also overcame some of the perceived design flaws in the captured trains by splitting the artillery car in two, leaving only one gun on each car

reducing loss of firepower if an artillery car was knocked out. Modern anti-aircraft guns and tank-hunter cars were also added making a formidable self-contained fighting unit.

During 1944 the Germans ran two Polish trains. These were Armoured Trains 21 and 22. Armoured Train 21 was armed with a mix of 10cm and 7.5cm guns. Armoured Train 22 was armed with 7.5cm guns. During 1944 these trains also received German BP44 Panzer IV turret tank-hunter cars. Armoured Train 23, while made up of ex-Czech cars also received some Polish cars and Russian two-turret artillery cars similar to the Polish models.

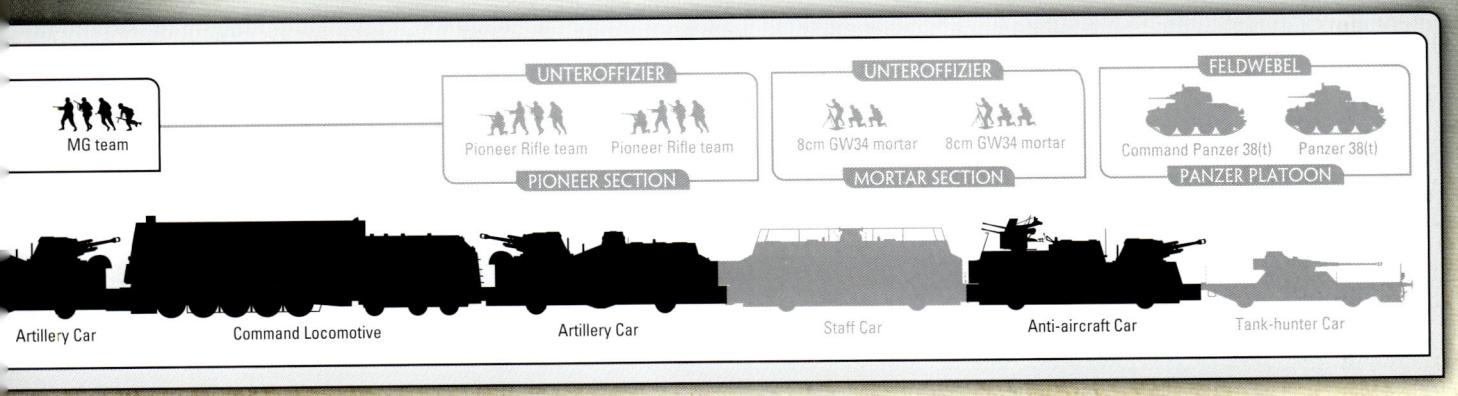

MG team

UNTEROFFIZIER
Pioneer Rifle team — Pioneer Rifle team
PIONEER SECTION

UNTEROFFIZIER
8cm GW34 mortar — 8cm GW34 mortar
MORTAR SECTION

FELDWEBEL
Command Panzer 38(t) — Panzer 38(t)
PANZER PLATOON

Artillery Car — Command Locomotive — Artillery Car — Staff Car — Anti-aircraft Car — Tank-hunter Car

ARMOURED TRAIN RULES

Armoured trains cost a fortune to build and operate, and they have a set way of doing things to maximise efficiency.

The main parts of any Armoured Train are the Locomotive, Artillery Cars, Infantry Cars (and the Infantry Platoons they carry), and supporting Tank platoons.

The Locomotive and each Car or Wagon in an Armoured Train count as a separate Tank team, with the whole train counting as a single platoon.

An Armoured Train must always operate in the order shown in the Intelligence Briefing. It cannot uncouple cars or change their order, but it may travel in either direction and may enter the table facing either direction.

ARTILLERY CARS

The main function of an armoured train is to provide heavy fire support for other troops using its artillery cars. These can range from cars mounting multiple heavy artillery pieces to tank-hunter cars mounting tank turrets, and even unarmed artillery staff cars to improve the performance of other artillery cars.

The Artillery Cars of an Armoured Train are divided into Artillery Batteries, usually of one Artillery Car per battery, although some have multiple cars in a battery.

INFANTRY CARS

Like the marines of a battleship, many armoured trains carry an infantry platoon in a specially-equipped assault car for its own protection and to clear obstacles from the track. The infantry platoon travels safe under the thick armour of the infantry car, only dismounting when they are needed.

An Armoured Train with an Infantry Car always deploys the Infantry Platoon as Passengers in the Infantry Car. The Infantry Car is a Tank team, but can Mount and Dismount the Infantry Platoon as if it is a Transport team. The Infantry Platoon operates as a separate platoon with its own Platoon Command team. Treat the Armoured Train and the Infantry Platoon as two separate platoons when calculating the number of platoons held in Ambush or Reserve, but both arrive together as a single platoon from Reserves.

SUPPORTING TANK PLATOONS

Many armoured trains have supporting tanks ready to dismount to deal with attacks on the train.

Supporting Tank Platoons operate as separate platoons with their own command teams. Treat the Armoured Train and the Supporting Tank Platoon as two separate platoons when calculating the number of platoons held in Ambush or Reserve, but both arrive together as a single platoon from Reserves.

The tanks of a Supporting Tank Platoon deploy evenly at each end of the train as shown in the Intelligence Briefing and move with the Armoured Train as part of it. Until they leave the train, Panzer 38(t) tanks of the Supporting Tank Platoon have Front and Side Armour of 3.

The tanks may leave the train at end of any Movement Step, moving up to an additional 4"/10cm in any direction. The tanks may also leave the train to Counterattack if it is assaulted. In this case simply move the vehicles from their current positions. Once the Tank teams of the Supporting Tank Platoon have left the train, they remain separate for the rest of the game.

ARMOURED TRAIN DEPLOYMENT

ARMOURED TRAIN DEPLOYMENT

Armoured trains must deploy in the order shown in diagram on pages 122 and 123

Panzergrenadier Platoon deploys in Assault or Infantry Car

One Panzer 38(t) from the Panzer Platoon deploys at each end of the Armoured Train

Panzer 38(t) — Tank-hunter car — Artillery Car — Assault or Infantry Car — Locomotive — Artillery Car — Tank-hunter car — Panzer 38(t)

Armoured trains can only move along railway lines, so their deployment options are usually very limited.

An Armoured Train that would normally be deployed on the table is held off the table until the start of its first Movement Step, when it moves onto the table. In the turn an Armoured Train moves onto the table, it always moves far enough to bring the entire train and its Supporting Tank Platoons onto the table.

The Armoured Train will move on from the point where the railway line crosses the player's Deployment Area, or if it does not do so, at the point the Railway Line crosses a table edge closest to the player's Deployment Area.

Armoured Trains cannot be held in Ambush, they are simply too big and obvious, but they may be held in Reserves.

PLACING RAILWAY LINES

PLACING RAILWAY LINES

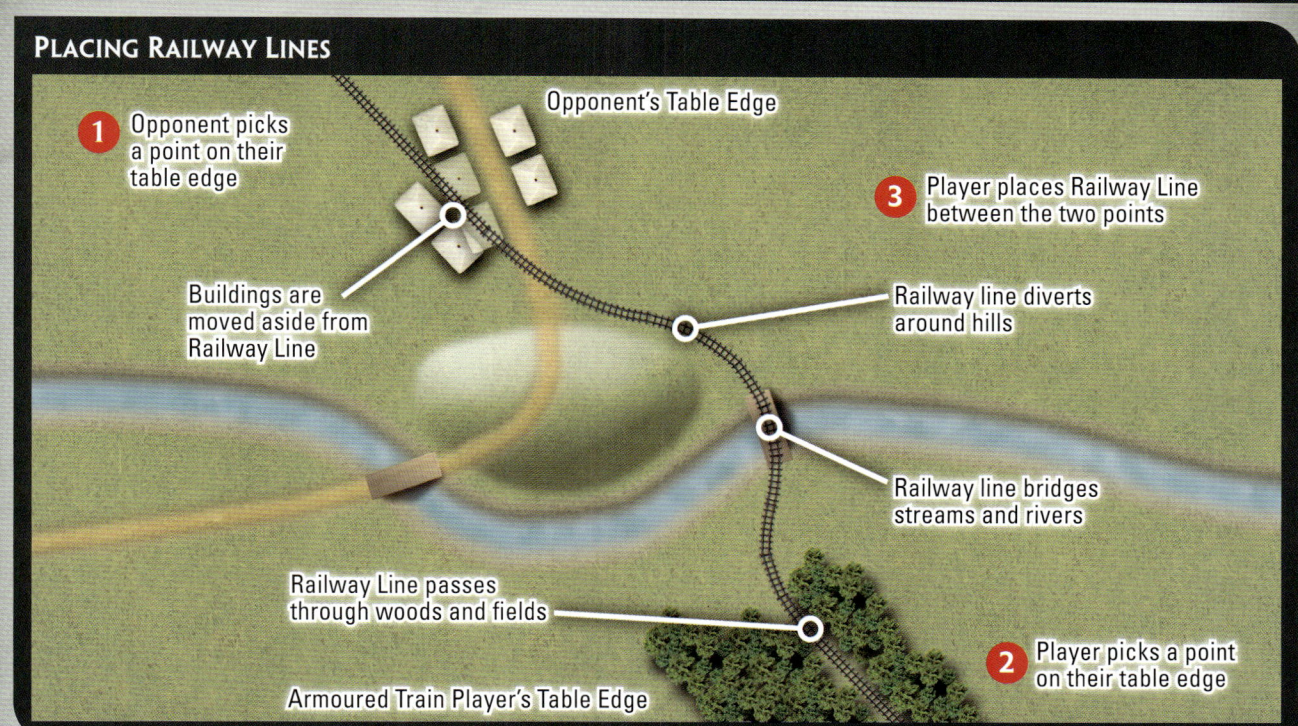

1 Opponent picks a point on their table edge

Opponent's Table Edge

3 Player places Railway Line between the two points

Buildings are moved aside from Railway Line

Railway line diverts around hills

Railway line bridges streams and rivers

Railway Line passes through woods and fields

2 Player picks a point on their table edge

Armoured Train Player's Table Edge

Without railway lines, there are no armoured trains. Since most gaming tables don't have tracks laid ready for an armoured train to use, we use the following rules to place the tracks before the game begins.

If the table has been set up with Railway Lines, use these for the Armoured Train. Otherwise, at the start of the game before any Objectives or Fortifications are placed on the table, the players place a Railway Line as follows:

1. *The player with the Armoured Train chooses two opposite table edges.*

2. *The opposing player picks a point on one of those table edges.*

3. *The Armoured Train player picks a point on the other table edge.*

4. *The Armoured Train player then places a Railway Line running in a straight line between these two points. The line cuts through woods, fields and other flat ground, and rivers or streams are assumed to be crossed with culverts or bridges. Any buildings in the way should be moved aside by the minimum necessary. The railway line is diverted by the minimum necessary to go around hills (even gentle gradients are avoided by railway lines).*

If both players have Armoured Trains, there is only one Railway Line and the attacking player starts by picking which table edges the Railway Line will run between.

DEMOLISHING RAILWAY LINES

Railway lines are remarkably hard to destroy, especially as the train carries spare rails and wooden sleepers or ties to allow its track-repair team to repair minor damage.

The only way to damage a Railway Line during a game is for a Pioneer team to place charges on the tracks. A Pioneer team must start the turn adjacent to the Railway Line and not be Pinned Down to place charges. The team cannot move in the Movement Step, although they do count as moving and cannot claim to be Concealed in the open or Gone to Ground.

Instead of Shooting, roll a Skill Test at the end of the Shooting Step.

If they pass the Skill Test, the section of Railway Line adjacent to the team becomes Difficult Going to trains.

Otherwise the railway line is undamaged.

Only the Locomotive needs to take a Bog Check when crossing damaged Railway Lines, but it does so whenever the train reaches the damaged section, or if the train starts the turn straddling a damaged section of railway line. If a train Bogs Down, only the Locomotive is considered Bogged Down.

ARMOURED TRAIN MOVEMENT

The key to the success of armoured trains is their operational mobility. They can move rapidly to trouble spots and deliver effective support when they arrive. In a country with few modern roads, this is essential.

Armoured Trains can only move on railway lines. Provided that the Locomotive is still operational they can move up to 16"/40cm forward or backward along the railway line each turn.

If any part of an Armoured Train leaves the table, the whole Armoured Train is considered to have left the table and cannot return.

MOVING NEAR ENEMY TEAMS

Armoured trains are unstoppable behemoths. They are not worried by mere soldiers on the track and will push aside any enemy tank that attempts to block their passage. Of course, they are a little more reluctant to ram a friendly tank!

Armoured Trains may not move through friendly teams. Any friendly team on the tracks must move off the tracks before the train can pass.

Armoured trains are not hindered by enemy teams except other Armoured Trains. Any other enemy team is moved aside by the opposing player by the smallest distance necessary for the train to pass, retaining its current facing. The Armoured Train player may move any wreck on the tracks aside to let the train pass.

MOVING THROUGH TEAMS

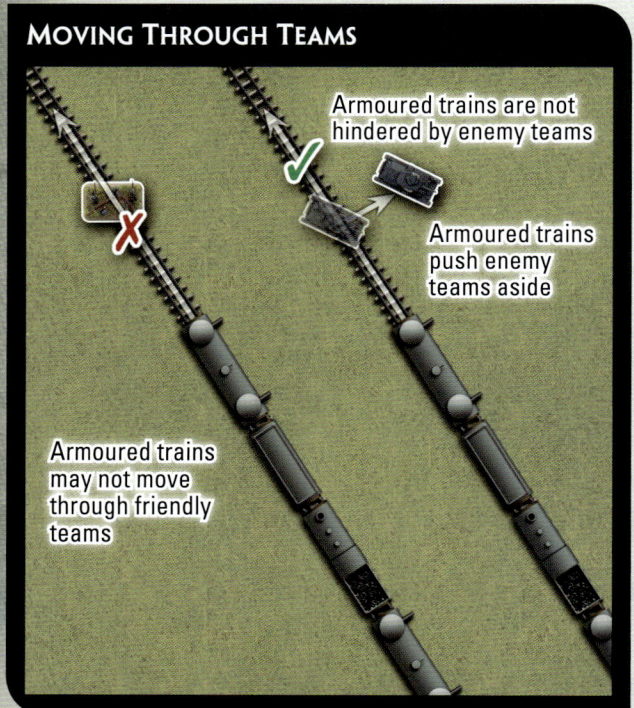

Armoured trains are not hindered by enemy teams

Armoured trains push enemy teams aside

Armoured trains may not move through friendly teams

ARMOURED TRAIN MOVEMENT

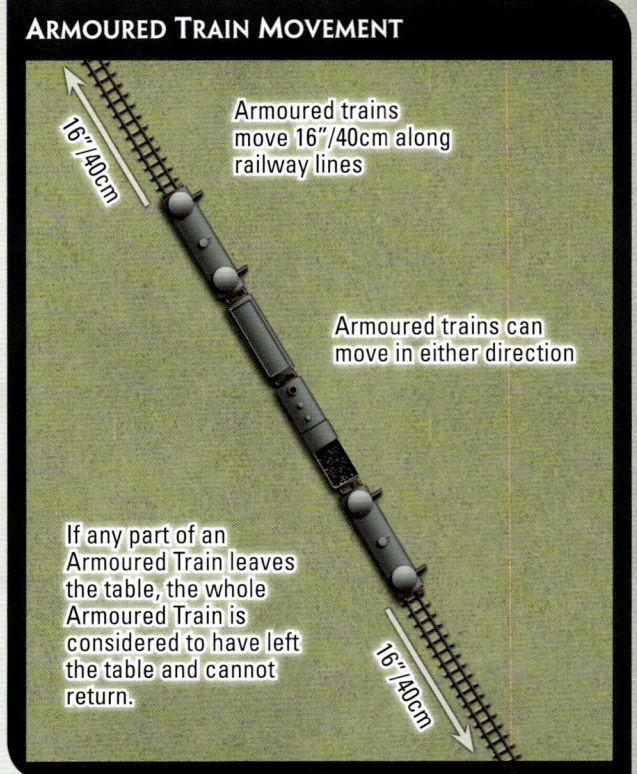

Armoured trains move 16"/40cm along railway lines

Armoured trains can move in either direction

If any part of an Armoured Train leaves the table, the whole Armoured Train is considered to have left the table and cannot return.

16"/40cm

16"/40cm

NO MOVING THROUGH TRAINS

Unlike a platoon of tanks or infantry, an armoured train is a single long entity with no gaps. There is nowhere for tanks to drive through and infantry are not foolish enough to attempt to move between the wagons risking getting crushed if the train should move.

Teams cannot move through an Armoured Train unless the Locomotive has been Destroyed. Once the Locomotive is Destroyed, it and other Destroyed cars in the train become Very Difficult Going. Armoured Trains cannot move through other Armoured Trains, even when the other Armoured Train is Destroyed.

NO MOVING THROUGH ARMOURED TRAINS

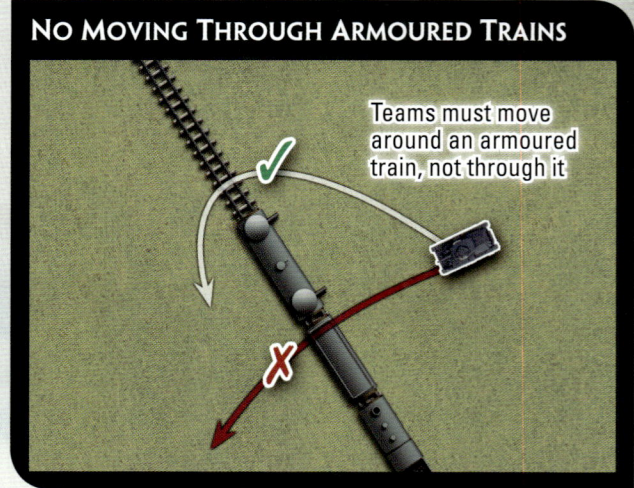

Teams must move around an armoured train, not through it

SHOOTING WITH ARMOURED TRAINS

SHOOTING FROM ARMOURED TRAINS

By comparison with a tank moving across country, a train provides a relatively smooth ride.

Armoured Trains do not suffer any penalty for shooting on the move, retaining their full ROF when shooting while moving.

Almost all guns (aside from machine-guns) on an Armoured Train are mounted in Deck Turrets. As such, their Field of Fire is restricted by the superstructure and other turrets of the Artillery Car they are mounted on, and possibly by other cars in the train.

ARTILLERY BOMBARDMENTS FROM TRAINS

The predictable movement of a train allows it to fire a bombardment on the move. The huge stocks of ready ammunition mean they can do so at a very high rate of fire.

When firing an Artillery Bombardment, count each weapon firing as two weapons firing.

Armoured Trains can fire Artillery Bombardments while moving.

All of the Artillery Batteries in an Armoured Train can combine to fire a single bombardment instead of firing as separate batteries.

TRAIN MACHINE-GUNS

A train's machine-guns are usually crewed by separate machine-gun sections leaving the gun crews free to fire their own guns.

The cars of an Armoured Train fire all of their machine-guns at ROF 2, even when other weapons are firing, including if they are firing an Artillery Bombardment.

The field of fire of a train MG mounted on the side of a car includes everything to the side of a line drawn along the side of the car. The field of fire of a train MG mounted on the end of a car includes everything to the front of a line drawn across the end of the car.

PASSENGER-FIRED MACHINE-GUNS

Some trains use the infantry platoon to crew the machine-guns. While this saves on manpower, it does mean that the train can't fire its machine-guns once the infantry dismount.

Each Passenger-fired Train MG requires an Infantry or Man-packed Gun team Mounted in an Infantry Car to fire it. The team does not need to be in the same car as the MG, and each team can fire any machine-gun in the train (although only one machine-gun per team at any one time).

FIELDS OF FIRE

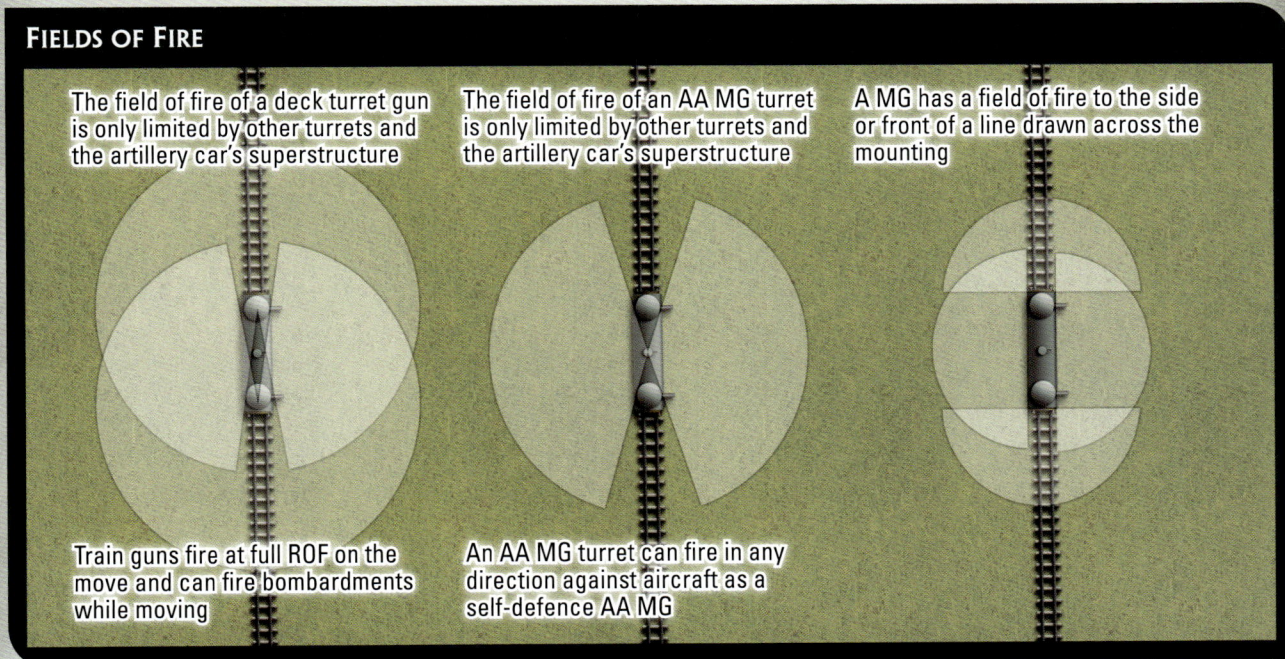

The field of fire of a deck turret gun is only limited by other turrets and the artillery car's superstructure

The field of fire of an AA MG turret is only limited by other turrets and the artillery car's superstructure

A MG has a field of fire to the side or front of a line drawn across the mounting

Train guns fire at full ROF on the move and can fire bombardments while moving

An AA MG turret can fire in any direction against aircraft as a self-defence AA MG

SHOOTING AT ARMOURED TRAINS

Armoured trains are remarkably tough unless hit with heavy-calibre guns. Even if the locomotive is disabled and the train halted, the gun cars will still fight on in place.

> *Armoured Trains are shot at in the same way as any platoon of tanks. If a car is Destroyed, it ceases to function along with all of its weapons, but otherwise has no effect on the train. The train moves with the Destroyed car still in place.*

DESTROYED LOCOMOTIVE

The one part of the train that is truly critical is the locomotive. Without it, nothing moves!

> *If the Locomotive is Destroyed, the train can no longer move. If the train fails a Platoon Morale Check after the Locomotive is Destroyed, the whole train is Destroyed in place. If the train fails a Platoon Morale Check while the Locomotive is still operational, the whole train is removed from the table instead.*
>
> *Once the Locomotive is Destroyed, a Destroyed Locomotive or Artillery or Infantry Car is Very Difficult Going. It still Conceals any team on the far side of it.*

HIT ALLOCATION

While some gunners just shoot at whatever part of the train they are looking at, smart gunners pick out the most dangerous cars and target them first.

> *The Gun Tank rule (see the rulebook) can be used to distinguish between different types of car and the Locomotive.*

SHOOTING ACROSS TRAINS

No one is going to risk taking a shot that might hit their own armoured train, but the enemy doesn't care.

> *In the same way that teams cannot shoot through any other type of Friendly team, Friendly teams cannot shoot through an Armoured Train.*
>
> *Enemy teams can shoot through an Armoured Train, but teams at least half obscured by the Armoured Train count as being Concealed.*

SHOOTING ACROSS ARMOURED TRAINS

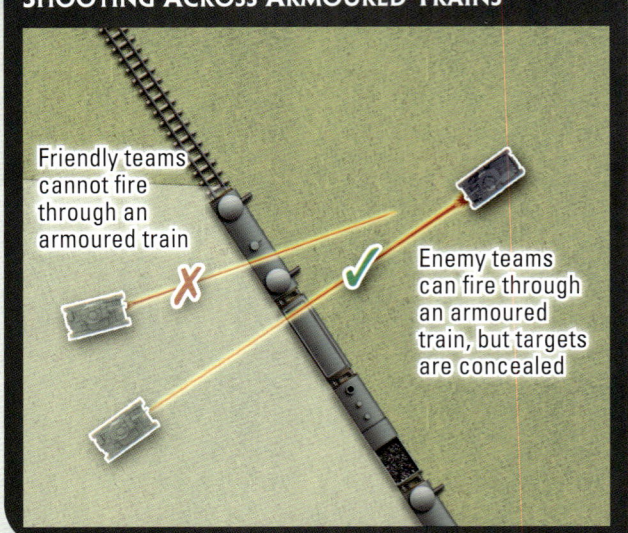

Friendly teams cannot fire through an armoured train

Enemy teams can fire through an armoured train, but targets are concealed

ARMOURED TRAINS IN ASSAULTS

Armoured trains are difficult to assault with machine-guns bristling from every aperture. These fearsome beasts are not worried by mere infantrymen and can ignore them at will.

Armoured Trains cannot Launch an Assault or Charge into Contact, and they do not move when Counterattacking.

If the Locomotive is operational when an Armoured Train Breaks Off from an assault, the train may move through or past any enemy teams without hindrance, and the train does not need to be more than 4"/10cm from enemy teams for the assault to end. It will not surrender, and Locomotives and Assault and Infantry Cars are unaffected by proximity to the Assaulting Platoon.

If the Locomotive is Bailed Out, Bogged Down or Destroyed when an Armoured Train Breaks Off from an assault, the Locomotive and any Assault or Artillery Car within 4"/10cm of the Assaulting Platoon are captured and Destroyed as usual.

Any Infantry Platoon or Supporting Tank Platoon is a separate platoon from the Armoured Train. If the Infantry Platoon is still mounted in the Infantry Car when the Infantry Car is Destroyed in an assault, the whole Infantry Platoon will be Destroyed. Infantry and Supporting Tank Platoons can Dismount to Counterattack if they are assaulted.

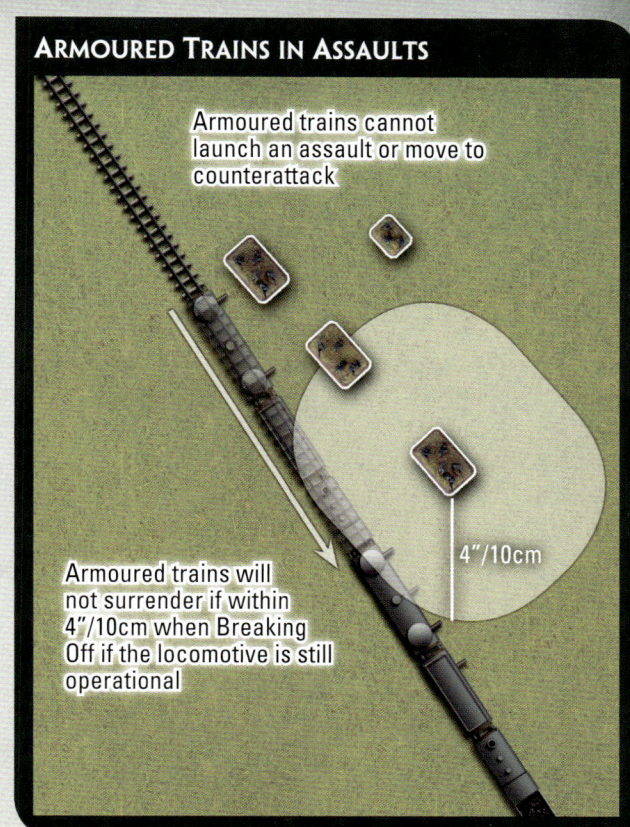

ARMOURED TRAINS IN ASSAULTS

Armoured trains cannot launch an assault or move to counterattack

4"/10cm

Armoured trains will not surrender if within 4"/10cm when Breaking Off if the locomotive is still operational

SS-Kavallerie fight to prevent Soviets breaking through their Street Barricade.

A counterattack by Feldherrnhalle Panzer IV/70 (V) tanks pushes the enemy back.

THE BATTLES FOR HUNGARY

Legend:
- Defensive Lines
- Axis Moves
- Soviet Advance

THE BATTLE FOR HUNGARY 1944

Map labels: SLOVAKIA, 1st Panzer Army, GALICIA, Fourth Ukrainian Front, AUSTRIA (GERMANY), Ibolyság, Army Group South, 8th Army, 1st Hungarian Army, Carpathian Mountains, ROMANIA, Börzsöny Hills, Esztergom, BUDAPEST, Hatvan, Debrecen, 2nd Hungarian Army, 8th Army, 13 October, Vál, 6th Army, Lake Velence, Székesfehérvár, HUNGARY, Lake Balaton, Kecskemét, R. Tisza, 3rd Hungarian Army, Second Ukrainian Front, Torda, 1 October, 1-13 October, R. Maros, Transylvania, ROMANIA, Carpathian Mountains (Transylvanian Alps), 1 December, Mohács, Pécs, Danube, Third Ukrainian Front, N, CROATIA, 2nd Panzer, YUGOSLAVIA

0 Miles — 100
0 KM — 100 — 200

FIGHTING FOR TRANSYLVANIA

The Soviet successes in the Ukraine and Romania led the Red Army to the very doorstep of Hungary in September 1944. The frontlines were drawn along the Carpathian Mountains in northeastern Hungary and on the Hungarian-Romanian border in Transylvania. Striking before the Red Army could cross Romania, the Hungarians took the opportunity to attack the Romanians on the weakly-held Transylvanian border, but rapidly arriving Soviet forces reinforced the Romanians and soon forced the Hungarians to withdraw back into Hungary. The Red Army launched a massive offensive on 6 October and the battle for Hungary had begun.

DEBRECEN

On 12 October, German and Hungarian forces started to withdraw back towards the Tisza River. The Soviet plan aimed to cut the withdrawing Axis armies off in eastern Hungary, to stop them reaching the Tisza River. The Soviets launched attacks on 17 October to cut off the withdrawing German Eighth Army and Hungarian First and Second Armies. Fighting centred around Debrecen (Hungary's third largest city) and along the Tisza River.

The Axis troops held on to Debrecen until 20 October. Savage street-fighting by both Hungarian and German infantry and assault guns held off the Soviets until the withdrawing German and Hungarians passed through.

Fast-moving Soviet cavalry and mechanised forces thrust past Debrecen into the rear of the German Eighth Army and captured Nyíregyháza on 22 October. The Germans counter-attacked on 23 October from Polgár on the Tisza River with armoured forces and from the northwest with mixed armour

and infantry. Nyíregyháza was recaptured on 29 October and the Soviet thrust was broken up resulting in the near destruction of a Soviet cavalry-mechanised group.

Hungarian troops held the main crossing of the Tisza at Polgár until 31 October. This allowed the German Eighth Army and Hungarian Second Army to withdraw across the river.

After capturing Debrecen on 20 October and pushing the Axis back to the Tisza River, the Soviets paused to take stock. Many of the Soviet units were worn down and the Tank and Mechanised Corps were in serious need of rest and refitting. However, the Soviet plans called for the offensive to continue on to the next objective, the Hungarian capital Budapest.

THE DRIVE TO BUDAPEST

A new Soviet offensive began on 29 October in the south against the Hungarian Third Army. This hastily formed army was made up of reserve and training units and quickly crumbled before the Soviet assault. The Red Army captured Kecskemét on 31 October, and by 1 November had advanced up to 30 miles (48 km) all along its front. This allowed the mechanised troops to get behind the German Sixth Army's flank, opening the way towards Budapest.

Mindful of the near destruction of the cavalry-mechanised group at Nyíregyháza, the Soviets made sure their spearheads were not overexposed. They kept sufficient numbers of tanks back to protect the flanks from the inevitable German counterattacks.

In the north, the Soviet 7th Guards Army had pushed across the Tisza and established bridgeheads after three days of fighting. On 4 November, the Soviet 7th Guards Army broke through the Hungarian 20th Infantry Division, took Cegléd, and advanced to within 10 miles of Budapest, before it was stopped by German counterattacks. Soon after Hitler's reinforcements arrived in Budapest. *8. SS-Kavallerie, 22. SS-Kavallerie,* and *Feldherrnhalle* divisions were engaged in various counterattacks during November around the city.

The German Sixth Army abandoned the Tisza line in early November and retreated west to new positions between Budapest and Miskolc.

During early November the Soviets regrouped their forces to prepare for an offensive on a much broader front. The Third Ukrainian Front entered Hungary from the south. The Second Ukrainian Front, which included the Romanian First and Fourth Armies, was east and northeast of Budapest.

THIRD UKRAINIAN FRONT

On 7 November the Third Ukrainian Front attempted to cross the west bank of the Danube south of Mohács, but determined German defence held them until 22 November.

The Third Ukrainian Front began a quick advance to the northwest of Pécs and north along the Danube in early December. German Panzer units were rushed south to reinforce this flank. German Panzer counterattacks finally stopped the Soviets' thrust at the Lake Balaton-Lake Velence line on 8 December after ferocious fighting in the rapidly deteriorating weather conditions. The Hungarian winter had arrived.

13. PANZERDIVISION AT KARCAG

The Soviet invasion forced the Germans to send reinforcements to Hungary. Many of these units were panzer or panzergrenadier formations ordered to stop the Soviet thrust towards Debrecen and keep open the withdrawal route of the Axis forces in eastern Hungary.

The *13. Panzerdivision* was assigned to the *III Panzerkorps* during the German counterattack towards Debrecen. On 8 October they were ordered to secure Karcag southwest of Debrecen. The division sent the *I Bataillon/66. Panzergrenadierregiment* towards Karcag supported by a company of 17 Panthers.

This *kampfgruppe* immediately ran into elements of the Soviet 7th Mechanised Corps. A desperate struggle ensued for the domination of ground. The Panthers engaged thirty Soviet T-34/85 tanks, winning the ferocious firefight, but with just six of their own tanks remaining. The Soviets entered Karcag in the late afternoon, which was already occupied by the *66. Panzergrenadierregiment.* Fighting for the town continued into the evening with Panzergrenadiers holding off the Soviets despite the latter's support from IS-2 heavy tanks. By nightfall there were just four Panthers in action, and as the Soviets penetrated to the south of the town, the Germans were forced to retreat to avoid being encircled.

In mid October *13. Panzerdivision* moved north to take part in the counterattack that eventually stopped the Red Army thrust north of Debrecen.

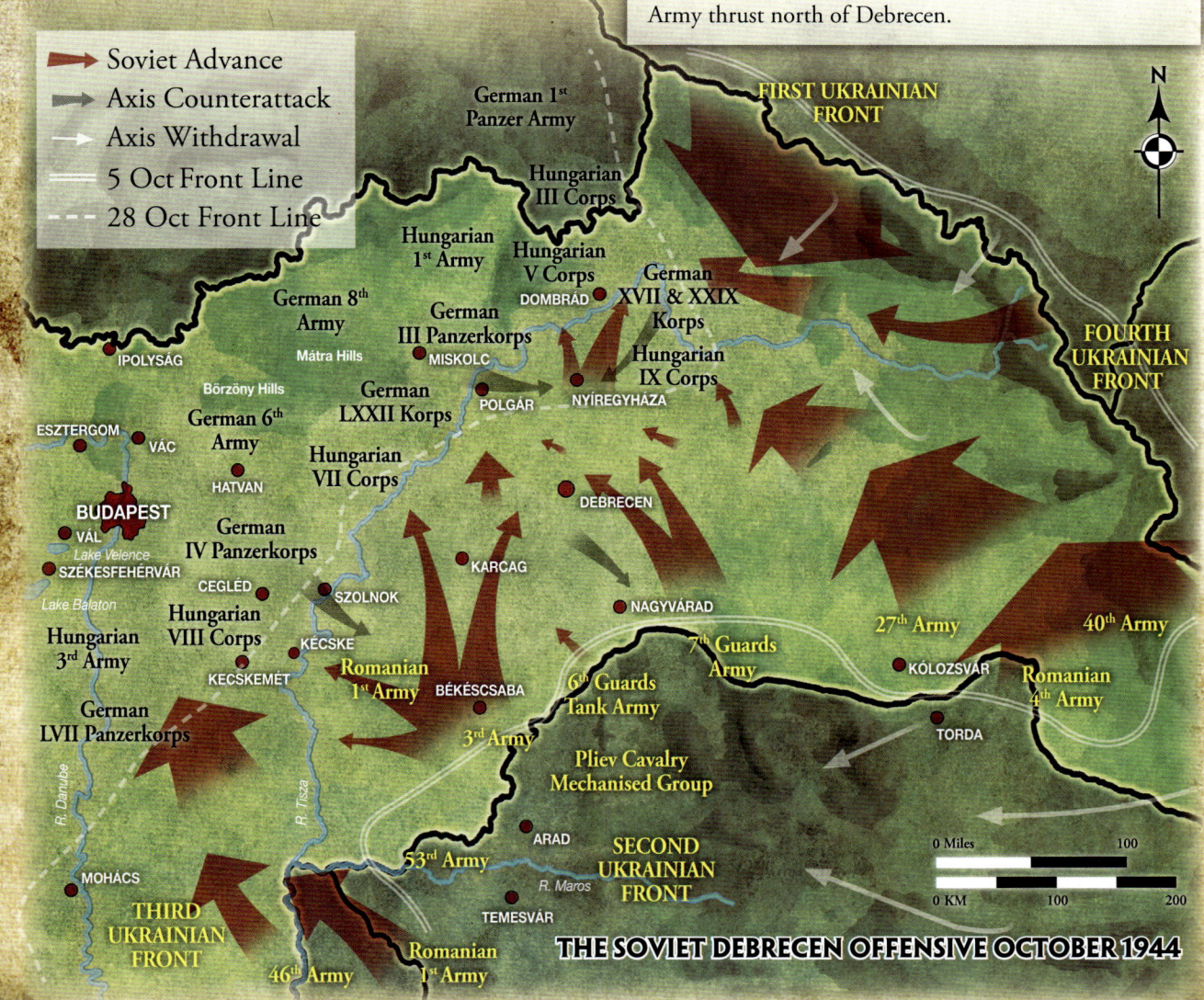

THE SOVIET DEBRECEN OFFENSIVE OCTOBER 1944

SECOND UKRAINIAN FRONT

The Second Ukrainian Front launched a major offensive on 11 November. Resistance was stronger than expected, forcing the Soviets to fight for every inch of ground. Counterattacks were conducted by all the available Panzer troops. After a week of fighting the Soviets had reached Miskolc, which they did not take until 4 December. The Second Ukrainian Front then pushed into the Mátra Hills, northeast of Budapest.

On 5 December, the Pliev Cavalry-Mechanised Group, 6th Guards Tank Army, and 7th Guards Army attacked in the Hatvan sector, at the junction of the German Sixth and Eighth Armies. The area, defended by three German divisions, including the *13. Panzerdivision*, was hit by over 500 Soviet tanks and assault guns on the first day. The Soviets punched through the frontlines and drove armour through the gaps.

German ounterattacks stopped the advance at Érd, on the southwest outskirts of Budapest on 8 December. The 6th Guards Tank Army then thrust north of Budapest and into the Börzsönyi Hills on 11 December.

FINAL ENCIRCLEMENT

On 20 December the 6th Guards Tank Army crushed the German *357. Infanteriedivision* and sped south out of the Börzsönyi Hills past Ipolyság. The Third Ukrainian Front attacked north on both sides of Lake Velence. By 22 December the Soviets had seized Vál and Budapest was in danger of encirclement.

The jaws of the trap closed on 24 December. The 18th Tank Corps of the Third Ukrainian Front and the vanguards of the 6th Guards Tank Army of the Second Ukrainian Front linked up at Esztergom and Budapest was encircled.

THE SIEGE OF BUDAPEST

Budapest was defended by 33,000 German and 37,000 Hungarian troops. From 4 December these troops came under the *IX SS-Gebirgskorps*. The city straddled the Danube River, with the hilly suburbs of Buda on the western side and Pest occupying the eastern bank.

Through December the Soviets made many attempts to penetrate into the city. They selected sectors held by the Hungarians because Germany's allies tended to be less well equipped. However, Hungarian armour, German *Feldherrnhalle,* and SS counterattacks threw back most of the Soviet thrusts.

By 9 December the Soviets had established a bridgehead on Csepel Island south of the city. The Soviets had formed a tight ring around the city by 10 December and began to pound it with heavy artillery.

Fighting died down between 17 and 23 December, but news arrived on 24 December that the city had been encircled. The Soviets tried to batter their way into Budapest on 25 December, but fierce fighting by *Feldherrnhalle* troops held them back.

The Soviets made gains on 28 December and on 30 December broke through Hungarian 10th Infantry Division lines near Csömör. The Germans concentrated flak with the *13. Panzerdivision* around Csömör to protect the New Racecourse airfield.

The counterattacks by *13. Panzerdivision* were not able to take back the airfield fully, forcing an Axis retreat to a shorter defensive line.

The battle to take the city was taking longer than the Soviets anticipated. Once the encirclement was completed the city was expected to fall in three to four days. However, the city's defenders were fighting with great determination to keep the Red Army out. The Soviets asked for the Budapest defenders' surrender on 29 December, but the offer was refused and the savage street-fighting continued about the city.

The Soviets and Romanians pressed the defences and by 31 December they had German troops pushed back to within half a mile of the Danube.

On the night of 31 December/1 January, the Red Army made a concerted attempt to throw the Axis forces out of Pest with a front-wide assault on the eastern portion of the city. A massive artillery barrage was unleashed all along the front followed by assaults. The *13. Panzerdivision* was pushed west out of Csömör and the defenders were forced to commit their last reserves in an attempt to hold Pest.

The Soviets made more attacks between 1 and 15 January, but the defenders threw them back. Each time the Germans and Hungarians were forced to shorten their lines, decreasing the defensive pocket. Pest's days were numbered. Axis Forces in Pest were finally ordered to withdraw across the Danube to Buda on 16 January. The evacuation lasted until daylight on 18 January. However, many Hungarian troops and civilians were trapped when the last bridge was blown by the Germans.

OPERATIONS KONRAD I-III

While the defenders of Budapest were fighting for their lives, several attempts were made to relieve the city. These were spearheaded by the *3. 'Totenkopf' SS-Panzerdivision* and *5. 'Wiking' SS-Panzerdivision* transferred from Army Group Centre. Three attempts were made, but each failed after heavy fighting. The final offensive ended on 27 January with the city still encircled.

END OF THE SIEGE

By 19 January food and ammunition were in short supply with only occasional airdrops getting through. Combat troop numbers were also low, the last members of the *IX SS-Gebirgskorps* staff were thrown into the battle as infantry on 27 January and the *Feldherrnhalle Panzerdivision* had to merge its Panzergrenadier regiments.

Fighting continued with savagery and on 3 February the infantry defenders destroyed eight Soviet tanks during the day's combat.

On 5 February the Soviets took their first hill in the west of Buda further tightening their hold on the decreasing pocket. A Red Army attack between Castle Hill and the Budapest citadel to split the defending forces was met with fierce resistance and failed. However, the end was near.

The German command decided to breakout on the night of 11 February. It began at 2000 hours (8pm). 10,000 seriously wounded were left behind, heavy equipment was destroyed and 16,000 men prepared to breakout with what light equipment they could carry.

THE SIEGE OF BUDAPEST JANUARY TO FEBRUARY 1945

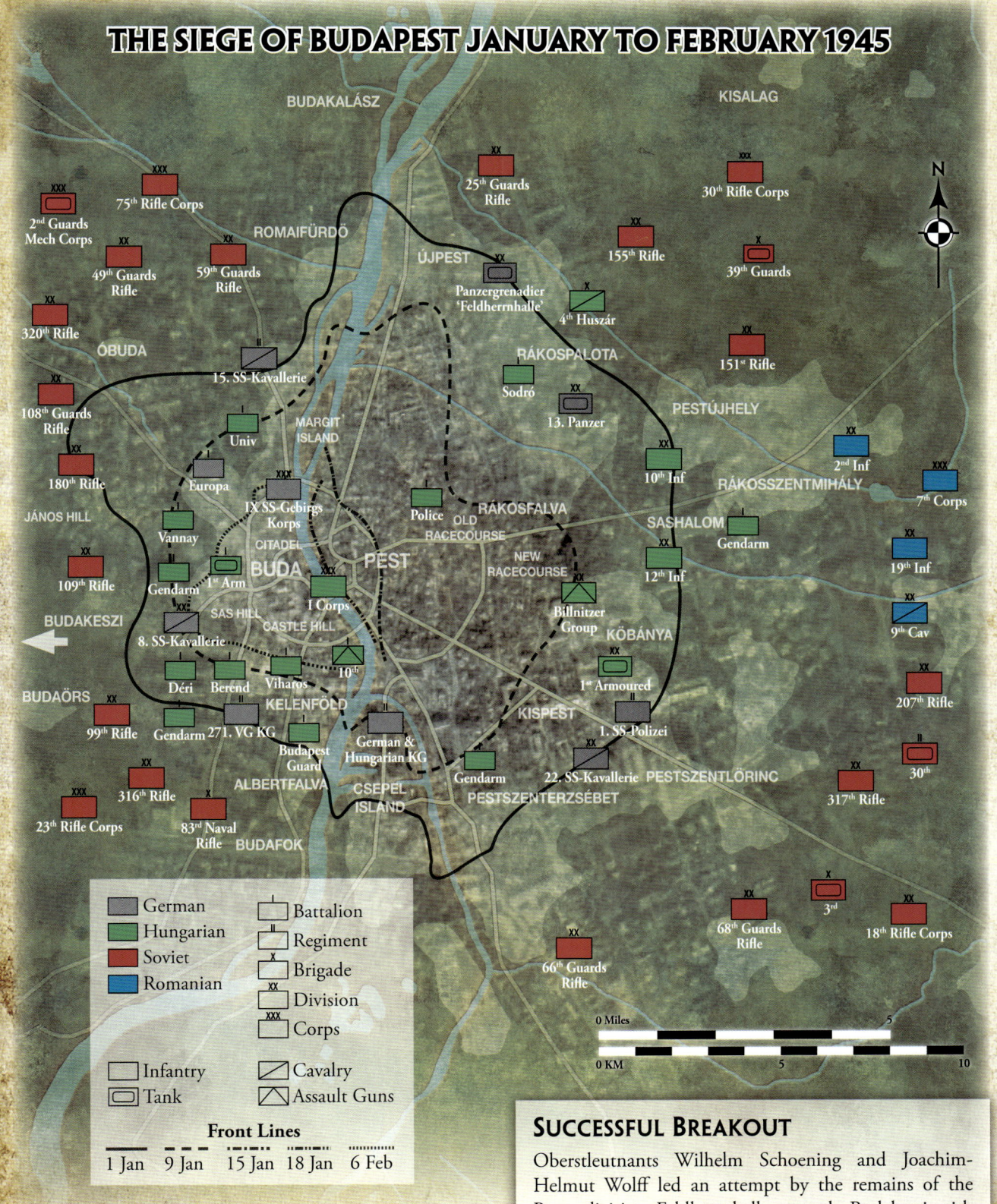

Legend:

German		Battalion
Hungarian		Regiment
Soviet		Brigade
Romanian		Division
		Corps
Infantry		Cavalry
Tank		Assault Guns

Front Lines

1 Jan   9 Jan   15 Jan   18 Jan   6 Feb

The Soviets opened fire with artillery as soon as it started. Lead elements from *13. Panzerdivision* and *8. SS-Kavalleriedivision* were slaughtered under the heavy barrage. A second wave did little better.

Various other groups of Hungarians and Germans made it out in small numbers, but the majority were either killed or taken prisoner. On 13 February the remaining defenders surrendered to the Soviets. After an epic struggle, the city of Budapest had been lost.

SUCCESSFUL BREAKOUT

Oberstleutnants Wilhelm Schoening and Joachim-Helmut Wolff led an attempt by the remains of the Panzerdivision Feldherrnhalle towards Budakeszi with about 600 men. They overran the Soviet blocking forces and made their way to woods using ravines and gullies for cover. By 14 February almost half the group had reached the main German lines. This was the largest single force to make it out of the Budapest encirclement.

FELDHERRNHALLE PANZERKAMPFGRUPPE
Feldherrnhalle Armoured Battle Group

(Tank Company)

You must field one platoon from each box shaded black and may field one platoon from each box shaded grey.

DIVISIONAL SUPPORT PLATOONS

COMBAT PLATOONS

ARMOUR

Feldherrnhalle Panzer Platoon — 137

ARMOUR

Feldherrnhalle Panzer Platoon — 137

ARMOUR

Feldherrnhalle Panzer Platoon — 137

Feldherrnhalle Gepanzerte Panzergrenadier Platoon — 138

INFANTRY

Feldherrnhalle Gepanzerte Panzergrenadier Platoon — 138

WEAPONS PLATOONS

INFANTRY

Panzer Pioneer Platoon — 74

RECONNAISSANCE

Panzer Scout Platoon — 75

ANTI-AIRCRAFT

Feldherrnhalle Anti-aircraft Gun Platoon — 138

ALLIED PLATOONS

✚ 🦅

Hungarian and Luftwaffe Platoons in your force are Allies and follow the Allies rules in the rulebook.

ARMOUR

Schwere Panzer Platoon — 71

Tank-hunter Platoon — 163

Hungarian Rohamágyús Platoon ✚ — 215

Anti-tank Gun Platoon — 165

Heavy Anti-tank Gun Platoon — 165

INFANTRY

Feldherrnhalle Gepanzerte Panzergrenadier Platoon — 138

Feldherrnhalle Panzergrenadier Platoon — 141

Aufklärungs Platoon — 93

Hungarian Önkéntes Puskás Platoon ✚ — 221

INFANTRY

Gepanzerte Panzerpionier Platoon — 87

Panzerpionier Platoon — 89

RECONNAISSANCE

Light Panzerspäh Platoon — 91

Half-tracked Panzerspäh Platoon — 91

Heavy Panzerspäh Platoon — 92

ARTILLERY

Motorised Artillery Battery — 167

Motorised Heavy Artillery Battery — 167

Armoured Artillery Battery — 168

Armoured Heavy Artillery Battery — 168

Hungarian Artillery Battery ✚ — 226

ARTILLERY

Rocket Launcher Battery — 169

Motorised Artillery Battery — 167

ANTI-AIRCRAFT

Anti-aircraft Gun Platoon — 171

Hungarian Anti-aircraft Platoon ✚ — 226

Luftwaffe Anti-aircraft Gun Platoon — 173

ANTI-AIRCRAFT

Heavy Anti-aircraft Gun Platoon — 171

Luftwaffe Heavy Anti-aircraft Gun Platoon — 173

AIRCRAFT

Air Support — 172

FELDHERRNHALLE

The title *'Feldherrnhalle'* was given to *Heer* (German Army) units who drew their recruits from the SA (*Sturmabteilung*) or 'Brown Shirts', the original paramilitary organisation of the Nazi Party, who were later usurped by the SS. The specific title *'Feldherrnhalle'* refers to the *Feldherrnhalle* (Field Marshals' Hall) in Munich where an attempted Nazi coup was crushed by the Bavarian state police on 9 November 1923. It resulted in Hitler's arrest and imprisonment.

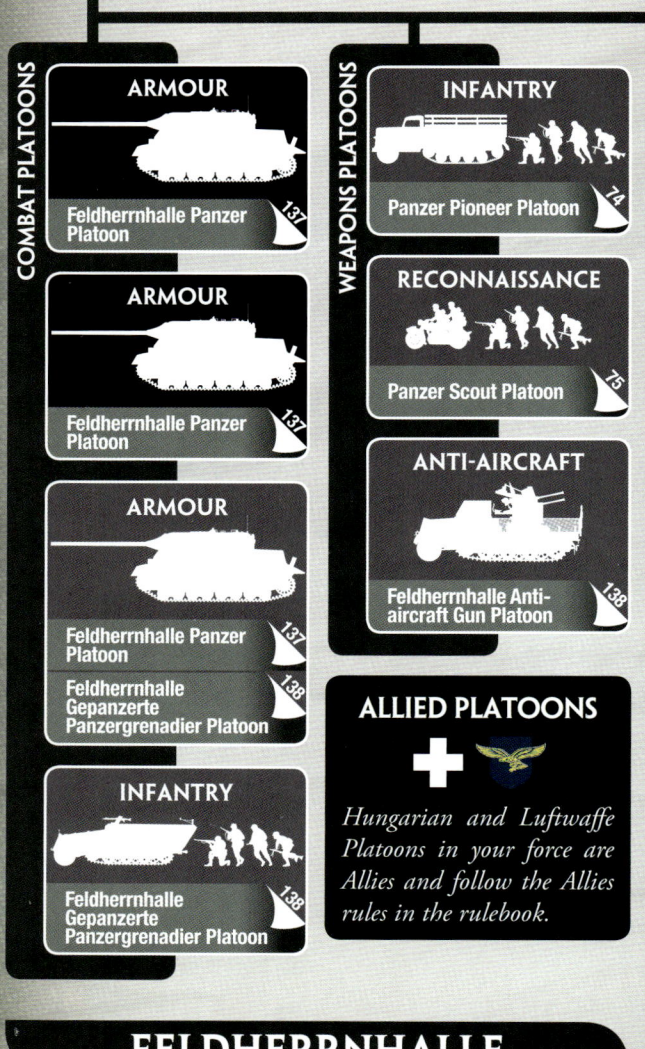

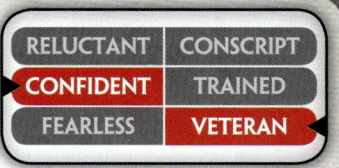

MOTIVATION AND SKILL

The Panzertruppen *(armoured troops) of the* 13. Panzerdivision *and* Panzerdivision Feldherrnhalle *are commanded by veterans of many hard battles on the eastern front. Thrown into the Hungarian cauldron as reinforcements while rebuilding they fight with the hard-nosed determination for which the German Panzer troops are renowned. A* Feldherrnhalle *Panzerkampfgruppe is rated as* **Confident Veteran.**

RELUCTANT	CONSCRIPT
CONFIDENT	TRAINED
FEARLESS	**VETERAN**

HEADQUARTERS

FELDHERRNHALLE PANZERKAMPFGRUPPE HQ

HEADQUARTERS

Company HQ with:

2 Panzer IV/70 (A)	270	points
1 Panzer IV/70 (A)	135	points
2 Panzer IV/70 (V)	300	points
1 Panzer IV/70 (V)	150	points
2 Panther G	375	points
1 Panther G	190	points

OPTION

- Add an Sd Kfz 9 (18t) recovery half-track for +5 points, a Bergepanzer III recovery vehicle for +10 points, or a Bergepanther recovery vehicle for +15 points.

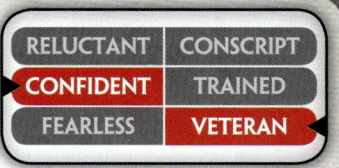

HAUPTMANN

HAUPTMANN

Company Command tank

2iC Command tank

COMPANY HQ

UNTEROFFIZIER

Recovery vehicle

RECOVERY SECTION

FELDHERRNHALLE PANZERKAMPFGRUPPE HQ

You must field at least one Panzer Platoon equipped with the same model of tank as the Company HQ.

The *13. Panzerdivision* and *Panzerdivision Feldherrnhalle* Panzer units were armed with the new Panzer IV/70 and Panther G tanks. These became part of the units when both divisions merged with a *Feldherrnhalle* Panzer brigade each.

These well equipped, but inexperienced units provided the division veterans with the raw materials needed to complete their refit. New tanks in hand they rushed to the front in time for the battles at Debrecen.

COMBAT PLATOONS

FELDHERRNHALLE PANZER PLATOON

PLATOON

4 Panzer IV/70 (A)	540	points
3 Panzer IV/70 (A)	405	points
4 Panzer IV/70 (V)	600	points
3 Panzer IV/70 (V)	450	points
4 Panther G	750	points
3 Panther G	560	points

The Panzer IV/70 tank combined the excellent gun of the Panther tank with the tested hull of the Panzer IV tank. With additional armour this made for a formidable fighting vehicle.

The *Feldherrnhalle* divisions also used the Panther tank, one of the best fighting vehicles to see service in the war. Its heavy

LEUTNANT

LEUTNANT

Command Tank

HQ SECTION

UNTEROFFIZIER

Tank

PANZER SECTION

UNTEROFFIZIER

Tank

PANZER SECTION

UNTEROFFIZIER

Tank

PANZER SECTION

FELDHERRNHALLE PANZER PLATOON

armour and excellent manoeuvrability made it devastating in an attack.

FELDHERRNHALLE GEPANZERTE PANZERGRENADIER PLATOON

PLATOON

HQ Section with:

3 Panzergrenadier Squads	220 points
2 Panzergrenadier Squads	155 points

OPTIONS

- Replace the Command MG team with a Command Panzerfaust SMG team for +10 points.
- Replace Sd Kfz 251/1 half-track in HQ Section with a Sd Kfz 251/10 (3.7cm) half-track at no cost.

Feldherrnhalle Gepanzerte Panzergrenadier Platoons may use the Mounted Assault special rule.

When a panzer division divides into battle groups during a battle the armoured panzergrenadiers always support the tanks, dealing with troublesome infantry as needed.

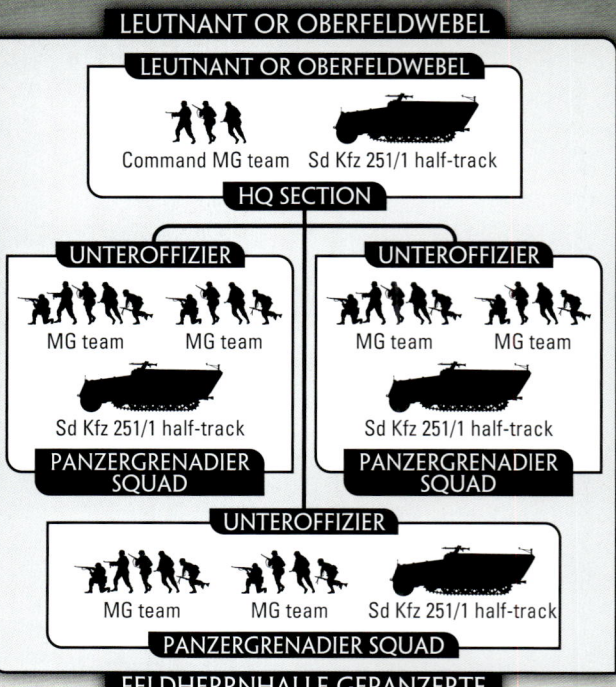

LEUTNANT OR OBERFELDWEBEL

LEUTNANT OR OBERFELDWEBEL

Command MG team   Sd Kfz 251/1 half-track

HQ SECTION

UNTEROFFIZIER

MG team   MG team

Sd Kfz 251/1 half-track

PANZERGRENADIER SQUAD

UNTEROFFIZIER

MG team   MG team

Sd Kfz 251/1 half-track

PANZERGRENADIER SQUAD

UNTEROFFIZIER

MG team   MG team   Sd Kfz 251/1 half-track

PANZERGRENADIER SQUAD

FELDHERRNHALLE GEPANZERTE PANZERGRENADIER PLATOON

WEAPONS PLATOONS

FELDHERRNHALLE ANTI-AIRCRAFT GUN PLATOON

PLATOON

3 Sd Kfz 7/1 (Quad 2cm)	120 points
2 Sd Kfz 7/1 (Quad 2cm)	80 points
3 Armoured Sd Kfz 7/1 (Quad 2cm)	150 points
2 Armoured Sd Kfz 7/1 (Quad 2cm)	100 points
3 Möbelwagen (3.7cm)	165 points
2 Möbelwagen (3.7cm)	110 points
3 Wirbelwind (Quad 2cm)	165 points
2 Wirbelwind (Quad 2cm)	110 points

One of the few vulnerabilities of your panzers is an air strike. Providing ample anti-aircraft support will assure your panzers are available to bring death and destruction down upon enemy armour.

LEUTNANT

LEUTNANT

Command Anti-aircraft tank

ANTI-AIRCRAFT SECTION

UNTEROFFIZIER

Anti-aircraft tank

ANTI-AIRCRAFT SECTION

UNTEROFFIZIER

Anti-aircraft tank

ANTI-AIRCRAFT SECTION

FELDHERRNHALLE ANTI-AIRCRAFT GUN PLATOON

'FELDHERRNHALLE' DIVISION

In October 1944 Hitler rushed *13. Panzerdivision* and *Panzergrenadierdivision Feldherrnhalle* to Hungary to stop the Red Army's invasion.

HUNGARIAN CAMPAIGN

The *13. Panzerdivision* first fought in the region of Püspökladany where they crushed Soviet tank forces and took the town. Soviet attacks intensified and the town was lost. The *13. Panzerdivision* withdrew on 9 October.

A *Kampfgruppe* of *Panzergrenadierdivision Feldherrnhalle* was mobilised on 11 October. They were deployed in the gap created when the *13. Panzerdivision* was forced to withdraw.

COUNTERATTACK AT DEBRECEN

By mid-October the *Feldherrnhalle* units were withdrawn north to the Tisza River to cover the retreat of the German 8th Army near Polgár. Soviet mobile units made a massive push past Debrecen, outpacing their infantry support.

The Germans quickly responded on 23 October and counterattacked with their available panzer troops to cut off the over-exposed armour and cavalry. The attack caught the Soviets by surprise and much of the Soviet Pliev Cavalry-Mechanised Group was cut off and destroyed. This allowed the Axis armies to complete their retreat across the Tisza River.

DEFENDING THE ROAD TO BUDAPEST

13. Panzerdivision remained with the *III Panzerkorps* and moved to the southeast of Budapest, while *Kampfgruppe Panzergrenadierdivision Feldherrnhalle* and *109. Panzer Brigade* moved to Hatvan where they were merged to complete the formation of *Panzerdivision Feldherrnhalle*.

Both divisions were back in action as the Soviets swept from the south and east across the Tisza and towards Budapest in early November. They spent November holding off the Soviet advance on Budapest to the north and east of the city. By the 23 November elements of *Panzerdivision Feldherrnhalle* were already fighting in the outskirts of Budapest. *Panzerdivision Feldherrnhalle* units defending the Danube River north of Budapest were forced to retreat to the outskirts of Budapest when Soviet tanks broke through the Hungarian units on their flank and took Vác. The Soviets advanced from the northeast to eventually link up with units coming from the south and encircle Budapest.

THE SIEGE OF BUDAPEST

The bulk of both divisions became encircled inside the city. During the fighting most of the armour from the two divisions was grouped together to act as a mobile reserve. The divisions' Panzergrenadiers fought side-by-side in the city's northeast, often counterattacking Soviet assaults and reinforcing Hungarian positions. *Panzerdivision Feldherrnhalle* troops were forced to abandon Újpest on 9 January 1945 when Soviet units penetrated towards Pestújhely, threatening their right flank (see map on page 135). At this point bitter and savage street fighting for the city proper began.

Slowly the pressure from the Soviets and Romanians wore the defenders down. By 17 January the *Feldherrnhalle* panzergrenadiers were fighting for the city centre of Pest with their backs to the Danube River. The order was finally given to evacuate to Buda on the western side of the river.

The house-to-house fighting continued on the other side of the river with the Soviets determined to wipe out the pocket before any of the German relief attempts could break through to the city. The *Feldherrnhalle* took charge of the defence in the south of Buda from 18 January. The hilly suburbs of Buda proved a hard nut to crack for the Soviet assault troops.

On 11 February a desperate breakout began. The largest and most successful group to make it back to the German lines was a 600 men from *Panzerdivision Feldherrnhalle*.

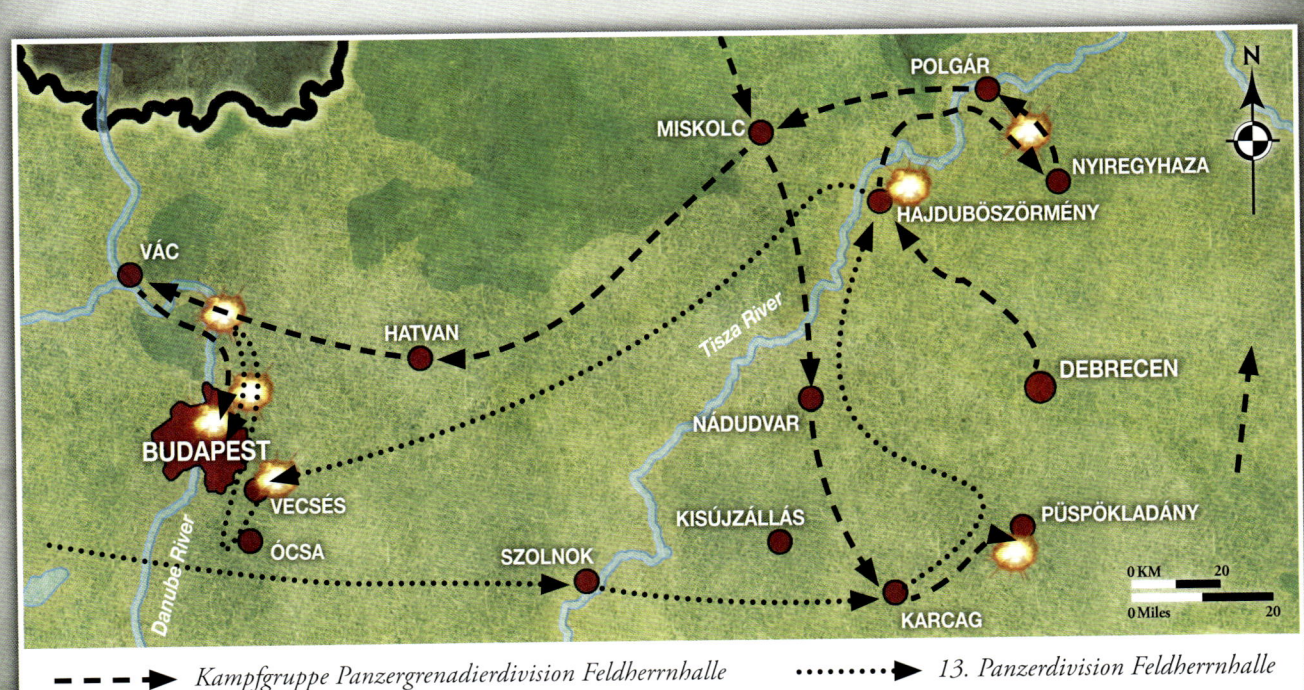

- - - ➤ *Kampfgruppe Panzergrenadierdivision Feldherrnhalle*   ·····➤ *13. Panzerdivision Feldherrnhalle*

FELDHERRNHALLE PANZERGRENADIERKOMPANIE
FELDHERRNHALLE MOTORISED INFANTRY COMPANY

(INFANTRY COMPANY)

HEADQUARTERS

Feldherrnhalle Panzer-grenadierkompanie HQ — 141

You must field one platoon from each box shaded black and may field one platoon from each box shaded grey.

DIVISIONAL SUPPORT PLATOONS

COMBAT PLATOONS

INFANTRY
Feldherrnhalle Panzergrenadier Platoon — 141

INFANTRY
Feldherrnhalle Panzergrenadier Platoon — 141

INFANTRY
Feldherrnhalle Panzergrenadier Platoon — 141

MACHINE-GUNS
Feldherrnhalle Heavy Platoon — 142

ALLIED PLATOONS

Hungarian and Luftwaffe Platoons in your force are Allies and follow the Allies rules in the rulebook.

WEAPONS PLATOONS

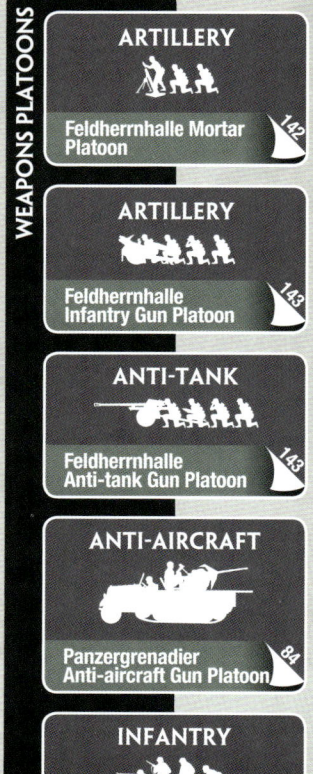

ARTILLERY
Feldherrnhalle Mortar Platoon — 142

ARTILLERY
Feldherrnhalle Infantry Gun Platoon — 143

ANTI-TANK
Feldherrnhalle Anti-tank Gun Platoon — 143

ANTI-AIRCRAFT
Panzergrenadier Anti-aircraft Gun Platoon — 84

INFANTRY
Panzerpionier Platoon — 89

REGIMENTAL SUPPORT PLATOONS

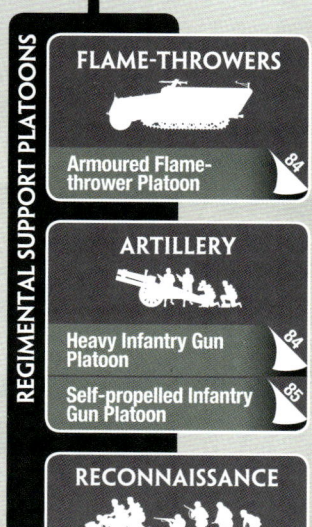

FLAME-THROWERS
Armoured Flame-thrower Platoon — 84

ARTILLERY
Heavy Infantry Gun Platoon — 84
Self-propelled Infantry Gun Platoon — 85

RECONNAISSANCE
Panzer Scout Platoon — 75

ARMOUR
Feldherrnhalle Panzer Platoon — 137
Hungarian Rohamágyús Platoon — 215

ARMOUR
Feldherrnhalle Panzer Platoon — 137
Tank-hunter Platoon — 163
Anti-tank Gun Platoon — 165

INFANTRY
Feldherrnhalle Gepanzerte Panzergrenadier Platoon — 138
Aufklärungs Platoon — 93
Hungarian Önkéntes Puskás Platoon — 221

INFANTRY
Gepanzerte Panzerpionier Platoon — 87
Panzerpionier Platoon — 89

RECONNAISSANCE
Light Panzerspäh Platoon — 91
Half-tracked Panzerspäh Platoon — 91
Heavy Panzerspäh Platoon — 92

FORTIFICATIONS
Street Fortifications — 157

ARMOUR
Motorised Artillery Battery — 167
Motorised Heavy Artillery Battery — 167
Armoured Artillery Battery — 168
Armoured Heavy Artillery Battery — 168
Hungarian Artillery Battery — 226

ARTILLERY
Rocket Launcher Battery — 169
Motorised Artillery Battery — 167

ANTI-AIRCRAFT
Anti-aircraft Gun Platoon — 171
Hungarian Anti-aircraft Platoon — 226
Luftwaffe Anti-aircraft Gun Platoon — 173

ANTI-TANK
Heavy Anti-aircraft Gun Platoon — 171
Luftwaffe Heavy Anti-aircraft Gun Platoon — 173
Heavy Anti-tank Gun Platoon — 165

AIRCRAFT
Air Support — 172

MOTIVATION AND SKILL

The motorised panzergrenadiers of 13. Panzerdivision *and* Panzerdivision Feldherrnhalle *fight with great bravery and skill in Hungary. Their role within the division makes them excellent troops on attack and defence. In Budapest they will be tested in some of the fiercest street fighting of the war. A Feldherrnhalle Panzergrenadierkompanie is rated as* **Confident Veteran.**

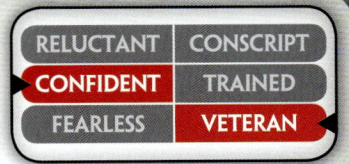

RELUCTANT	CONSCRIPT
CONFIDENT ◄	TRAINED
FEARLESS	VETERAN

HEADQUARTERS

FELDHERRNHALLE PANZERGRENADIERKOMPANIE HQ

HEADQUARTERS

| Company HQ | 45 points |

OPTIONS

- Replace either or both Command SMG teams with Command Panzerfaust SMG teams for +10 points per team.
- Add Anti-tank section for +25 points.
- Add Motorcycle and sidecar and Kfz 15 field car for Company HQ and Kfz 70 truck for Anti-tank section for +5 points for the Headquarters.
- Add up to three Sniper teams for +50 points per team.

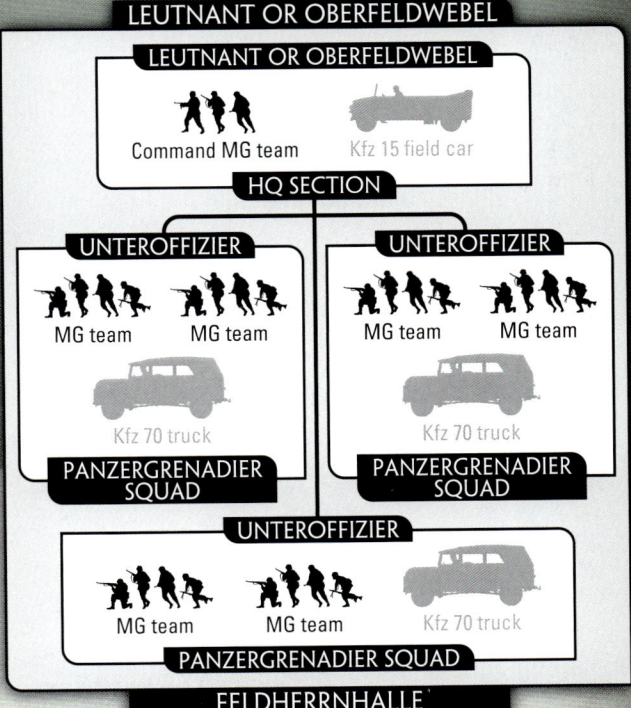

HAUPTMANN

HAUPTMANN

Company Command SMG team

Motorcycle and sidecar

2iC Command SMG team

Kfz 15 field car

COMPANY HQ

UNTEROFFIZIER

Panzerschreck team

Kfz 70 truck

ANTI-TANK SECTION

FELDHERRNHALLE PANZERGRENADIERKOMPANIE HQ

While fighting in Hungary the *Feldherrnhalle* panzergrenadiers fought in various battles.

On the Tisza River they counterattacked to drive back the Soviets and Romanians and then around Budapest fought running defensive battles against Soviet mobile troops. In the savage street fighting inside the city they fought some of their bloodiest battles. Finally they made last desperate breakout attempts to escape the encirclement in February 1945.

COMBAT PLATOONS

FELDHERRNHALLE PANZERGRENADIER PLATOON

PLATOON

HQ Section with:

| 3 Panzergrenadier Squads | 180 points |
| 2 Panzergrenadier Squads | 130 points |

OPTIONS

- Replace the Command MG team with a Command Panzerfaust SMG team for +10 points.
- Add Kfz 15 field car and Kfz 70 trucks for +5 points for the platoon.

STREET BRAWLERS

Feldherrnhalle Company Command, 2iC Command and Platoon Command Infantry teams are Street Brawlers.

If there are no enemy Tank teams or Bunkers within 2"/5cm of a Street Brawler, the Street Brawler hits on a roll of 2+ in Assaults.

LEUTNANT OR OBERFELDWEBEL

LEUTNANT OR OBERFELDWEBEL

Command MG team

Kfz 15 field car

HQ SECTION

UNTEROFFIZIER

MG team   MG team

Kfz 70 truck

PANZERGRENADIER SQUAD

UNTEROFFIZIER

MG team   MG team

Kfz 70 truck

PANZERGRENADIER SQUAD

UNTEROFFIZIER

MG team   MG team

Kfz 70 truck

PANZERGRENADIER SQUAD

FELDHERRNHALLE PANZERGRENADIER PLATOON

FELDHERRNHALLE HEAVY PLATOON

PLATOON

HQ Section with:

2 Machine-gun Sections	135 points
1 Machine-gun Section	70 points
No Machine-gun Sections	10 points

OPTIONS

- Add a Mortar Section for +65 points.
- Add Kfz 15 field car, Kübelwagen jeep and Kfz 70 trucks for +5 points for the platoon.

A Feldherrnhalle Heavy Platoon must have a Mortar Section if it has no Machine-gun sections.

A Feldherrnhalle Heavy Platoon may make Combat Attachments to Feldherrnhalle Panzergrenadier Platoon.

The Heavy Platoon provides the company's immediate heavy weapons support. Machine-guns and mortars can provide covering fire, or can be attached directly to the panzer-grenadiers in more direct support.

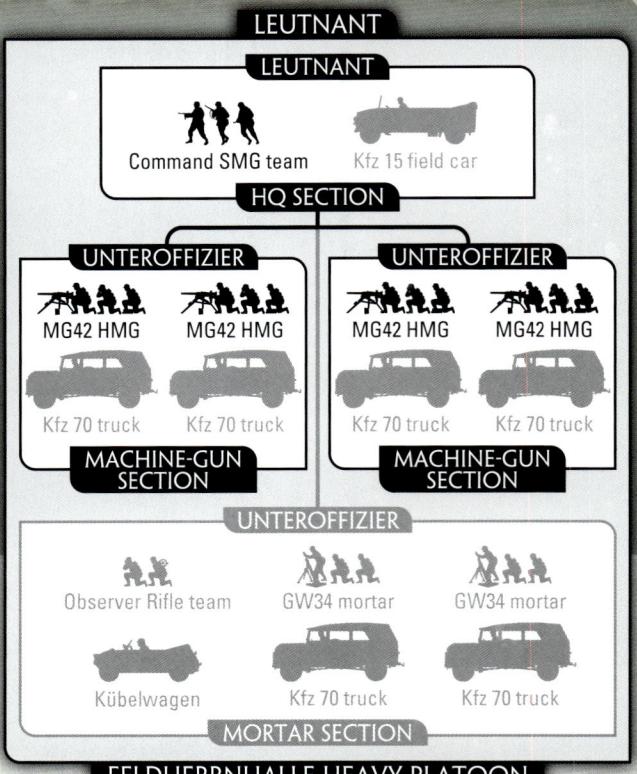

WEAPONS PLATOONS

FELDHERRNHALLE MORTAR PLATOON

PLATOON

HQ Section with:

3 Mortar Sections	180 points
2 Mortar Sections	125 points
1 Mortar Section	65 points

OPTIONS

- Add Kfz 15 field car, Kübelwagen jeeps, and 3-ton trucks for +5 points for the platoon.
- Upgrade the 8cm GW34 mortars to 12cm sGW43 mortars for +20 points per Mortar Section.

A Feldherrnhalle Mortar Platoon upgraded to 12cm sGW43 mortars may not have more than two Mortar Sections.

Mortars are ideal for fighting in the urban environment where open spaces are limited and big batteries of guns cannot be deployed. The mortars are close at hand, ready to provide support when called on, unlike the artillery who are often deployed kilometres away in large parks or sports grounds.

Out on the open plains their fast response time and ease of relocation also make them a vital support weapon for the fast-moving panzergrenadiers.

FELDHERRNHALLE
INFANTRY GUN PLATOON

PLATOON

HQ Section with:

2 7.5cm leIG18	70 points

OPTION

- Add Kfz 15 field car and Kfz 70 trucks for +5 points for the platoon.

The small and easily handled 7.5cm leIG18 infantry gun is an excellent weapon in a street battle, whether it is providing direct or indirect fire support to an attack, rolling up with an advance or defending a street intersection from advancing communists or Romanian traitors.

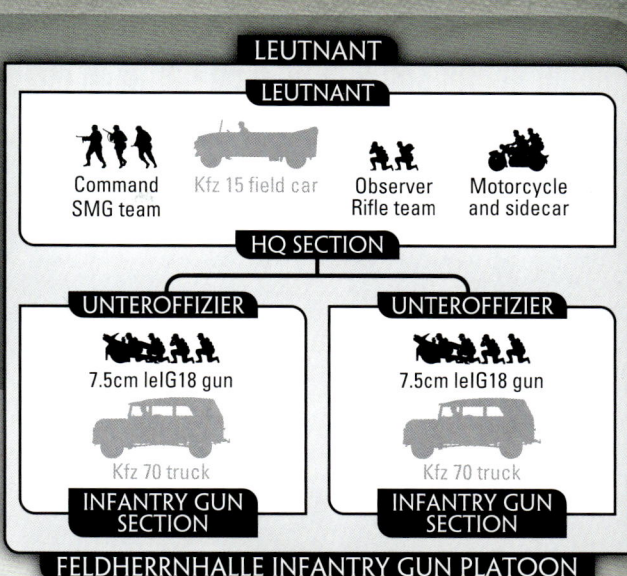

LEUTNANT
LEUTNANT
Command SMG team · Kfz 15 field car · Observer Rifle team · Motorcycle and sidecar
HQ SECTION
UNTEROFFIZIER — 7.5cm leIG18 gun — Kfz 70 truck — INFANTRY GUN SECTION
UNTEROFFIZIER — 7.5cm leIG18 gun — Kfz 70 truck — INFANTRY GUN SECTION
FELDHERRNHALLE INFANTRY GUN PLATOON

FELDHERRNHALLE
ANTI-TANK GUN PLATOON

PLATOON

HQ Section with:

3 7.5cm PaK40	155 points
2 7.5cm PaK40	105 points

OPTION

- Add Kfz 15 field car and Kfz 70 trucks for +5 points for the platoon.

The excellent 7.5cm PaK40 anti-tank gun has become the standard tank-busting weapon in the arsenal of the panzergrenadiers.

Ambushing from a wood or lurking at the end of a street, the PaK40 is equally devastating against an unsuspecting enemy vehicle. It also works well against entrenched enemy guns and infantry.

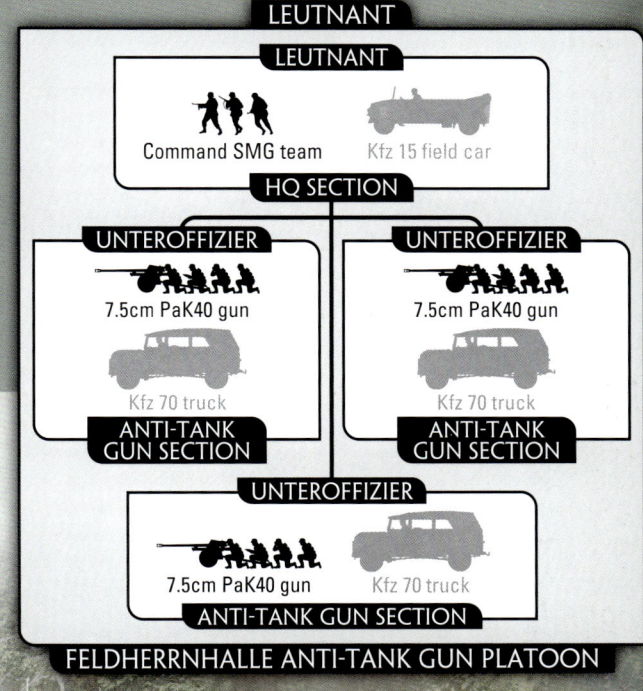

LEUTNANT — Command SMG team · Kfz 15 field car — HQ SECTION — UNTEROFFIZIER 7.5cm PaK40 gun / Kfz 70 truck / ANTI-TANK GUN SECTION (x3) — FELDHERRNHALLE ANTI-TANK GUN PLATOON

PANZERJÄGERKOMPANIE
TANK-HUNTER COMPANY

(TANK COMPANY)

HEADQUARTERS

HEADQUARTERS

Panzerjägerkompanie HQ — 145

You must field one platoon from each box shaded black and may field one platoon from each box shaded grey.

DIVISIONAL SUPPORT PLATOONS

COMBAT PLATOONS

ARMOUR

Panzerjäger Platoon — 145

ARMOUR

Panzerjäger Platoon — 145

ARMOUR

Panzerjäger Platoon — 145

ALLIED PLATOONS

✚ 🦅

Hungarian and Luftwaffe Platoons in your force are Allies and follow the Allies rules in the rulebook.

ANTI-TANK

Anti-tank Gun Platoon — 165

INFANTRY

Grenadier Platoon — 27

Pionier Platoon — 61

Hungarian Puskás Platoon ✚ — 217

INFANTRY

Grenadier Platoon — 27

ARTILLERY

Artillery Battery — 166

Heavy Artillery Battery — 166

Hungarian Artillery Battery ✚ — 226

ARTILLERY

Rocket Launcher Battery — 169

Artillery Battery — 166

ANTI-AIRCRAFT

Anti-aircraft Gun Platoon — 171

Hungarian Anti-aircraft Platoon ✚ — 226

Luftwaffe Anti-aircraft Gun Platoon 🦅 — 173

ANTI-TANK

Heavy Anti-aircraft Gun Platoon — 171

Luftwaffe Heavy Anti-aircraft Gun Platoon 🦅 — 173

Heavy Anti-tank Gun Platoon — 165

AIRCRAFT

Air Support — 172

144

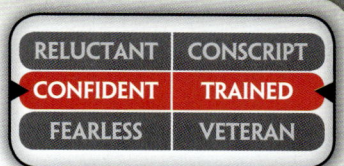
MOTIVATION AND SKILL

In the second half of 1944 as the production of the Hetzer tank-hunter got into full swing many reforming tank-hunter and assault gun battalions were issued these weapons instead of Marders and StuGs. They were quickly filled with new recruits under the leadership of old hands and sent to the front. A Panzerjägerkompanie is rated as **Confident Trained**.

RELUCTANT	CONSCRIPT
CONFIDENT	**TRAINED**
FEARLESS	VETERAN

HEADQUARTERS

PANZERJÄGERKOMPANIE HQ
HEADQUARTERS

Company HQ with:

2 Hetzer	130 points
1 Hetzer	65 points

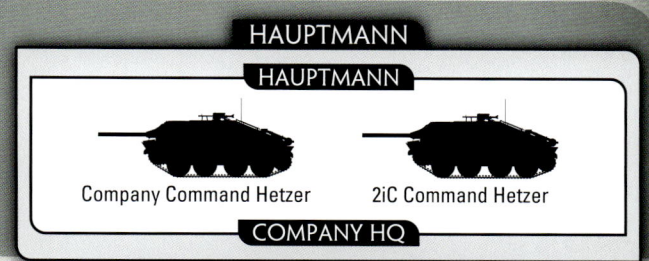

HAUPTMANN
HAUPTMANN
Company Command Hetzer · 2iC Command Hetzer
COMPANY HQ
PANZERJÄGERKOMPANIE HQ

Both assault gun and tank-hunter units began equipping with Hetzers in 1944. A number of these units were quickly committed to the fighting in Hungary supporting infantry against the Soviet and Romanian attack.

Many of these units had gained the title *abteilung* (detachment), but though this might imply they were battalion strength most were in fact only company sized units.

COMBAT PLATOONS

PANZERJÄGER PLATOON
PLATOON

4 Hetzer	260 points
3 Hetzer	195 points

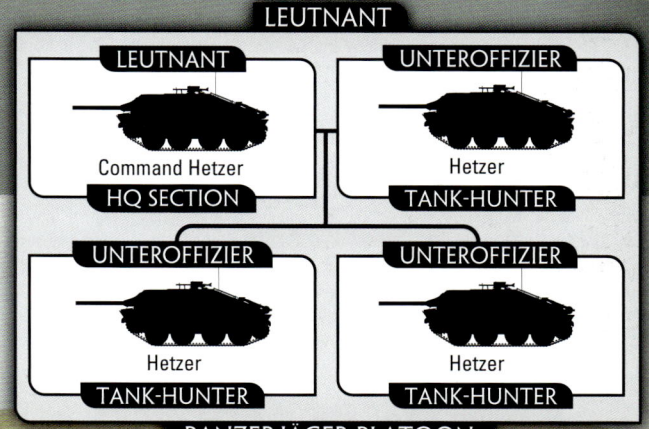

LEUTNANT
LEUTNANT · UNTEROFFIZIER
Command Hetzer · Hetzer
HQ SECTION · TANK-HUNTER
UNTEROFFIZIER · UNTEROFFIZIER
Hetzer · Hetzer
TANK-HUNTER · TANK-HUNTER
PANZERJÄGER PLATOON

The *StuG* (assault gun) companies armed with Hetzer tank-hunters usually had three platoons of three vehicles, while the Hetzer tank-hunters in *Panzerjäger* (tank-hunter) companies were organised in platoons of four.

Despite the usual support role of these units, often the need for armour to mount counterattacks was great and they were sent in en mass to push back the enemy attack.

SS-KAVALLERIESCHWADRON
SS Cavalry Squadron

(Infantry Company)

HEADQUARTERS

SS-Kavallerieschwadron HQ — 147

You must field one platoon from each box shaded black and may field one platoon from each box shaded grey.

You must choose to field your force from either 8. SS-Kavalleriedivision Florian Geyer (marked ⬛), or from 22. SS-Freiwilligen-Kavalleriedivision Maria Theresia (marked ⬛). Whichever SS-Division you chose, you may only take platoons and options marked with your division's symbol or no symbol.

DIVISIONAL SUPPORT PLATOONS

COMBAT PLATOONS

INFANTRY

SS-Kavallerie Platoon — 147

INFANTRY

SS-Kavallerie Platoon — 147

INFANTRY

SS-Kavallerie Platoon — 147

ALLIED PLATOONS

✚ (Luftwaffe symbol)

Hungarian and Luftwaffe Platoons in your force are Allies and follow the Allies rules in the rulebook.

WEAPONS PLATOONS

MACHINE-GUNS

SS-Kavallerie Machine-gun Platoon — 148

ARTILLERY

SS-Kavallerie Mortar Platoon — 148

REGIMENTAL SUPPORT PLATOONS

ARTILLERY

SS-Kavallerie Infantry Gun Platoon — 148

ANTI-TANK

SS-Kavallerie Anti-tank Gun Platoon — 149

INFANTRY

SS-Kavallerie Pioneer Platoon — 149

ARMOUR

SS-Assault Gun Platoon — 150

⬛ SS-Security Tank Platoon — 150

ARMOUR

SS-Assault Gun Platoon — 150

Hungarian Rohamágyús Platoon ✚ — 215

INFANTRY

Hungarian Önkéntes Puskás Platoon ✚ — 221

INFANTRY

Hungarian Önkéntes Puskás Platoon ✚ — 221

ARTILLERY

SS-Horse Artillery Battery — 150

Motorised SS-Artillery Battery — 178

Motorised Heavy SS-Artillery Battery — 178

Hungarian Artillery Battery ✚ — 226

ARTILLERY

Motorised SS-Artillery Battery — 178

ANTI-AIRCRAFT

Self-propelled SS-Anti-aircraft Gun Platoon — 181

Luftwaffe Light Anti-aircraft Gun Platoon — 173

Hungarian Anti-aircraft Platoon ✚ — 226

ANTI-AIRCRAFT

Heavy SS-Anti-aircraft Gun Platoon — 180

Luftwaffe Heavy Anti-aircraft Gun Platoon — 173

AIRCRAFT

Air Support — 172

FORTIFICATIONS

Street Fortifications — 157

MOTIVATION AND SKILL

The 8. SS-Kavalleriedivision Florian Geyer has spent most of their time on the eastern front on security duties, their move to Hungary has given them an opportunity to prove themselves in battle. The 22. SS-Freiwilligen-Kavalleriedivision Maria Theresia was formed around a core of old hands from Florian Geyer SS-Kavalleriedivision *with many new recruits from Hungary's ethnic German population. An SS-Kavallerieschwadron is rated* **Fearless Trained.**

RELUCTANT	CONSCRIPT
CONFIDENT	**TRAINED**
FEARLESS	VETERAN

An SS-Kavallerieschwadron uses the Florian Geyer and Maria Theresia Waffen-SS special rules on page 189 as well as all the normal German special rules on pages 183 to 187.

HEADQUARTERS

SS-KAVALLERIESCHWADRON HQ

HEADQUARTERS

Company HQ	40 points

OPTIONS

- Replace either or both Command SMG teams with Command Panzerfaust SMG teams for +10 points per team.
- Anti-tank Section with:

2 Panzerschreck teams	+50 points
1 Panzerschreck team	+25 points

SS-KAVALLERIESCHWADRON HQ

HAUPTSTURMFÜHRER

HAUPTSTURMFÜHRER
Company Command SMG team · 2iC Command SMG team
COMPANY HQ

UNTERSCHARFÜHRER
Panzerschreck team
Panzerschreck team
ANTI-TANK SECTION

During the siege of Budapest both *SS-Kavallerie* (SS cavalry) divisions were trapped inside the city. They fought furiously against the red onslaughts on foot along side the Hungarians to save their glittering capital. With most of their horses lost during the fighting on the Hungarian plains, those few horses inside the city soon became additional food as the siege lengthened into 1945.

Like all good modern cavalry they easily adapted to fighting on foot, making use of the good defensive positions created by Soviet artillery bombardments.

COMBAT PLATOONS

SS-KAVALLERIE PLATOON

PLATOON

HQ Section with:

3 Kavallerie Squads	145 points
2 Kavallerie Squads	100 points

OPTIONS

- Replace the Command Rifle/MG team with a Command Panzerfaust SMG team for +10 points.
- Replace up to one Rifle/MG team per squad with a Panzerfaust Rifle/MG team for +10 points per Squad.

In Budapest the *SS-Kavallerie* men fight dismounted. The men of the *8. SS-Kavalleriedivision Florian Geyer* are tough and fight with great courage. The newly raised *22. SS-Freiwilligen-Kavalleriedivision Maria Theresia*, are made up of Hungarian citizens, many of who know the streets of Budapest well.

UNTERSTURMFÜHRER

UNTERSTURMFÜHRER
Command Rifle/MG team
HQ SECTION

UNTERSCHARFÜHRER
Rifle/MG team · Rifle/MG team
KAVALLERIE SQUAD

UNTERSCHARFÜHRER
Rifle/MG team · Rifle/MG team
KAVALLERIE SQUAD

UNTERSCHARFÜHRER
Rifle/MG team · Rifle/MG team
KAVALLERIE SQUAD

SS-KAVALLERIE PLATOON

WEAPONS PLATOONS

SS-KAVALLERIE MACHINE-GUN PLATOON

PLATOON

HQ Section with:

2 Machine-gun Sections	115 points
1 Machine-gun Section	60 points

SS-Kavallerie Machine-gun Platoons may make Combat Attachments to SS-Kavallerie Platoons.

The high rate-of-fire of the MG-42 machine-gun is the cornerstone of any good urban defence. The machine-gun platoon makes a killing field of any street.

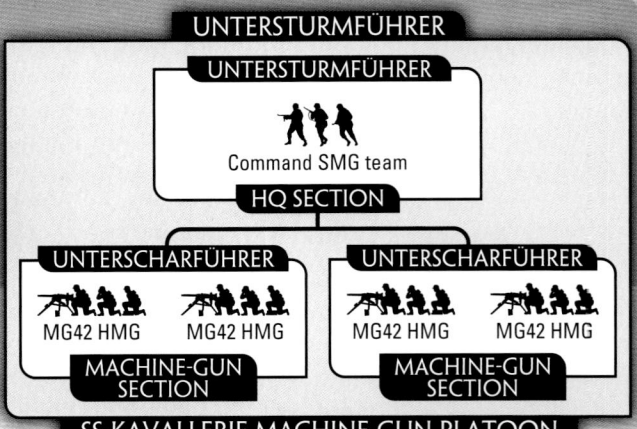

SS-KAVALLERIE MORTAR PLATOON

PLATOON

HQ Section with:

3 Mortar Sections	160 points
2 Mortar Sections	115 points
1 Mortar Section	60 points

Good light artillery is critical in pinning or blinding enemy positions or strongpoints with smoke, especially in street fighting. SS-Mortar platoons are excellent in this role due to the speed with which they respond to calls for fire.

With plenty of observer teams they can engage any target across the whole battlefield much faster than the big guns of the artillery. The 8cm GW34 mortar can pin down enemy attacks as well as deliver covering smoke just as well as any artillery battery.

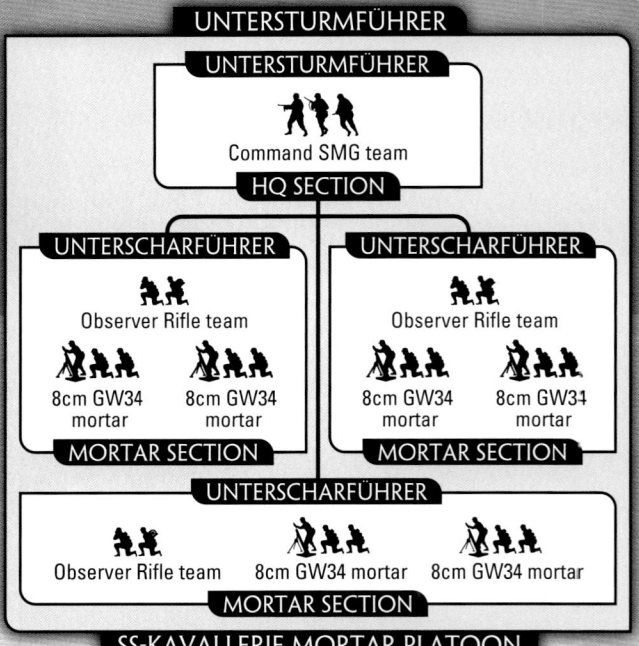

REGIMENTAL SUPPORT PLATOONS

SS-KAVALLERIE INFANTRY GUN PLATOON

PLATOON

HQ Section with:

2 7.5cm leIG18	60 points

OPTIONS

- Add Kfz 15 field car and 3-ton trucks for +5 points for the platoon.
- Replace 3-ton trucks with RSO tractors at no cost.

Though not overpowering, the 7.5cm leIG18 gun provides flexibility to your company. Adding an infantry gun platoon provides an answer to a number of field problems. It can provide smoke and artillery support for advancing infantry

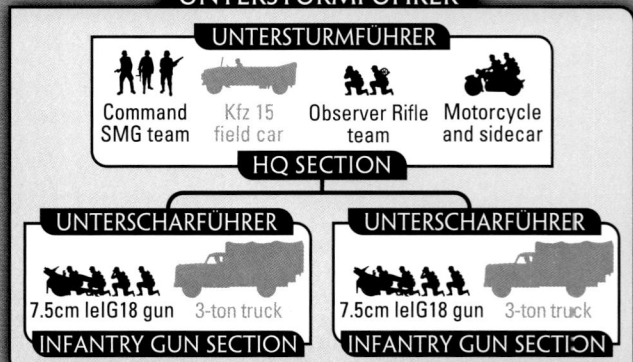

while also protecting the front lines against assaults. It can dig out enemy machine-guns and anti-tank guns as well as provide some anti-tank capability against assaulting tanks.

SS-KAVALLERIE ANTI-TANK GUN PLATOON

PLATOON

HQ Section with:

3 7.5cm PaK40	135 points
2 7.5cm PaK40	90 points

OPTION

- Add Kübelwagen jeep and 3-ton trucks for +5 points for the platoon.

SS-Kavallerie Anti-tank Gun Platoons may make Combat Attachments to SS-Kavallerie Platoons.

The 7.5cm PaK40 gun has become the standard anti-tank gun of the SS-Kavallerie divisions. The hard hitting PaK40 will destroy almost any tank the enemy cares to put in range of them.

Place your anti-tank assets in good cover behind your front lines, then spring your ambush when the enemy tanks close to overrun your position. Your concentrated fire will stop the assault in its tracks.

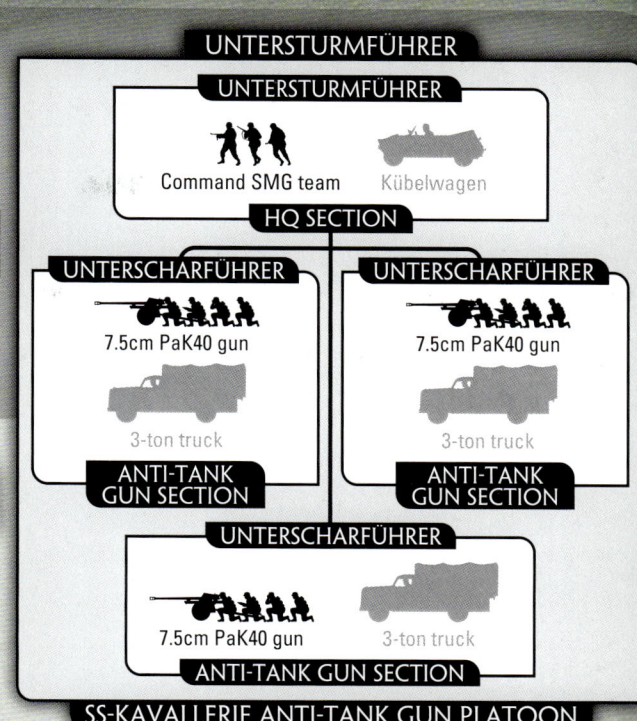

SS-KAVALLERIE ANTI-TANK GUN PLATOON

SS-KAVALLERIE PIONEER PLATOON

PLATOON

HQ Section with:

3 Pioneer Squads	190 points
2 Pioneer Squads	135 points

OPTIONS

- Replace Command Pioneer Rifle/MG team with Command Pioneer Panzerknacker SMG team for +5 points or Command Pioneer Panzerfaust SMG team for +10 points.
- Equip one Pioneer Rifle/MG team with a Goliath demolition carrier in addition to its normal weapons for +30 points.
- Add Kfz 15 field car and 3-ton trucks for +5 points for the platoon.
- Add Pioneer Supply 3-ton truck for +25 points, or Pioneer Supply Maultier for +30 points.

You may replace up to one Pioneer Rifle/MG team per Pioneer Squad with a Flame-thrower team at the start of the game before deployment.

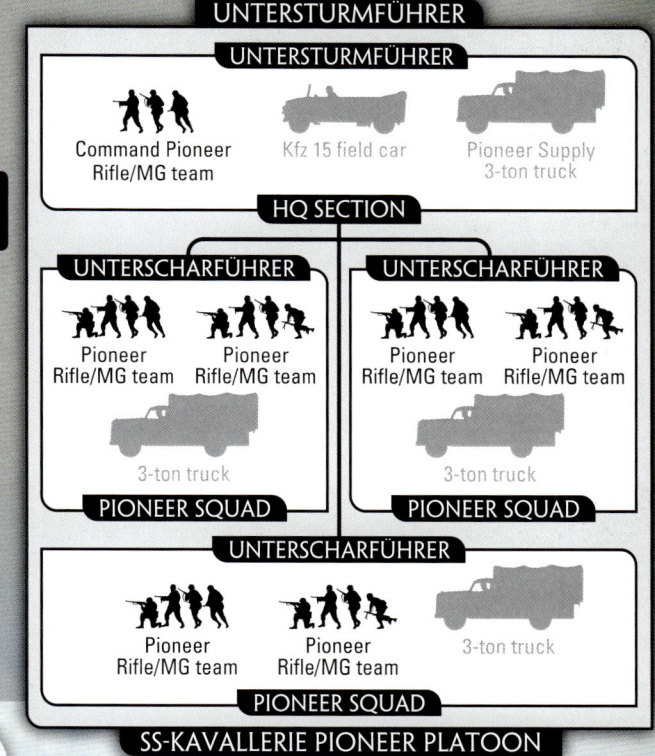

SS-KAVALLERIE PIONEER PLATOON

SS-Assault Gun Platoon

Platoon

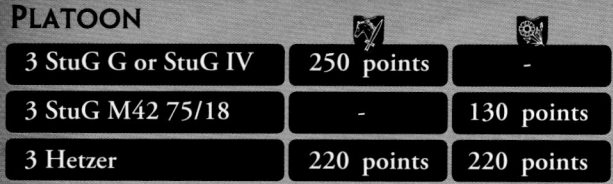

3 StuG G or StuG IV	250 points	-
3 StuG M42 75/18	-	130 points
3 Hetzer	220 points	220 points

8. SS-Kavalleriedivision Florian Geyer had its own assault gun battery in support. Their *Panzerjägerabteilung* (tank-hunter battalion) received 15 Hetzer tank-hunters in September 1944.

22. SS-Freiwilligen-Kavalleriedivision Maria Theresia fielded a few Italian Semovente 75/18 (StuG M42) in their *panzerjäger* (tank-hunter) battalion alongside their small numbers of Hetzers. *22. SS-Freiwilligen-Kavalleriedivision Maria Theresia*

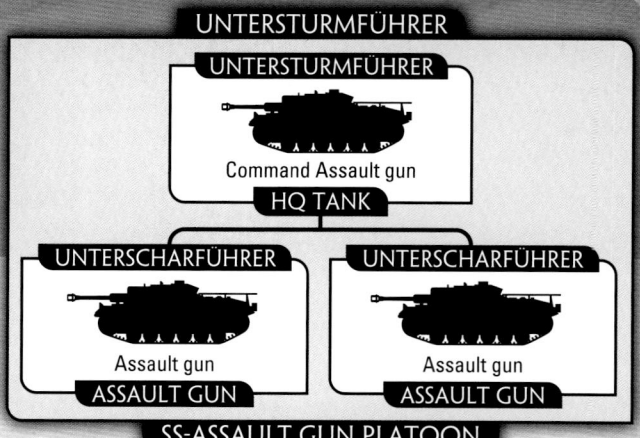

got seven Hetzer tank-hunters to supplement their Italian tanks in October 1944.

SS-Security Tank Platoon

Platoon

5 Panzer M14/41	200 points
4 Panzer M14/41	160 points
3 Panzer M14/41	120 points

The *22. SS-Freiwilligen-Kavalleriedivision Maria Theresia* was initially intended for anti-partisan duties in Yugoslavia. For this work they were issued with Italian made M14/41 tanks.

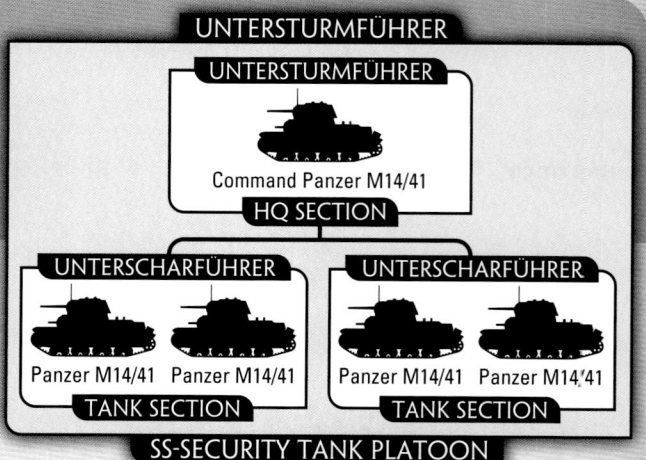

SS-Horse Artillery Battery

Platoon

HQ Section with:

4 7.5cm FK18	130 points
2 7.5cm FK18	80 points

Option

- Add cavalry wagon and cavalry limbers at no cost.

A SS-Horse Artillery Batteries uses the Horse Artillery special rule (see the rulebook).

The two SS cavalry divisions in Budapest had specialist horse artillery. Trained for quick deployment, these small guns proved ideal for reinforcing defensive positions against Soviet assaults.

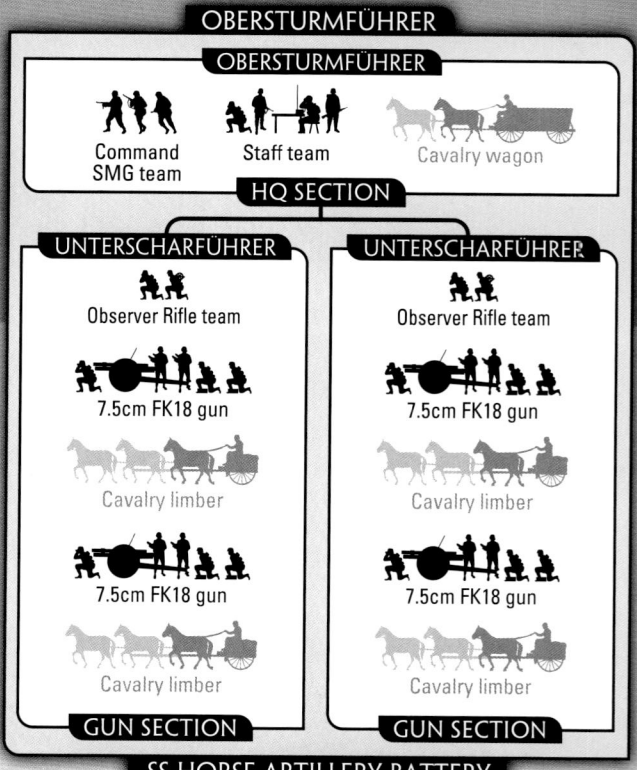

8. SS-KAVALLERIEDIVISION FLORIAN GEYER

The division was formed in 1942 from the SS Cavalry Brigade with the addition of about 9000 ethnic German Romanians from the Yugoslavian border areas. It was named 'Florian Geyer' in March 1944 after a Franconian knight who led the Black Company in the Peasants War (1524-25).

RUSSIAN FRONT

The newly raised division fought on the eastern front in the Rzhev and Orel sectors until April 1943. The division then moved to Bobruisk for anti-partisan and security duties until September 1943, before returning to the frontlines. They moved south to fight during the retreat to the Dnepr River. In October 1943 they reorganised in Hungary before being posted to Croatia to take part in further anti-partisan duties. They returned in April 1944 and took part in the fighting for Transylvania in September. After fighting on the Hungarian Plain they were assigned to the *IX SS-Gebirgskorps* (9th SS-Mountain Corps) defending Budapest.

BUDAPEST

Initially the division fought in the east of Budapest, on the outskirts of Pest, but as the threat from the encircling Soviets from the west became greater they were ordered to move to Buda on the western bank of the Danube River on 24 December 1944. They were immediately in action as Soviet forces started to push into Buda from the south. They fought alongside a number of Hungarian volunteer battalions and assault guns. A number of counterattacks were launched on 1 January 1945, but strong Red Army forces repelled these. Savage fighting for the city continued street by street.

The *8. SS-Kavalleriedivision Florian Geyer* formed the backbone of the defence in Buda until the arrival of the *Feldherrnhalle* units on 18 January. They were engaged in heavy fighting around the Károly Király Barracks and Farkasrét Cemetery hill counterattacking several Soviet attacks between 12 and 15 January. One heavy Soviet attack took the hill, but a hasty counterattack quickly swept the Red Army troops back. The division was supported by their own Hetzer tank-hunters, heavy FlaK troops and Hungarian assault guns and heavy anti-aircraft guns.

Fighting continued into February. The defenders were forced to give ground in small increments, fighting for every scrap of open ground and hill. They fought until 11 February, when a breakout attempt was made by the defenders. Only about 170 SS cavalry men were able to escape to German lines to the west of Budapest.

22. SS-FREIWILLIGEN-KAVALLERIEDIVISION MARIA THERESIA

In December 1943 a second SS cavalry division was created. It was formed around the 17th SS-Cavalry Regiment from *8. SS-Kavalleriedivision Florian Geyer*. The regiment was sent to Kisbér, Hungary to begin forming the new division. The rest of the division's manpower came from ethnic German Hungarians, most of who were transferred from the Hungarian Army. It was titled 'Maria Theresia' after the 18th Century Austro-Hungarian Empress. From her the division also received its symbol, the cornflower, which was Empress Maria Theresia's favourite flower. The division's first new regiment was ready for combat in June 1944.

FIRST ACTION

The 52nd SS-Cavalry Regiment formed a *kampfgruppe* (battle group) to send to the front in August. They fought around Arad in Transylvania attached to the *III Panzerkorps*, before being encircled as the Red Army launched its assault towards Debrecen. Secondary forces assaulted the encircled group and split it in two. One group launched a counterattack and managed to open an escape route towards friendly lines. However, the other group found it impossible to break out as a fighting unit, so *SS-Hauptsturmführer* Ameiser ordered his men to strip down to light equipment and weapons, and breakout. He then led his men on an epic 200-mile retreat that ended on 30 October with 48 survivors at the town of Dunaföldvár. For this he was awarded the Knight's Cross.

BUDAPEST

The rest of the division was ready for action by November 1944 and these units were sent to Budapest to take part in the defence of the city's perimeter. By late December they found themselves trapped inside as the Soviets encircled the city from the west.

As the Soviet encirclement closed they fought in the south-west of Pest, alongside the Hungarian troops of Group Billnitzer and 1st Parachute Battalion. On 7 January 1945 the division fought alongside the assault guns of Group Billnitzer in nine counterattacks to retake the suburb of Köbánya, before finally abandoning it in the evening. Following this they fought hard for the Köbánya-alsó Railway Station, the nearby railway workers housing and the railway embankment itself. The heavy fighting took its toll and by mid-January the continuous fighting had worn the division down to a combat strength of about 800 men. The Soviets pushed the division out of Kispest on 8 January.

On 18 January the remnants of the division retreated across the Danube to Buda and joined *8. SS-Kavalleriedivision*. They held off the Soviet assaults until 11 February 1945.

170 SS-Cavalrymen made it to friendly lines following the last ditch desperate breakout on 11 February 1945. The remnants of both SS cavalry divisions not trapped in the city were used to create *37. SS-Kavalleriedivision Lützow*.

FORTIFICATIONS

The rules for Fortification and Bunkers can be found in *Das Book* and are available as a free PDF download from the Flames of War website (*http://www.flamesofwar.com*).

FORTIFICATION PLACEMENT

Fortifications are deployed as either Area Defences or as part of Fortified Platoons.

FORTIFIED COMPANY

Soldiers manning fortifications hunkered down behind barbed wire and minefields, sheltering in bunkers, waiting for the enemy to attack. In static warfare, neither side will launch an all-out attack. Instead they aggressively patrol No Man's Land, particularly at night.

> *A Fortified Company will always be the defender against a Tank, Mechanised, or Infantry Company.*
>
> *When two Fortified Companies face off, they automatically play the No Man's Land mission (see www.flamesofwar.com) rather than normal missions.*

FORTIFICATION PLACEMENT

Bunkers and fortifications take a lot of time and effort to build and rarely come as a surprise to the enemy.

You must place all Fortifications before the opposing player places any Objectives unless the mission specifies otherwise.

> *Entrenchments and Bunkers must be placed in your Deployment Area. Obstacles may be placed in either your Deployment Area or in No Man's Land.*
>
> *Entrenchments and Bunkers cannot be placed in a stream, river, swamp, lake or other water feature.*

BUNKER DEPLOYMENT

Bunkers are usually built with mutual support in mind and are usually not placed right next to each other.

> *Bunkers may not be placed:*
> - *within 2"/5cm of another Bunker, nor*
> - *on a road, track, bridge, railway line, blocking a ford, or in any other way obstructing a constructed route.*
>
> *Bunkers can be placed in the ground floor of Buildings. If the Building is large enough, place the Bunker in the Building, otherwise, declare the whole Building to be a Bunker. Use Openings in the Building as the Bunker's Firing Slits.*
>
> *Objectives may not be placed in, on or under Bunkers.*

REMOVE EMPTY GUN PITS

Sometimes additional gun pits are dug as decoy positions to confuse the enemy as to where the artillery is actually positioned. The truth of the matter quickly becomes clear once the fighting begins.

> *Remove any unoccupied Gun Pits and Tank Pits from the table at the end of deployment before the game begins.*

AREA DEFENCES

Bunkers must be placed at least 2"/5cm apart and may not be placed on roads, Bridges, or Railway Lines.

Bunkers and entrenchments may not be placed in water obstacles, although other Obstacles can be.

Not Placing your Fortifications

Sometimes Fortified Platoons may be operating away from their fortifications.

You may choose to not place any or all of your Fortifications in a game. If you do this the Fortifications are not used during the game.

Area Defences

Area defences are fortifications (usually obstacles) used to create a continuous barrier across the battlefield, slowing down and channelling the enemy's advance. Area defences are dispersed across the frontage being defended to strengthen weak areas.

Fortifications for Area Defence are deployed as described above. A force with two or more Bunkers in its Area Defences is a Fortified Company and will Always Defend.

Fortified Platoons

Fortified platoons occupy strongpoints. These heavily-fortified positions are difficult to overrun and dominate the ground around them with fire. Fortified platoons occupy strongpoints to defend an area, using firepower to deny the surroundings to the enemy.

Fortified Platoons place their fortifications at the same time as other Fortifications before objectives are placed.

Mark a spot on the table (use a die or counter) as the centre of each Fortified Platoon's position. All of the platoon's Fortifications must be placed entirely within 12"/30cm of the marked point, but may not be placed within 6"/15cm of any Fortification from another Fortified Platoon or of either side table edge.

Deploying the Company HQ

The company commander usually chooses to place themselves in a central position within easy reach of all of the other fortified platoons or safely within the crucial position.

The Company HQ of a Fortified Company may deploy as part of another Fortified Platoon, amalgamating their Fortifications as a single position. If it does not do this, it deploys in its own fortifications which are placed as if it were a Fortified Platoon on its own.

Deploying Fortified Platoons

Strongpoints are assigned to a particular unit to defend, and woe betide any commander who abandons his position before a battle.

When the teams of a Fortified Platoon are Deployed, they must be placed in the Trench Lines, Gun Pits, and Tank Pits that come with their platoon. Only teams from the Fortified Platoon may be deployed in its Entrenchments. If there are insufficient Entrenchments for the platoon, any teams that do not fit in the Entrenchments may not be used in the game.

If a Fortified Platoon elects not to place any of its Fortifications, it is no longer a Fortified Platoon. However, such a platoon never benefits from the Prepared Positions special rule as they have not had time to dig alternative positions.

Fortified Platoons cannot be deployed in Ambush or Reserves of any sort unless it elects not to deploy any of its Fortifications.

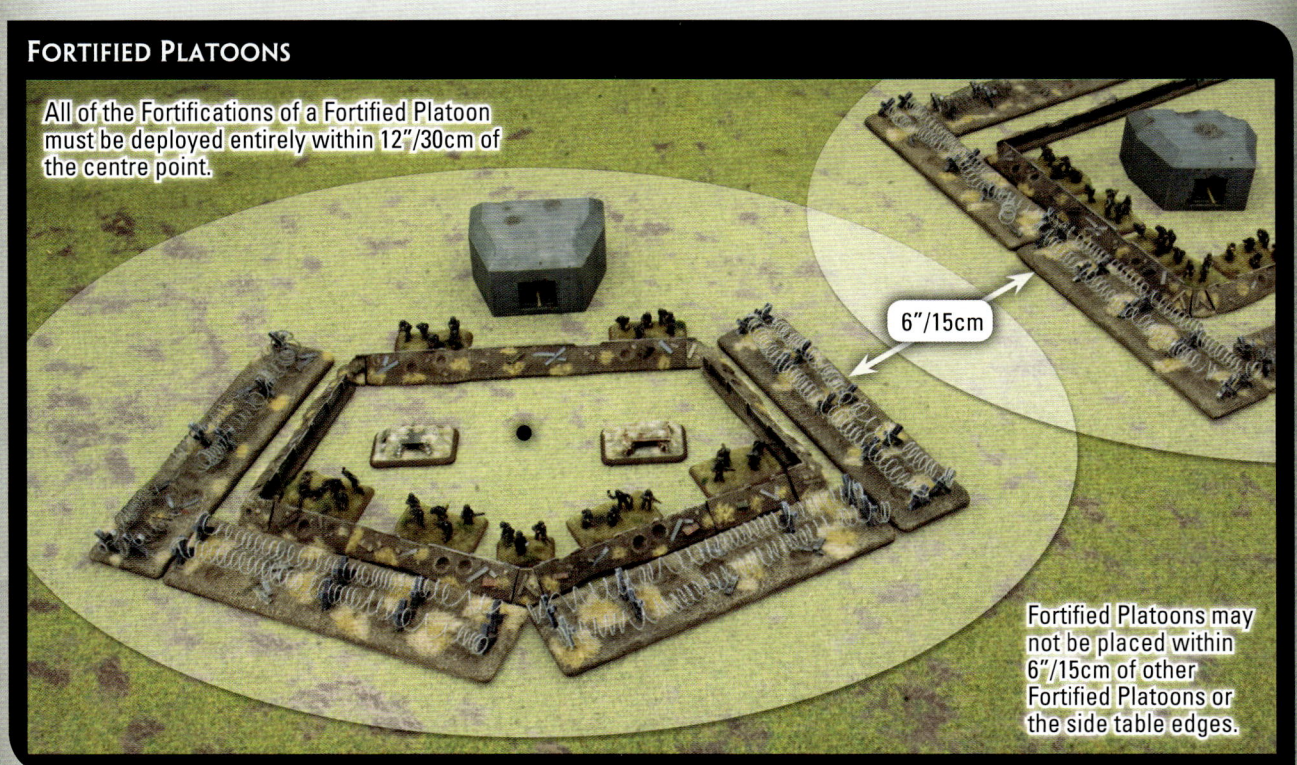

FORTIFIED PLATOONS

All of the Fortifications of a Fortified Platoon must be deployed entirely within 12"/30cm of the centre point.

6"/15cm

Fortified Platoons may not be placed within 6"/15cm of other Fortified Platoons or the side table edges.

BOOBY TRAPS SPECIAL RULES

Retreating troops often booby trapped areas before withdrawing. The pursuing troops usually had little warning of booby traps and relied on their skill to detect and neutralise any they encountered. Those unwary enough not to notice them suffered accordingly.

> *Booby Traps are Dangerous Obstacles. As such platoons need to pass a Motivation Test before they can cross or move off them.*
>
> *Unlike other Obstacles, a Booby Trap occupies the area of a small base (1¼"/32mm x 1"/25mm).*
>
> *A Booby Trap only affects the first team to move across or off the Booby Trap base. After that the Booby Trap is expended and removed.*

PLACING BOOBY TRAPS

Booby traps are always carefully hidden and come as a surprise to the enemy.

> *Booby Traps do not need to be deployed before the game begins. Instead, they may be placed at the same time as Ambushing platoons during the Starting Step.*
>
> *Booby Traps may be placed anywhere in your own Deployment Area or No man's Land, including under enemy troops. However, you may not place Booby Traps under teams already on Booby Traps, nor under Dug In teams or those that deployed in Prepared Positions and have not yet moved.*

TRIPPING A BOOBY TRAP

Because booby traps are carefully hidden, soldiers can find themselves in a dangerous situation without having realised it.

> *If a Booby Trap is placed under a team, the team's platoon will test as if crossing the Booby Trap when it first moves off it. Otherwise the Booby Trap goes off when a team first crosses it.*
>
> *Roll a Skill Test for a team that crosses a Booby Trap.*
> - *If the team passes the Skill Test, they are unharmed.*
> - *If they fail, they take a hit from the Booby Trap and must stop moving immediately.*
>
> *In either case, the Booby Trap is removed.*
>
> *If any team in a platoon is hit by a Booby Trap, whether the hit causes a casualty or not, the entire platoon is Pinned Down at the end of their movement.*

ARMOURED VEHICLES HIT BY BOOBY TRAPS

Booby traps are normally small and intended for infantry on foot rather than tanks, but some of them are much larger and can knock out a tank with a bit of luck.

> *Treat hits from Booby Traps on Armoured vehicles as a hit on the Top armour by a weapon with an Anti-tank rating of 3 and a Firepower rating of 5+.*

UNARMOURED VEHICLES HIT BY BOOBY TRAPS

Booby traps are usually powerful enough to wreck a truck, but a lucky driver might find themselves shocked but unharmed.

> *Unarmoured vehicles that are hit by a Booby Trap save on a roll of 5+.*

INFANTRY AND GUNS HIT BY BOOBY TRAPS

Infantry are generally cautious in booby trapped areas, but occasionally they make a serious miscalculation and pay the price.

> *Roll a save for each Infantry or Gun team hit by a Booby Trap. An Infantry team survives on a roll of 3+, but is Destroyed otherwise. Gun teams survive on a roll of 5+.*

PASSENGERS HIT BY BOOBY TRAPS

With a little luck, the passengers in a vehicle or tank riders on a tank that hits a booby trap survive unharmed.

> *Passengers carried in or on a vehicle Destroyed by a Booby Trap must roll a 5+ save. If they pass the save, they Dismount Under Fire.*

GAPPING A BOOBY TRAP

Booby traps are designed to be tricky to locate and hard to disarm. One mistake and the clearing team is in serious trouble.

> *If a team fails its Skill Test to gap a Booby Trap, it is hit by mines as if it failed its Skill Test to cross the Booby Trap.*

ADDING FORTIFICATIONS TO YOUR FORCE

Some companies may add fortifications to enhance their defences, these can either be Field or Street Fortifications.

A Company with Field or Street Fortifications is a Fortified Company and follows the Fortification Placement rules on Pages 152 to 153. The Company Headquarters becomes a Fortified Company HQ.

Each type of Fortification has Strongpoints and Artillery Positions available, these provide defences, nests and bunkers for infantry and gun platoons.

Strongpoints or Artillery Positions are added to platoons in your company. Only one Strongpoint or Artillery positions may be added to any one platoon.

A Platoon with a Strongpoint or Artillery Positions is a Fortified Platoon and follows the rules on pages 152 to 153.

FIELD FORTIFICATIONS

A Company with Field Fortifications is a Fortified Company and follows the Fortification Placement rules on Pages 152 to 153. The Company Headquarters becomes a Fortified Company HQ.

If you choose to take Field Fortifications, you must field one fortification option from each box shaded black and may field a fortification option from each box shaded grey.

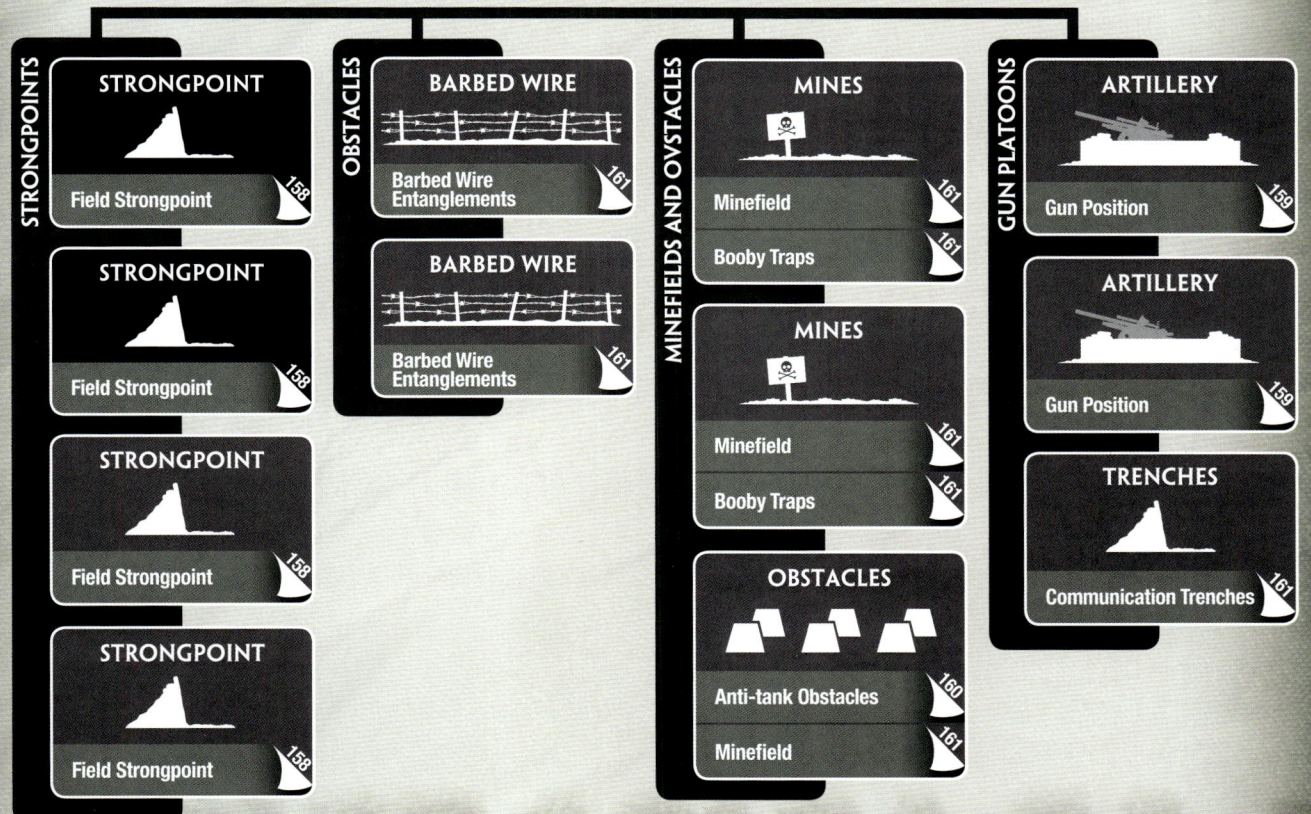

STRONGPOINTS	OBSTACLES	MINEFIELDS AND OVSTACLES	GUN PLATOONS
STRONGPOINT — Field Strongpoint 158	**BARBED WIRE** — Barbed Wire Entanglements 161	**MINES** — Minefield 161 / Booby Traps 161	**ARTILLERY** — Gun Position 159
STRONGPOINT — Field Strongpoint 158	**BARBED WIRE** — Barbed Wire Entanglements 161	**MINES** — Minefield 161 / Booby Traps 161	**ARTILLERY** — Gun Position 159
STRONGPOINT — Field Strongpoint 158		**OBSTACLES** — Anti-tank Obstacles 160 / Minefield 161	**TRENCHES** — Communication Trenches 161
STRONGPOINT — Field Strongpoint 158			

STREET BARRICADE SPECIAL RULES

During the many battles for cities on the Eastern Front the defenders used improvised barricades to block access to the streets and thoroughfares. These were used to slow the momentum of the attackers and as an anchor point for the defence. The barricade would slow or stop any enemy armour, making them vulnerable to attacks from the surrounding buildings.

A barricade was usually built to act as both an obstacle and as protection for the troops manning it. They took many different forms depending on the material available to construct it. Many were made from rubble and stonework salvaged from the surrounding streets and buildings. Felled lampposts and telegraph poles could also reinforce them. Sometimes cars, trucks, buses or city trams were used to build the barricade around. A knocked out tank made a particularly impressive barricade anchor, providing the defenders plenty of protection from rifle and machine-gun fire.

> Street Barricades are Tall Linear Obstacles and are Very Difficult Going. They are Impassable to Cavalry.
>
> Teams must stop their movement on reaching a Street Barricade. They must start their movement square up against a Street Barricade to cross it.
>
> Teams cannot end a Step sitting on a Street Barricade. Like a Linear Obstacle, teams must be on one side or the other and clearly either up against the Street Barricade or back from it.

Taking Cover behind a Barricade

The height and solidity of a street barricade gives the defenders a distinct advantage since the enemy can't see them moving about behind it.

> Street Barricades block Line of Sight in the same way as Tall Linear Obstacles, and provide Concealment and Bulletproof Cover against shooting from the other side for teams square up against them. As with Linear Obstacles, they provide no protection against Artillery Bombardments, or aircraft.

Assaulting a Barricade

The height and solidity of a street barricade gives the defenders a distinct advantage since the enemy can't see them moving about behind it.

> As with any Linear Obstacle, Assaulting teams that contact a Street Barricade can Roll to Hit and be hit by enemy teams on the other side of the Street Barricade.
>
> However, unlike other Linear Obstacles, teams may only Roll to Hit against teams on the other side of the Street Barricade if they started their Charge into Contact square up against the Street Barricade.

Gapping a Street Barricade

Street barricades are quite massive and difficult to demolish completely.

> Do not remove a Street Barricade when it is gapped. Instead, remove the middle piece, creating a gap in the Street Barricade rated as Difficult Going. The remainder of the Street Barricade continues to function as normal.

Bunker Busters Demolish Barricades

Assault tanks like the German Brummbär or the Soviet SU-152 have little to worry about from a street barricade. They simply settle down and pound it until the blow a gap right through it.

> Weapons rated as Bunker Buster can attempt to gap a Street Barricade. A Bunker Buster must not move in the Movement Step, and must be able to draw a Line of Sight to a Street Barricade and be within 16"/40cm of it to attempt to gap it.
>
> Roll a Skill Test for the Bunker Buster in the Shooting Step instead of shooting.
>
> • If the test is successful, the Street Barricade is gapped,
>
> • Otherwise, it remains intact.

STREET FORTIFICATIONS

A Company with Street Fortifications is a Fortified Company and follows the Fortification Placement rules on Pages 152 to 153. The Company Headquarters becomes a Fortified Company HQ.

If you choose to take Street Fortifications, you must field one fortification option from each box shaded black and may field a fortification option from each box shaded grey.

OBSTACLES

OBSTACLES
Street Barricade 160

OBSTACLES
Street Barricade 160

OBSTACLES
Street Barricade 160

OBSTACLES
Anti-tank Obstacles 160

BARBED WIRE

BARBED WIRE
Barbed Wire Entanglements 161

BARBED WIRE
Barbed Wire Entanglements 161

MINEFIELDS

MINES
Minefield 161
Booby Traps 161

MINES
Minefield 161
Booby Traps 161

STRONGPOINTS

STRONGPOINT
Street Strongpoint 159

STRONGPOINT
Street Strongpoint 159

STRONGPOINT
Street Strongpoint 159

STRONGPOINT
Street Strongpoint 159

ARTILLERY
Gun Position 159

ARTILLERY
Gun Position 159

STRONGPOINTS

FIELD STRONGPOINT

STRONGPOINT

Strongpoint	55 points

OPTIONS

- Add Trench Line for +5 points.
- Add second HMG Nest for +40 points.
- ✚ Add 5cm PaK38 Pillbox for +80 points or 7.5cm PaK40 Pillbox for +125 points.
- Replace up to one HMG Nest with a HMG Pillbox for +40 points.
- ✚ Replace up to one HMG Nest with a 2cm FlaK Nest for -15 points, or a Quad 2cm FlaK Nest for +5 points.
- ✚ Replace up to one HMG Nest with a 40mm Bofors AA Nest for +30 points.
- ✚ Replace up to one HMG Nest with a Panzer I turret for -15 points, Panzer II turret for -10 points, T-70 turret for +5 points, T-34 turret for +70 points or a Panther turret for +140 points.

Options marked (✚) are only available to German forces. Options marked (✚) are only available to Hungarian or Finnish forces. Unmarked options are available to any nation.

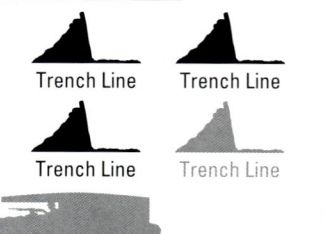

FORTIFICATIONS

HMG Nest	Trench Line	Trench Line
HMG Nest	Trench Line	Trench Line

Anti-tank Pillbox

STRONGPOINT

FIELD STRONGPOINT

A Field Strongpoint can be attached to any Platoon entirely made up of Infantry teams and/or Man-packed Gun teams in a company with Field Fortifications.

A Platoon with a Field Strongpoint attached is a Fortified Platoon, see page 153.

STREET STRONGPOINT

STRONGPOINT

Strongpoint	15 points

OPTION

- Add Trench Line for +5 points.

A Street Strongpoint can be attached to any Platoon entirely made up of Infantry teams and/or Man-packed Gun teams in a company with Street Fortifications.

FORTIFICATIONS

FORTIFICATIONS

Trench Line	Trench Line
Trench Line	Trench Line

STRONGPOINT

STREET STRONGPOINT

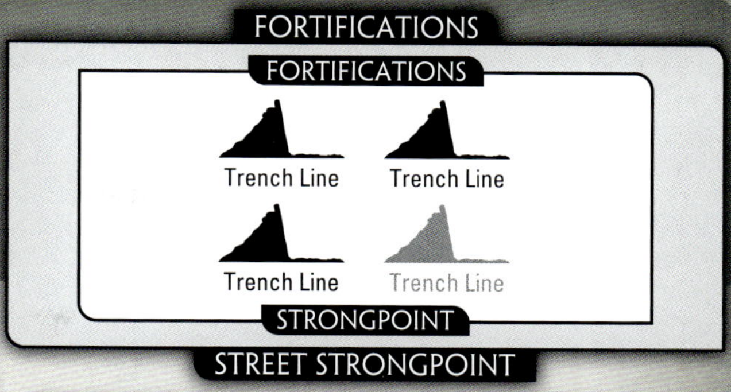

A Platoon with a Street Strongpoint attached is a Fortified Platoon, see page 153.

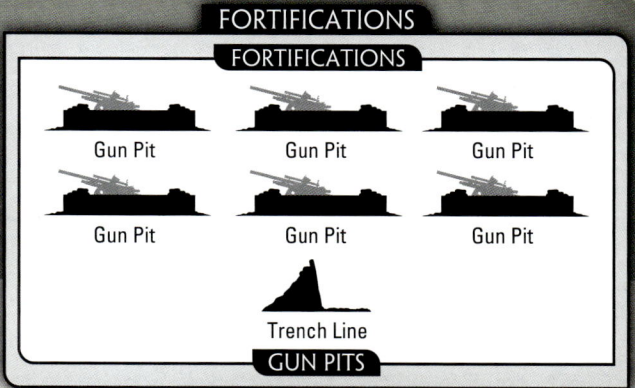

GUN POSITION

STRONGPOINT

Trench Line and:

6 Gun Pits	35 points
4 Gun Pits	25 points
2 Gun Pits	15 points

A Platoon with a Gun Position attached is a Fortified Platoon, see page 153.

FORTIFICATIONS

FORTIFICATIONS

Gun Pit	Gun Pit	Gun Pit
Gun Pit	Gun Pit	Gun Pit

Trench Line

GUN PITS

GUN POSITION

A Gun Position can be attached to any Platoon containing Gun teams in a company with Field or Street Fortifications.

OBSTACLES

ANTI-TANK OBSTACLE

FORTIFICATION

| 2 Anti-tank Obstacles | 200 points |
| 1 Anti-tank Obstacle | 100 points |

FORTIFICATION

FORTIFICATION

Anti-tank Obstacle   Anti-tank Obstacle

OBSTACLE

ANTI-TANK OBSTACLE

Anti-tank Obstacles are Area Defences, see page 153.

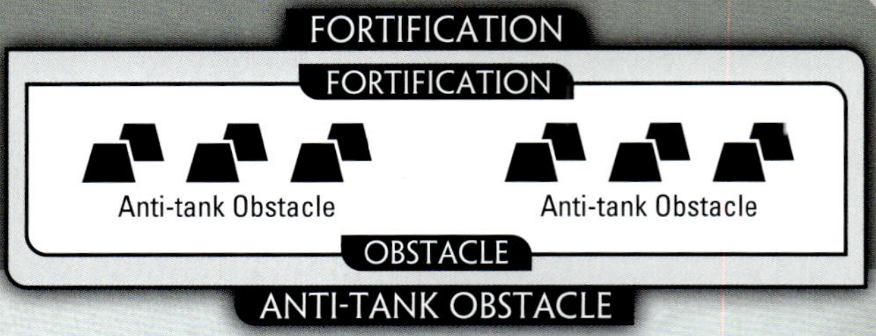

STREET BARRICADE

FORTIFICATION

| 2 Street Barricades | 60 points |
| 1 Street Barricade | 30 points |

FORTIFICATION

FORTIFICATION

Street Barricade   Street Barricade

OBSTACLE

STREET BARRICADE

Street Barricades are Area Defences, see page 153.

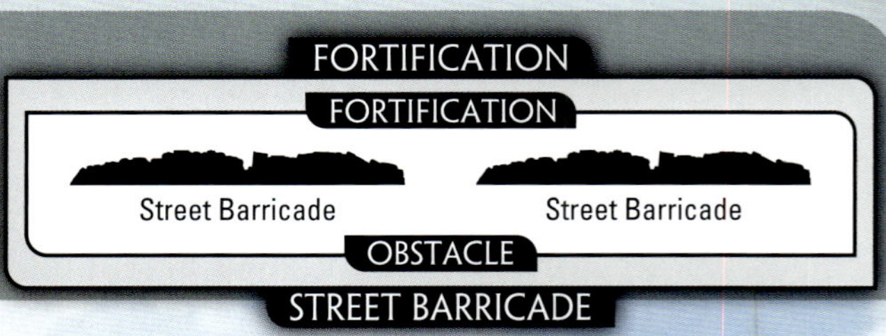

MINEFIELD

FORTIFICATION

| 2 Minefields | 100 points |
| 1 Minefield | 50 points |

FORTIFICATION
FORTIFICATION

Minefield   Minefield

MINES

MINEFIELD

Minefields are Area Defences, see page 153.

BOOBY TRAPS

FORTIFICATION

8 Booby Traps	80 points
7 Booby Traps	70 points
6 Booby Traps	60 points
5 Booby Traps	50 points
4 Booby Traps	40 points
3 Booby Traps	30 points
2 Booby Traps	20 points

FORTIFICATION
FORTIFICATION

Booby Trap   Booby Trap   Booby Trap   Booby Trap

Booby Trap   Booby Trap   Booby Trap   Booby Trap

BOOBY TRAPS

BOOBY TRAPS

Booby Traps are Area Defencess, see page 153.

BARBED WIRE ENTANGLEMENTS

FORTIFICATION

6 Barbed Wire Entanglements	60 points
4 Barbed Wire Entanglements	40 points
2 Barbed Wire Entanglements	20 points

FORTIFICATION
FORTIFICATION

Barbed Wire Entanglement   Barbed Wire Entanglement   Barbed Wire Entanglement

Barbed Wire Entanglement   Barbed Wire Entanglement   Barbed Wire Entanglement

WIRE

BARBED WIRE ENTANGLEMENTS

Barbed Wire Entanglements are Area Defences, see page 153.

COMMUNICATION TRENCHES

FORTIFICATION

9 Trench Lines	45 points
6 Trench Lines	30 points
3 Trench Lines	15 points

FORTIFICATION
FORTIFICATION

Trench Line   Trench Line   Trench Line

Trench Line   Trench Line   Trench Line

Trench Line   Trench Line   Trench Line

TRENCHES

TRENCH LINES

Trench Lines from the Communication Trenches option are Area Defences, see page 153.

HEER DIVISIONAL SUPPORT

MOTIVATION AND SKILL

Like the troops they support, the Divisional Support Platoons are experienced troops that know their worth. Divisional Support Platoons are rated as **Confident Veteran**.

Heer Divisional and Corps Support Platoons that support SS companies are Allies to the SS and follow the Allies rules in the Flames Of War rulebook.

RELUCTANT	CONSCRIPT
CONFIDENT	TRAINED
FEARLESS	**VETERAN**

RADIO-CONTROL TANK PLATOON

PLATOON

HQ Section with:

4 StuG G & 4 Borgward B IV	460	points
3 StuG G & 3 Borgward B IV	345	points
2 StuG G & 2 Borgward B IV	230	points

No gun in existence can deliver half a ton of explosives with the unerring accuracy of a Borgward BIV demolition carrier.

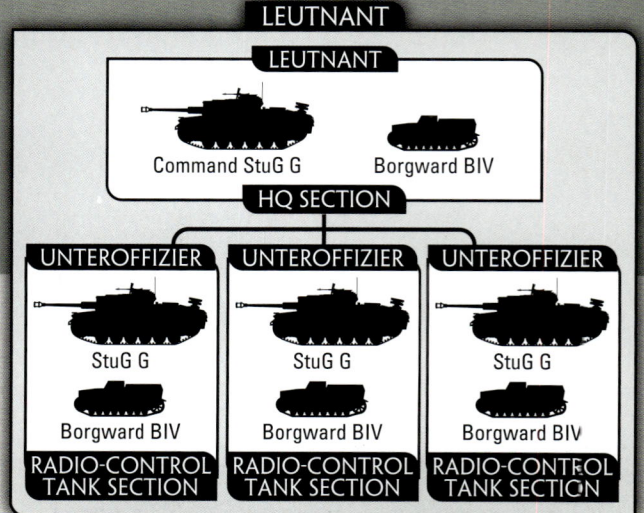

STURMPANZER PLATOON

PLATOON

4 Brummbär	280	points
3 Brummbär	210	points
2 Brummbär	140	points

Sturmpanzer Abteilung 218 saw service during the Warsaw uprising before fighting against the Soviet offensive of January 1945 where they were destroyed.

In Hungary *Sturmpanzer Abteilung 219* fought around the Velence Lake area. In January 1945 the battalion was assigned to *23. Panzerdivision* for the Budapest relief attempt. During March the battalion lost nearly all of its Brummbär assault tanks and was transferred to Czechoslovakia. At the end of March 1945 the 2nd company received ten Brummbär assault tanks and was again transferred to Hungary.

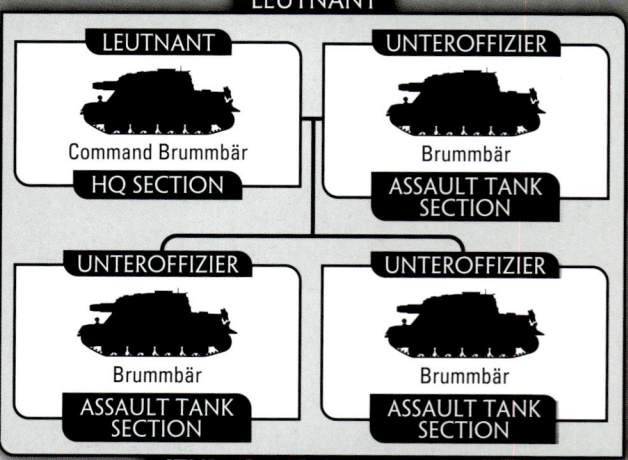

TANK HUNTER PLATOON

PLATOON

4 Marder II	260	points
3 Marder II	195	points
2 Marder II	130	points
4 Marder III H	260	points
3 Marder III H	195	points
2 Marder III H	130	points
4 Marder III M	255	points
3 Marder III M	190	points
2 Marder III M	125	points
4 Hetzer	340	points
3 Hetzer	255	points
2 Hetzer	170	points
4 Jagdpanzer IV	380	points
3 Jagdpanzer IV	285	points
2 Jagdpanzer IV	190	points
4 Hornisse	465	points
3 Hornisse	350	points
2 Hornisse	235	points
4 Elefant	1200	points
3 Elefant	900	points
2 Elefant	600	points

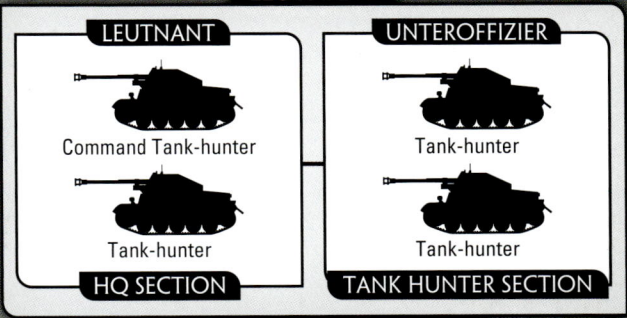

LEUTNANT

LEUTNANT	UNTEROFFIZIER
Command Tank-hunter	Tank-hunter
Tank-hunter	Tank-hunter
HQ SECTION	TANK HUNTER SECTION

TANK-HUNTER PLATOON

The life of a tank-hunter is one of fire and movement if he is to survive in the world of heavy and fast tanks. Being able to get off the first shot, on target, becomes your main concern.

Self-propelled anti-tank guns are an inexpensive way to get high calibre guns into the front lines. More mobile than normal anti-tank platoons, the tank-hunters can outmanoeuvre or ambush the enemy to bring devastating fire upon advancing enemy armoured formations.

The heavier tank-hunters mount the overlong 8.8cm PaK43, a weapon more than capable of knocking out any heavy tank at any distance.

Though tank-hunters can contribute immensely to your defence, to keep them in the field, you must be wary of their two biggest handicaps. Lighter tank-hunters have thin armour so they must avoid direct confrontation with armoured tanks. Heavy tank-hunters have the armour to stand toe to toe with tanks but lack the protection to ward off assaulting infantry.

VETERAN TANK-HUNTER PLATOON

PLATOON

4 7.5cm PaK40 auf RSO	225	points
3 7.5cm PaK40 auf RSO	170	points
2 7.5cm PaK40 auf RSO	115	points
4 Hetzer	340	points
3 Hetzer	255	points
2 Hetzer	170	points

SHOOT AND SCOOT:

Veteran Tank-hunter Platoons Stormtrooper on a roll of 2+.

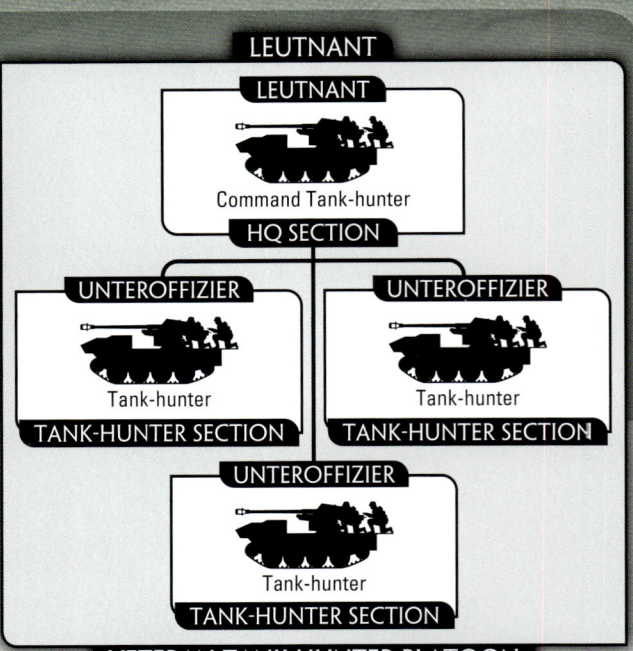

LEUTNANT

LEUTNANT

Command Tank-hunter

HQ SECTION

UNTEROFFIZIER
Tank-hunter
TANK-HUNTER SECTION

UNTEROFFIZIER
Tank-hunter
TANK-HUNTER SECTION

UNTEROFFIZIER
Tank-hunter
TANK-HUNTER SECTION

VETERAN TANK-HUNTER PLATOON

MANOEUVRE AND FIRE

Experienced tank-hunters would prepare firing and movement routes within concealing terrain to maximize their effectiveness against advancing enemy armour.

Veteran Tank-hunter Platoons may prepare alternate firing positions. At the start of the game choose a single piece of Concealing Area Terrain or Linear Obstacle that provides Concealment. If a Veteran Tank-hunter Platoon is placed from Ambush choose the terrain feature at the start of the

Starting Step in the turn the ambushing platoon is placed. The selected terrain feature must be at least partially within your deployment area.

As long as the whole platoon remains Concealed within or behind the selected terrain feature, they may shoot using their full ROF when they move, as if they had not moved. If the platoon leaves the selected terrain feature, or is no longer Concealed from enemy teams other than aircraft, they lose the benefit of the Manoeuvre and Fire special rule for the remainder of the game.

AREA TERRAIN

Tank-hunters may move freely within the area terrain and fire at their full rate of fire.

If any Tank-hunter from the platoon leaves the selected terrain feature, the platoon loses the benefit of the Manoeuvre & Fire special rule.

LINEAR TERRAIN

Tank-hunters can use Manoeuvre & Fire by remaining behind and right up against a Linear Obstacle

Linear Obstacles that are not continuous prevent Manoeuvre & Fire

HEAVY ANTI-TANK GUN PLATOON

PLATOON

HQ Section with:

4 8.8cm PaK43	420 points
3 8.8cm PaK43	315 points
2 8.8cm PaK43	210 points
1 8.8cm PaK43	105 points
4 8.8cm PaK43/41	405 points
3 8.8cm PaK43/41	305 points
2 8.8cm PaK43/41	205 points
1 8.8cm PaK43/41	105 points

OPTIONS

- Add Kfz 15 field car and RSO tractors for +5 points for the platoon.
- Replace all RSO tractors with Sd Kfz 7 half-tracks for +5 points for the platoon.

If you find yourself pitted against Soviet heavy armour then enlisting a few sections of 8.8cm PaK43 or PaK43/41 guns is exactly what the doctor ordered. They have the added punch needed to destroy those Soviet iron monsters.

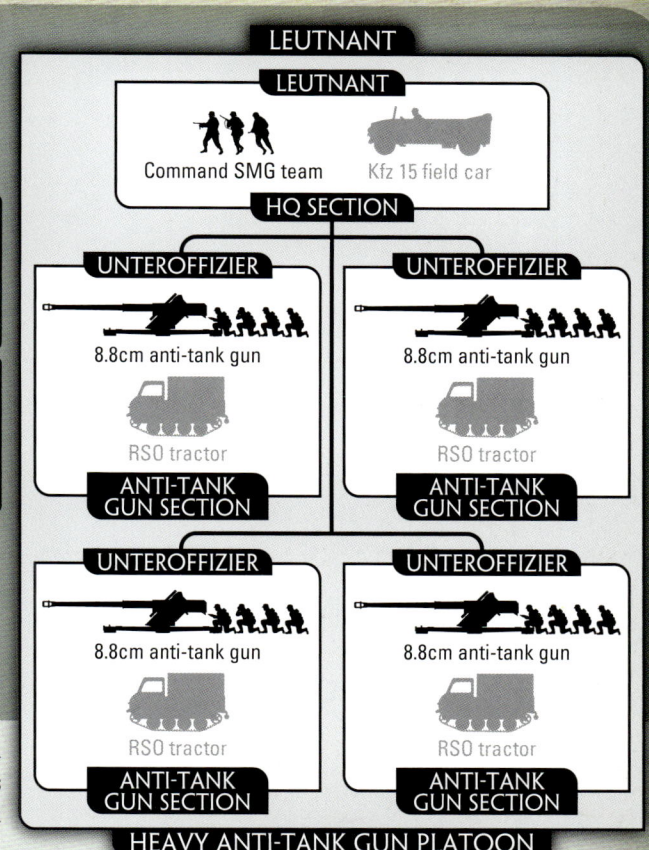

LEUTNANT

LEUTNANT

Command SMG team — Kfz 15 field car

HQ SECTION

UNTEROFFIZIER — 8.8cm anti-tank gun — RSO tractor — ANTI-TANK GUN SECTION

UNTEROFFIZIER — 8.8cm anti-tank gun — RSO tractor — ANTI-TANK GUN SECTION

UNTEROFFIZIER — 8.8cm anti-tank gun — RSO tractor — ANTI-TANK GUN SECTION

UNTEROFFIZIER — 8.8cm anti-tank gun — RSO tractor — ANTI-TANK GUN SECTION

HEAVY ANTI-TANK GUN PLATOON

ANTI-TANK GUN PLATOON

PLATOON

HQ Section with:

4 7.5cm PaK40	205 points
3 7.5cm PaK40	155 points
2 7.5cm PaK40	105 points
4 7.62cm PaK36(r)	165 points
3 7.62cm PaK36(r)	125 points
2 7.62cm PaK36(r)	85 points

OPTIONS

- Add Kfz 15 field car and 3-ton trucks to the platoon for +5 points for the platoon.
- Replace all 3-ton trucks with RSO tractors at no cost or with Sd Kfz 7 or 11 half-tracks for +5 points for the platoon.

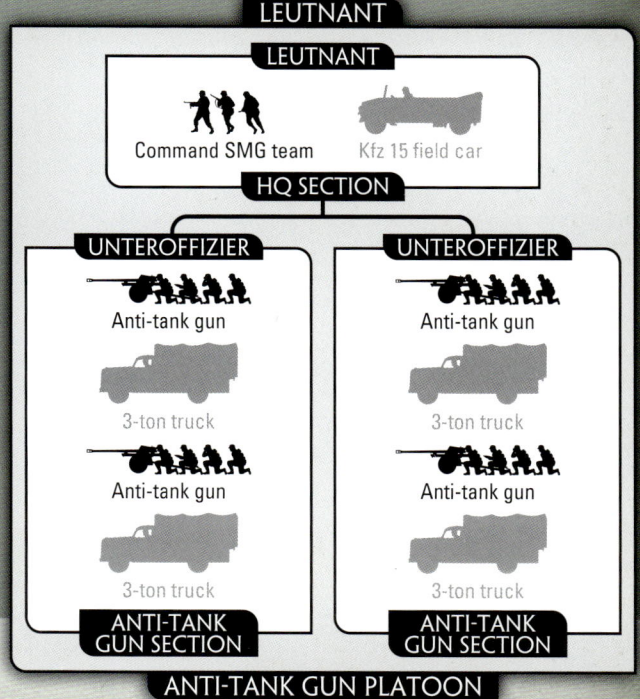

LEUTNANT

LEUTNANT

Command SMG team — Kfz 15 field car

HQ SECTION

UNTEROFFIZIER — Anti-tank gun — 3-ton truck — Anti-tank gun — 3-ton truck — ANTI-TANK GUN SECTION

UNTEROFFIZIER — Anti-tank gun — 3-ton truck — Anti-tank gun — 3-ton truck — ANTI-TANK GUN SECTION

ANTI-TANK GUN PLATOON

ARTILLERY BATTERY

PLATOON

HQ Section with:

4 10.5cm leFH18	210 points
3 10.5cm leFH18	155 points
2 10.5cm leFH18	115 points
4 10.5cm leFH18/40	215 points
2 10.5cm leFH18/40	120 points

OPTIONS

- Add horse-drawn wagon and limbers for +5 points for the battery.
- Replace all horse-drawn limbers and wagons with 3-ton trucks or RSO tractors at no cost.

You must purchase all of the guns from one Gun Section before adding any extra teams from the second Gun Section.

An Artillery Battery with three 10.5cm leFH18 howitzers may only contain one Observer Rifle team.

The 10.5cm leFH18 is the standard artillery piece of the German army. It has a heavy shell making its bombardments quite destructive.

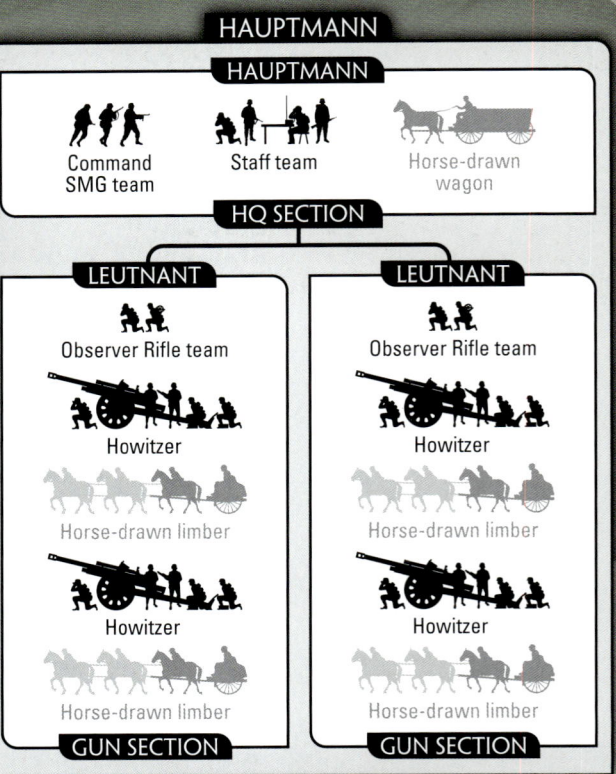

HEAVY ARTILLERY BATTERY

PLATOON

HQ Section with:

4 15cm sFH18	310 points
2 15cm sFH18	160 points
4 s10cm K18	395 points
2 s10cm K18	205 points
4 12.8cm K81	530 points
2 12.8cm K81	275 points

OPTIONS

- Add horse-drawn wagon and limbers for +5 points for the battery.
- Replace all horse-drawn limbers and wagons with 3-ton trucks or RSO tractors or Sd Kfz 7 half-tracks a no cost.

You must purchase all of the guns from one Gun Section before adding any extra teams from the second Gun Section.

Heavy Artillery Batteries may not be placed from Ambush within 16"/40cm of enemy teams.

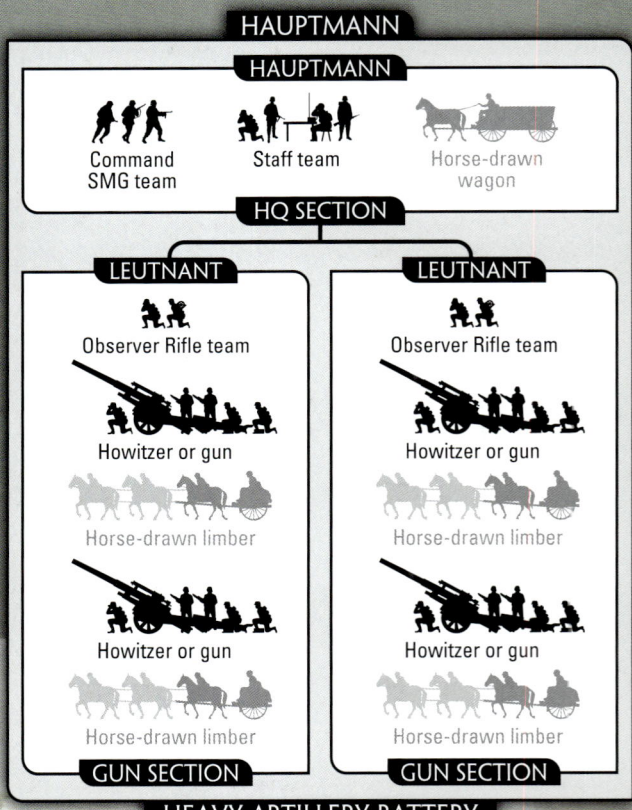

MOTORISED ARTILLERY BATTERY

PLATOON

HQ Section with:

4 10.5cm leFH18	210	points
2 10.5cm leFH18	115	points
4 10.5cm leFH18/40	215	points
2 10.5cm leFH18/40	120	points

OPTIONS

- Add Kfz 15 field car, Kfz 68 radio truck and Sd Kfz 11 half-tracks for +5 points for the battery.
- Replace all Kübelwagen jeeps with Sd Kfz 250 half-tracks for +5 points per half-track.
- Replace any or all Observer Rifle teams and their Kübelwagen jeep with Observer Panzer II OP tanks for +10 points per tank.

You must purchase all of the guns from one Gun Section before adding any extra teams from the second Gun Section.

Divisional towed artillery is an effective and cost efficient means of supporting your defence. The 10.5cm howitzer can deliver a high volume of fire with the ability to destroy the toughest enemy targets.

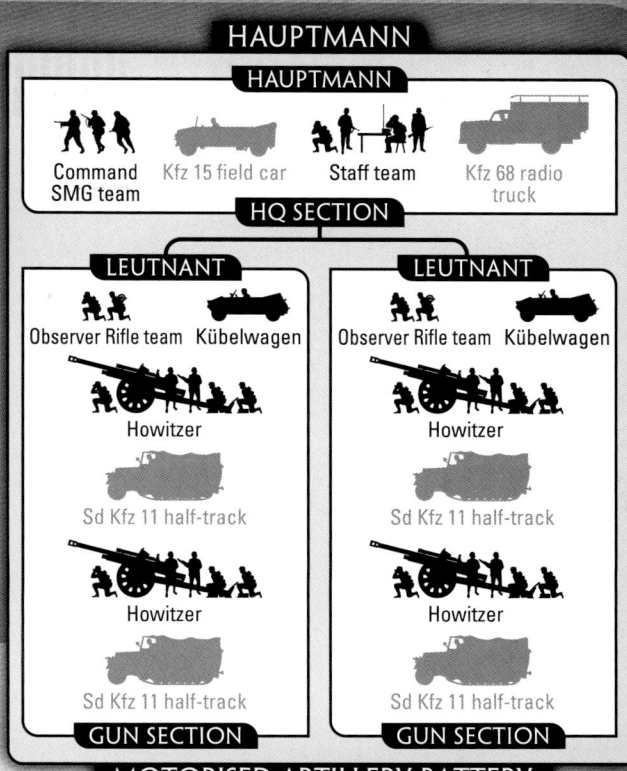

MOTORISED ARTILLERY BATTERY

Observer Panzer II OP tanks cannot launch assaults.

MOTORISED HEAVY ARTILLERY BATTERY

PLATOON

HQ Section with:

4 15cm sFH18	310	points
2 15cm sFH18	160	points
4 s10cm K18	395	points
2 s10cm K18	205	points
4 12.8cm K81	530	points
2 12.8cm K81	275	points

OPTIONS

- Add Kfz 15 field car, Kfz 68 radio truck and Sd Kfz 7 half-tracks for +5 points for the battery.
- Replace all Kübelwagen jeeps with Sd Kfz 250 half-tracks for +5 points per half-track.
- Replace any or all Observer Rifle teams and their Kübelwagen jeep with Observer Panzer II OP tanks for +10 points per tank.

You must purchase all of the guns from one Gun Section before adding any extra teams from the second Gun Section.

Observer Panzer II OP tanks cannot launch assaults.

MOTORISED HEAVY ARTILLERY BATTERY

Motorised Heavy Artillery Batteries may not be placed from Ambush within 16"/40cm of enemy teams.

ARMOURED ARTILLERY BATTERY

PLATOON

HQ Section with:

6 Wespe	435 points
4 Wespe	315 points
3 Wespe	235 points
2 Wespe	170 points

OPTIONS

- Replace any or all Observer Rifle teams and their Sd Kfz 250 with Observer Panzer III OP tanks for +10 points per tank.
- Add an Sd Kfz 9 (18t) recovery half-track for +5 points.

You must purchase all of the guns from one Gun Section before adding any extra teams from the second Gun Section.

Observer Panzer III OP tanks cannot launch assaults.

Providing concentrated artillery support to counterattacking Panzers will create the environment needed to halt the enemy advance and send them reeling back to their lines.

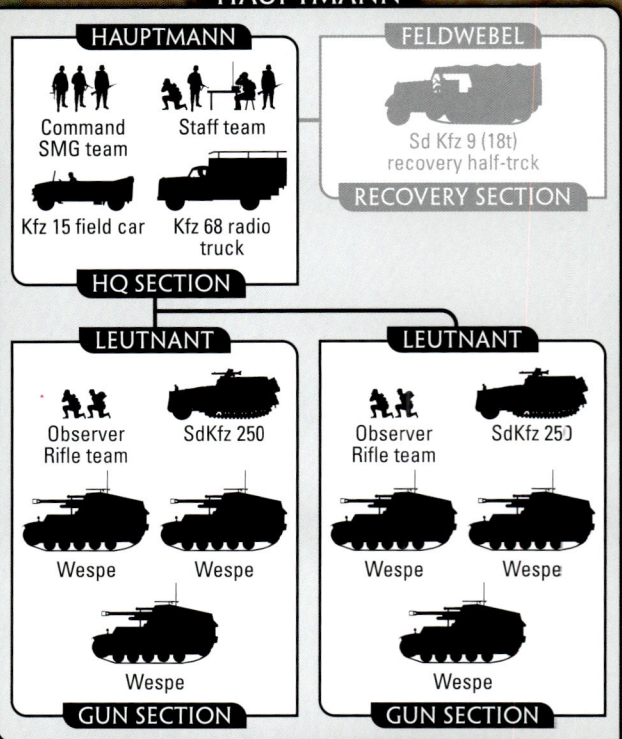

HAUPTMANN

HAUPTMANN

Command SMG team

Staff team

Kfz 15 field car

Kfz 68 radio truck

HQ SECTION

FELDWEBEL

Sd Kfz 9 (18t) recovery half-trck

RECOVERY SECTION

LEUTNANT

Observer Rifle team

SdKfz 250

Wespe

Wespe

Wespe

GUN SECTION

LEUTNANT

Observer Rifle team

SdKfz 250

Wespe

Wespe

Wespe

GUN SECTION

ARMOURED ARTILLERY BATTERY

ARMOURED HEAVY ARTILLERY BATTERY

PLATOON

HQ Section with:

6 Hummel	560 points
4 Hummel	405 points
3 Hummel	300 points
2 Hummel	210 points

OPTIONS

- Replace any or all Observer Rifle teams and their Sd Kfz 250 with Observer Panzer III OP tanks for +10 points per tank.
- Add an Sd Kfz 9 (18t) recovery half-track for +5 points.

You must purchase all of the guns from one Gun Section before adding any extra teams from the second Gun Section.

Observer Panzer III OP tanks cannot launch assaults.

HAUPTMANN

HAUPTMANN

Command SMG team

Staff team

Kfz 15 field car

Kfz 68 radio truck

HQ SECTION

FELDWEBEL

Sd Kfz 9 (18t) recovery half-trck

RECOVERY SECTION

LEUTNANT

Observer Rifle team

SdKfz 250

Hummel

Hummel

Hummel

GUN SECTION

LEUTNANT

Observer Rifle team

SdKfz 250

Hummel

Hummel

Hummel

GUN SECTION

ARMOURED HEAVY ARTILLERY BATTERY

ROCKET LAUNCHER BATTERY

PLATOON

HQ Section with:

6 15cm NW41 Launchers	200	points
4 15cm NW41 Launchers	145	points
3 15cm NW41 Launchers	105	points
2 15cm NW41 Launchers	75	points
6 21cm NW42 Launchers	240	points
4 21cm NW42 Launchers	175	points
3 21cm NW42 Launchers	125	points
2 21cm NW42 Launchers	90	points

OPTIONS

- Add a Kfz15 field car and Sd/Kfz 11 half-tracks to the battery for +5 points.
- Add an Anti-tank Section for +30 points.
- Replace 5cm PaK38 gun and Kfz 70 truck with 7.5cm PaK40 gun and Sd Kfz 11 half-track for +25 points.

You must purchase all of the guns from one Launcher Section before adding any extra teams from the second Launcher Section.

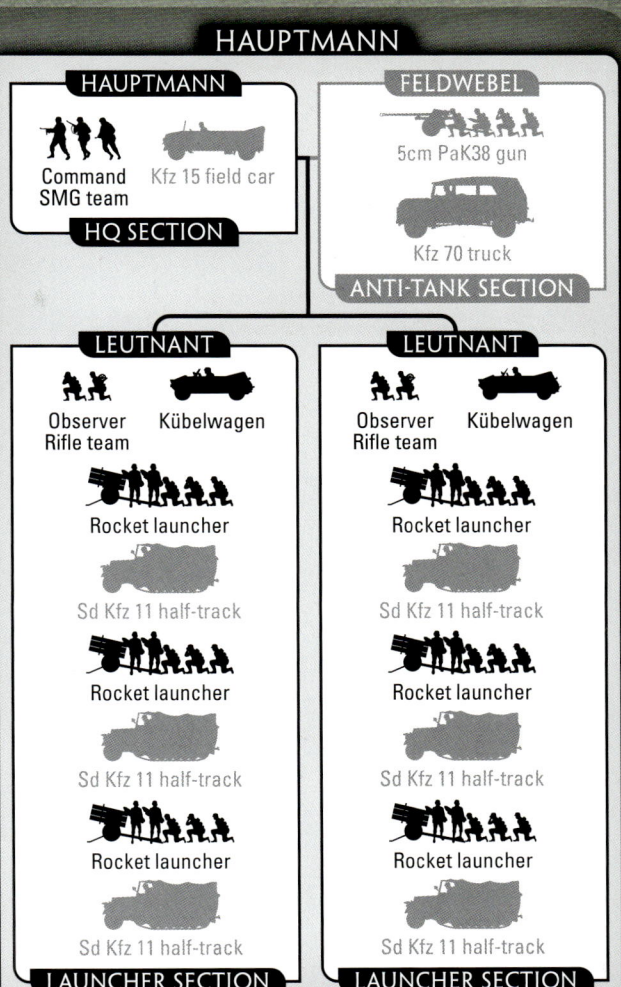

An economical way to deliver massive support in operations, the 15cm NW41 or 21cm NW42 rocket launcher batteries deliver crushing firepower at a moment's notice. Their ability to bring a constant reign of fire upon advancing enemy infantry can break their assault in one volley.

Though vulnerable to tank attack and easily identified for counterbattery fire by the smoke trails, their greater numbers provide you the firepower needed to successfully support your army with a potentially battle winning option.

Massing your rocket launchers can provide a much wider area of destruction in support of your mission. This option provides the most economical means of stopping massive enemy assaults in their tracks.

WALKÜRE PLATOON

PLATOON

HQ Section with:

3 Walküre Squads	120	points
2 Walküre Squads	85	points

OPTION

- Replace Command Rifle/MG team with a Command Panzerfaust SMG team for +10 points.

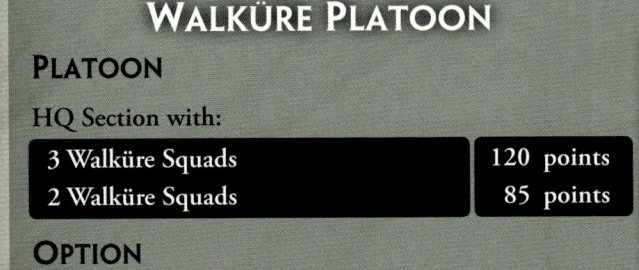

Walküre *Platoons are rated as* **Confident Trained.**

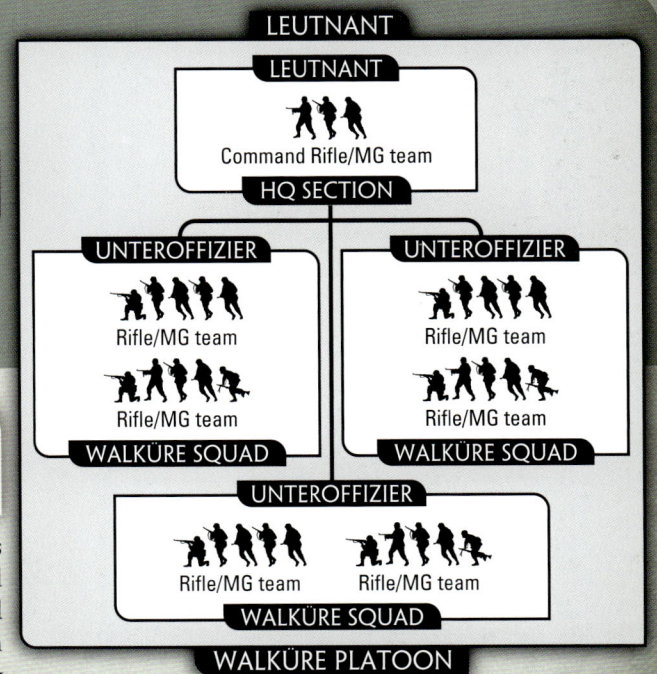

Walküre platoons provide excellent infantry reinforcements for developing a defence against an advancing mechanised enemy. Fully combat equipped, they can easily be integrated into existing formations to form a ready defence. Though fielded as full companies they were generally supported by more veteran formations. This allowed for a quick development of a solid defence backed by crack combat veterans able to stop, hold, and then counterattack enemy incursions.

ARMOURED ROCKET LAUNCHER BATTERY

PLATOON

HQ Section with:

8 Panzerwerfer 42	365	points
6 Panzerwerfer 42	305	points
4 Panzerwerfer 42	210	points
3 Panzerwerfer 42	165	points

OPTIONS

- Model Panzerwerfer 42 (Maultier) rocket launchers with 5 or more crew and count each rocket launcher as two weapons when firing a bombardment for +5 points per rocket launcher.
- Add Anti-tank Section for +30 points.
- Replace 5cm PaK38 gun with 7.5cm PaK40 gun for +20 points.

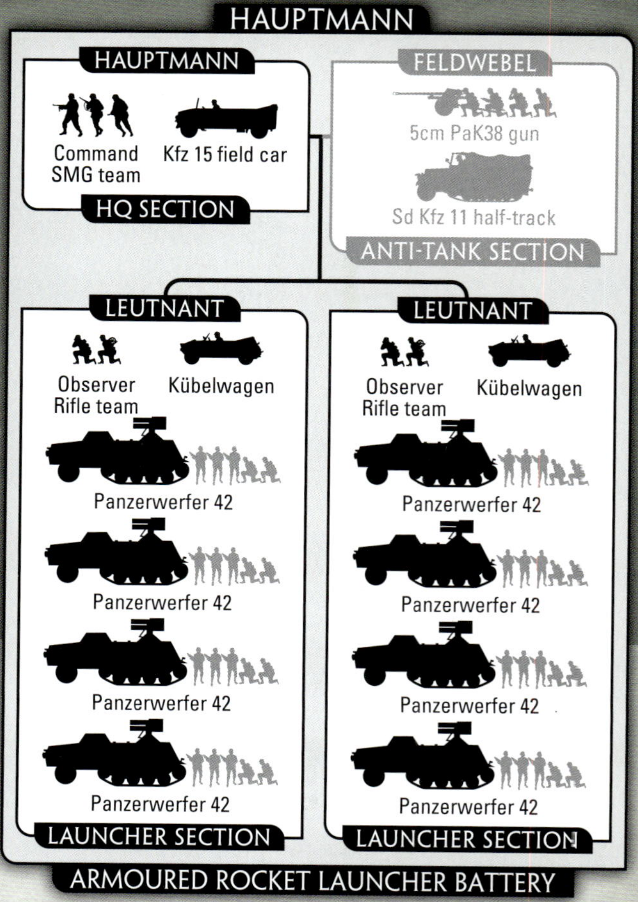

DEVASTATING BOMBARDMENT

Like the Soviet *Katyusha* rocket launchers, Panzerwerfer 42 rocket launchers are ready to signal the beginning of the offensive. Very little can withstand the fury of a full *Panzerwerfer* battery!

If the Bombarding platoon has nine to thirteen weapons firing, use a Devastating Bombardment Template to determine which teams are hit.

If the Bombarding platoon has fourteen or more weapons, use a Devastating Bombardment Template to determine which teams are hit and re-rolls failed To Hit rolls.

HAUPTMANN

HAUPTMANN

Command SMG team — Kfz 15 field car

FELDWEBEL

5cm PaK38 gun

Sd Kfz 11 half-track

ANTI-TANK SECTION

HQ SECTION

LEUTNANT

Observer Rifle team — Kübelwagen

Panzerwerfer 42

Panzerwerfer 42

Panzerwerfer 42

Panzerwerfer 42

LAUNCHER SECTION

LEUTNANT

Observer Rifle team — Kübelwagen

Panzerwerfer 42

Panzerwerfer 42

Panzerwerfer 42

Panzerwerfer 42

LAUNCHER SECTION

ARMOURED ROCKET LAUNCHER BATTERY

You must purchase all of the guns from one Launcher Section before adding any extra teams from the second Launcher Section.

Armoured Rocket Launcher Batteries use the Armoured Rocket Launcher special rule on page 186.

HEAVY ANTI-AIRCRAFT GUN PLATOON

PLATOON

HQ Section with:

2 8.8cm FlaK36	165 points
1 8.8cm FlaK36	85 points

OPTION

• Model 8.8cm FlaK36 guns with eight or more crew and increase their ROF to 3 for +10 points per gun.

The Allies have learned to respect this weapon and will often go out of their way to avoid it. When well positioned these weapons can halt an attack as it starts.

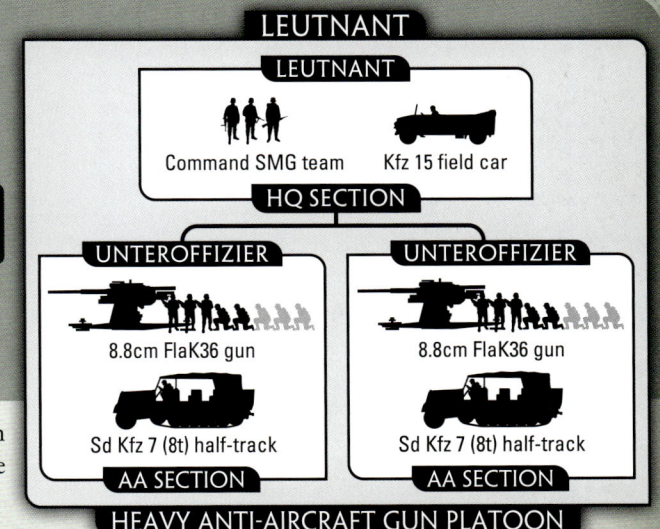

ANTI-AIRCRAFT GUN PLATOON

PLATOON

3 Sd Kfz 10/5 (2cm)	90 points
2 Sd Kfz 10/5 (2cm)	60 points
3 Armoured Sd Kfz 10/5 (2cm)	120 points
2 Armoured Sd Kfz 10/5 (2cm)	80 points
3 Sd Kfz 7/1 (Quad 2cm)	120 points
2 Sd Kfz 7/1 (Quad 2cm)	80 points
3 Armoured Sd Kfz 7/1 (Quad 2cm)	150 points
2 Armoured Sd Kfz 7/1 (Quad 2cm)	100 points
3 Sd Kfz 7/2 (3.7cm)	140 points
2 Sd Kfz 7/2 (3.7cm)	95 points
3 Armoured Sd Kfz 7/2 (3.7cm)	165 points
2 Armoured Sd Kfz 7/2 (3.7cm)	110 points

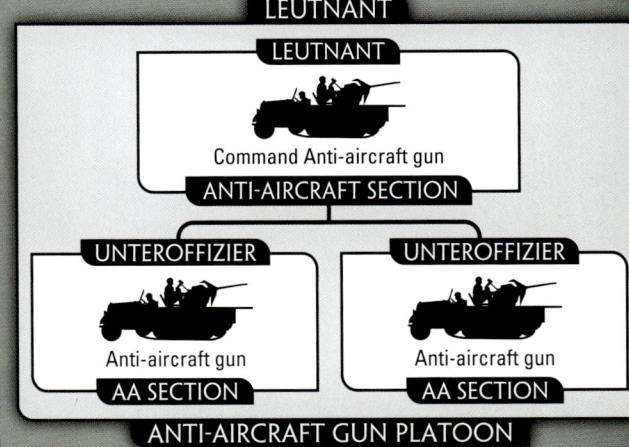

LUFTWAFFE DIVISIONAL SUPPORT

The Luftwaffe Flak Korps are well-trained in anti-aircraft work, but have little to no expertise in ground combat. They are normally kept behind the German front line, but as the Soviets rapidly advance keep finding themselves in the thick of the battle.

Luftwaffe divisional platoons are rated as **Reluctant Trained.**

RELUCTANT	CONSCRIPT
CONFIDENT	TRAINED
FEARLESS	VETERAN

AIR SUPPORT

SPORADIC AIR SUPPORT

Hs 129B	115 points
Hs 129B3	115 points
Ju 87D Stuka	100 points
Ju 87G Stuka	100 points

The Hs 129 was a formidable ground-attack aircraft nicknamed the *Panzerknacker* by the *Luftwaffe*. Armed with the 7.5cm main gun from the Panzer IV tank, the Hs 129B3 was able to knock out any tank in the world.

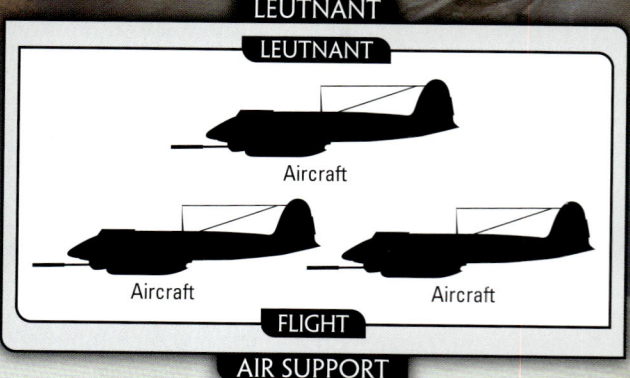

LEUTNANT
LEUTNANT
Aircraft
Aircraft
Aircraft
FLIGHT
AIR SUPPORT

LUFTWAFFE JÄGER PLATOON

PLATOON

HQ Section with:

3 Jäger Squads	90 points
2 Jäger Squads	65 points

OPTION

- Replace Command Rifle/MG teams with Command Panzerknacker SMG teams for +5 points.

The combat capable remnants of *10. Luftwaffe Felddivision* fought in a *Kampfgruppe* with *SS-Panzeraufklärungsabteilung 11* and *SS-Pionierbataillon 11* from 14 January onwards until they were "absorbed" by *11. SS-Panzergrenadierdivision Nordland* as replacements.

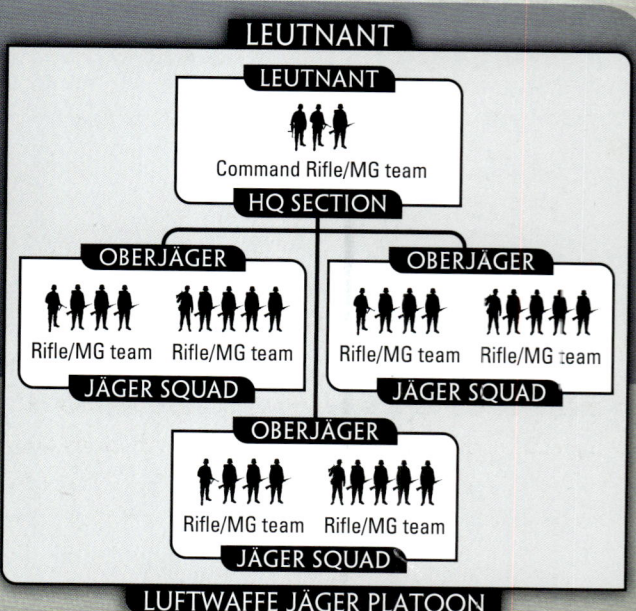

LEUTNANT
LEUTNANT
Command Rifle/MG team
HQ SECTION
OBERJÄGER
Rifle/MG team Rifle/MG team
JÄGER SQUAD
OBERJÄGER
Rifle/MG team Rifle/MG team
JÄGER SQUAD
OBERJÄGER
Rifle/MG team Rifle/MG team
JÄGER SQUAD
LUFTWAFFE JÄGER PLATOON

LUFTWAFFE HEAVY ANTI-AIRCRAFT GUN PLATOON

PLATOON

HQ Section with:

2 8.8cm FlaK36	95 points
1 8.8cm FlaK36	50 points

OPTION

- Model 8.8cm FlaK36 guns with eight or more crew and increase their ROF to 3 for +10 points per gun.

Flak battalions of the *Luftwaffe* have supported the defensive operations since the beginning of the war. Though not your primary operation you can, when called upon, bring the power of the 8.8cm gun to bear.

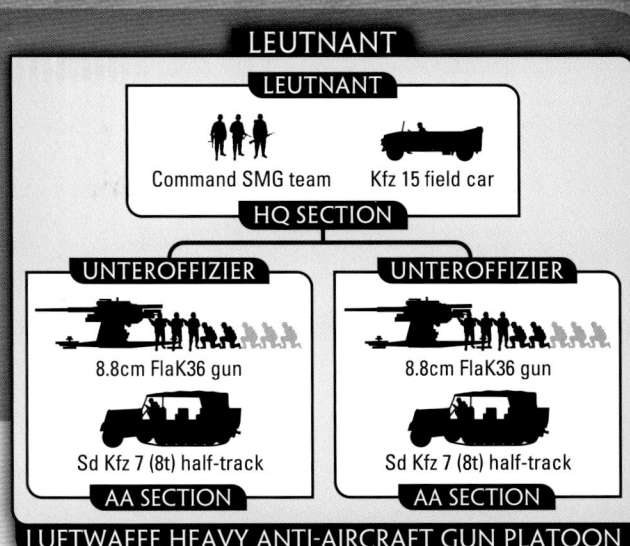

LUFTWAFFE LIGHT ANTI-AIRCRAFT GUN PLATOON

PLATOON

HQ Section with:

3 3.7cm FlaK43	65 points
3 2cm Flakvierling 38	60 points

OPTIONS

- Add Kfz 15 field car and 3-ton trucks for +5 points for the platoon.
- Replace all trucks with RSO tractors at no cost.
- Replace all 2cm Flakvierling 38 guns with Quad 2cm FlaK Nests and remove the Command SMG team for +25 points per gun.

A company containing Quad 2cm Flak Nests from a Luftwaffe Light Anti-aircraft Gun Platoon as its only fortifications is not a Fortified Company.

Quad 2cm Flak Nests follow the fortification placement rules for Nests on pages 152 to 153.

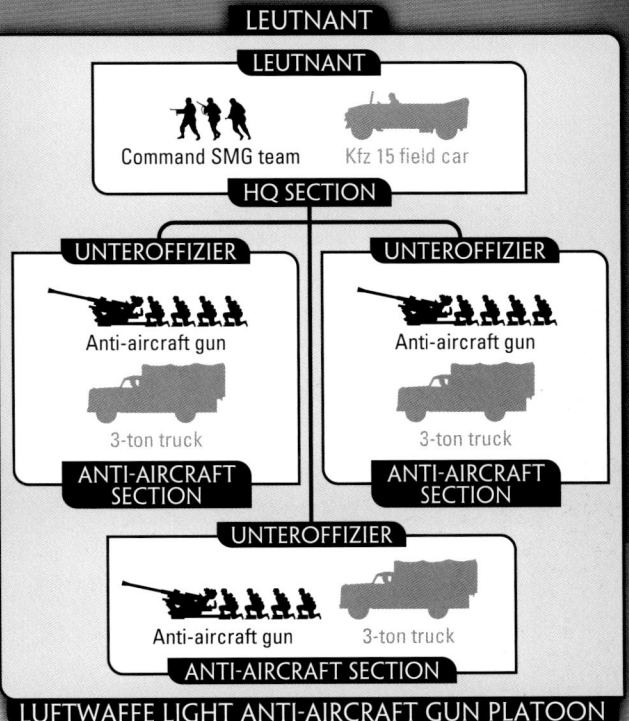

Field anti-aircraft guns to protect your soft rear-area assets from enemy fighter bombers.

LUFTWAFFE ANTI-AIRCRAFT GUN PLATOON

PLATOON

3 Sd Kfz 7/2 (3.7cm)	95 points
2 Sd Kfz 7/2 (3.7cm)	65 points
3 Armoured Sd Kfz 7/2 (3.7cm)	110 points
2 Armoured Sd Kfz 7/2 (3.7cm)	75 points

The high rate of fire of anti-aircraft guns effectively clears the skies of aircraft or brings much needed support to your dwindling infantry when fighting advancing enemy formations. Anti-aircraft half-tracks also afford some protection against enemy reconnaissance vehicles and infantry platoons.

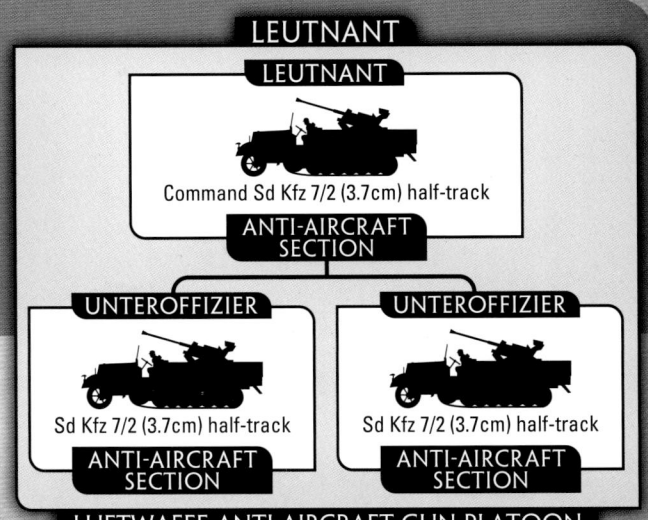

SS DIVISIONAL SUPPORT

MOTIVATION AND SKILL

V *Veteran SS-Divisions have vast experience of the harsh warfare of the Eastern Front. Divisional and Corps support platoons from 3. SS-Panzerdivision Totenkopf (marked* ⬛ *) and 4. SS-Freiwilligen-Panzergrenadier Brigade Nederland (marked* ⬛ *) are rated* **Fearless Veteran**, *unless otherwise noted, and select their support platoons from the column marked (* **V** *), or with no symbol.*

RELUCTANT	CONSCRIPT
CONFIDENT	TRAINED
FEARLESS	**VETERAN**

V **VETERAN SS**

RELUCTANT	CONSCRIPT
CONFIDENT	**TRAINED**
FEARLESS	VETERAN

T **TRAINED SS**

T *Volunteer SS-Divisions have many new recruits and replacement from all over Europe. Divisional and Corps support platoons from , 5. SS-Panzerdivision Wiking (marked* ⬛ *), 11. SS-Freiwilligen-Panzergrenadier Division Nordland (marked* ⬛ *), 8. SS-Kavalleriedivision Florian Geyer (marked* ⬛ *), 22. SS-Freiwilligen-Kavalleriedivision Maria Theresia (marked* ⬛ *) are rated* **Fearless Trained**, *unless otherwise noted, and select their support platoons from the column marked (* **T** *), or with no symbol.*

SS Divisional Support Platoons use the Waffen-SS special rules of the SS Company they are supporting (See pages 188 to 189) as well as all the normal German special rules on pages 183 to 187.

HEAVY SS-TANK PLATOON

PLATOON

PLATOON	**V**	**T**
5 Tiger I E	1200 points	-
4 Tiger I E	960 points	-
3 Tiger I E	720 points	-
2 Tiger I E	480 points	-
1 Tiger I E	240 points	-

Remember to roll for your Tiger Ace Skills before each game.

9. Kompanie 3. SS-Panzerregiment has been an integral part of the *3. SS-Panzerdivision Totenkopf* since May 1943. Originally containing ten Tiger heavy tanks and nine Panzer III tanks, the company upgraded to 14 Tiger heavy tanks in September 1943.

For the latter part of 1943 the company was overstrength and had 4 Tiger I platoons. The fourth platoon had five tanks and were numbered 941 to 945.

The Tigers of the company were readily used in battlegroups to support various offensive and defensive operations of the

UNTERSTURMFÜHRER

UNTERSTURMFÜHRER

Command Tiger I E

HQ SECTION

UNTERSCHARFÜHRER

Tiger I E

Tiger I E

HEAVY TANK SECTION

UNTERSCHARFÜHRER

Tiger I E

Tiger I E

HEAVY TANK SECTION

HEAVY SS-TANK PLATOON

3. SS-Panzerdivision Totenkopf. They were regularly assigned as platoons subordinate to a *Panzergrenadier regiment* or the *3. SS-Panzerjägerabteilung* (anti-tank battalion).

SS-TANK-HUNTER PLATOON

PLATOON

	V	T
4 Marder III M	285 points	220 points
3 Marder III M	215 points	165 points
2 Marder III M	145 points	110 points
4 StuG G	435 points	-
3 StuG G	325 points	-
2 StuG G	215 points	-
4 Jagdpanzer IV	435 points	335 points
3 Jagdpanzer IV	325 points	250 points
2 Jagdpanzer IV	215 points	165 points
3 Hetzer	285 points	220 points
2 Hetzer	190 points	145 points

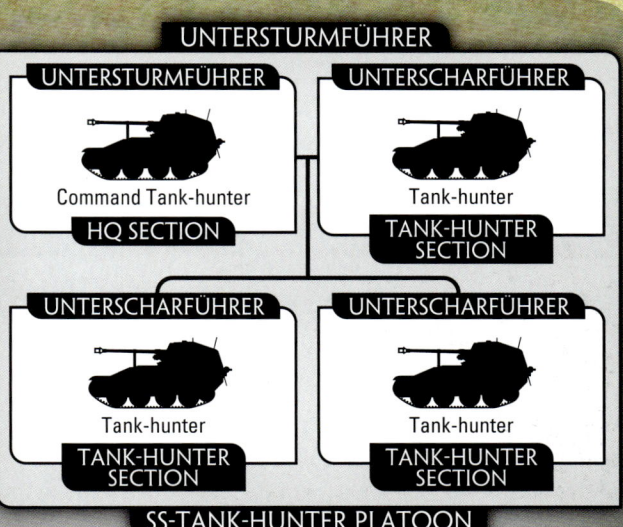

SS-AUFKLÄRUNGS PLATOON

PLATOON

HQ Section with:

	V	T
3 Aufklärungs Squads	260 points	195 points
2 Aufklärungs Squads	190 points	140 points

OPTIONS

- Replace Command MG team with a Command Panzerfaust SMG team for +10 points.
- Add an additional Sd Kfz 250 half-track to each squad for +10 points per half-track.

SS-Aufklärungs Platoons may use the Mounted Assault special rule.

Though originally a reconnaissance element in blitzkrieg, circumstances now have the *Aufklärungs* platoon providing much needed *Panzergrenadier* support for your counterattacking panzer platoons. Still capable of deep penetration behind enemy lines, the great numbers of enemy armour requires the intelligent use of this highly valued unit to ensure its maximum effectiveness.

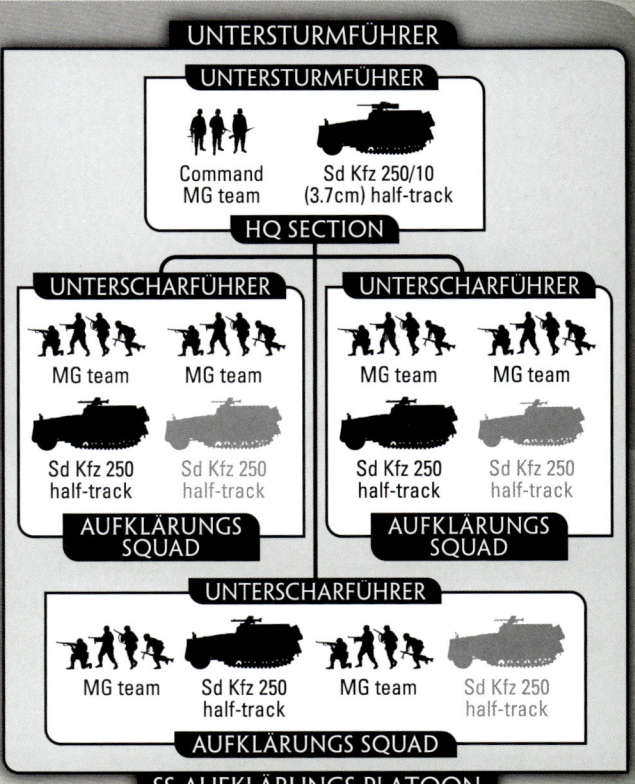

SS-Panzerspäh Platoon

PLATOON

	V	T
2 Panzerspäh Patrols	240 points	190 points
1 Panzerspäh Patrol	120 points	95 points

Panzerspäh Patrols of a SS-Panzerspäh Platoon operate as separate platoons, each with their own command team.

Panzerspäh Patrols are Reconnaissance Platoons.

Armoured car reconnaissance plays a key role in assessing the enemy's intentions. If he is advancing your armoured cars can reconnoitre your counterattack route while keeping any flanking infantry or light armoured transport at bay.

UNTERSTURMFÜHRER

UNTERSTURMFÜHRER

Command Sd Kfz 223 (radio)

Sd Kfz 222 (2cm)

Sd Kfz 222 (2cm)

PANZERSPÄH PATROL

UNTERSCHARFÜHRER

Command Sd Kfz 223 (radio)

Sd Kfz 222 (2cm)

Sd Kfz 222 (2cm)

PANZERSPÄH PATROL

SS-PANZERSPÄH PLATOON

HEAVY SS-Panzerspäh Platoon

PLATOON

	V	T
3 Panzerspäh Patrols	270 points	210 points
2 Panzerspäh Patrols	180 points	140 points
1 Panzerspäh Patrol	90 points	70 points

Panzerspäh Patrols of a Heavy SS-Panzerspäh Platoon operate as separate platoons, each with their own command team.

Panzerspäh Patrols are Reconnaissance Platoons.

When attacking, the armoured reconnaissance can find or even make a hole through the enemy's defensive line if he is foolish enough to leave his infantry unprotected by armour or anti-tank guns. But even so, your armoured cars are swift enough to discover the enemy's positions without staying around to absorb his firepower.

In counterattack, armoured reconnaissance can protect the flanks of your armour from hidden anti-tank or assaulting infantry. This allows your panzers to put their maximum effort at the point of attack.

UNTERSTURMFÜHRER

UNTERSTURMFÜHRER

Command Sd Kfz 231 (8-rad)

Sd Kfz 231 (8-rad)

PANZERSPÄH PATROL

UNTERSCHARFÜHRER

Command Sd Kfz 231 (8-rad)

Sd Kfz 231 (8-rad)

PANZERSPÄH PATROL

UNTERSCHARFÜHRER

Command Sd Kfz 231 (8-rad)

Sd Kfz 231 (8-rad)

PANZERSPÄH PATROL

HEAVY SS-PANZERSPÄH PLATOON

SS-PANZERGRENADIER PIONEER PLATOON

PLATOON

HQ Section with:

	V	T
3 Pioneer Squads	245 points	190 points
2 Pioneer Squads	175 points	135 points

OPTIONS

- Replace Command Pioneer Rifle/MG team with Command Pioneer Panzerknacker SMG team for +5 points or Command Pioneer Panzerfaust SMG team for +10 points.

- Equip one Pioneer Rifle/MG team with a Goliath demolition carrier in addition to its normal weapons for +30 points.

- Add Pioneer Supply 3-ton truck for +25 points, or Pioneer Supply Maultier for +30 points.

You may replace up to one Pioneer Rifle/MG team per Pioneer Squad with a Flame-thrower team at the start of the game before deployment.

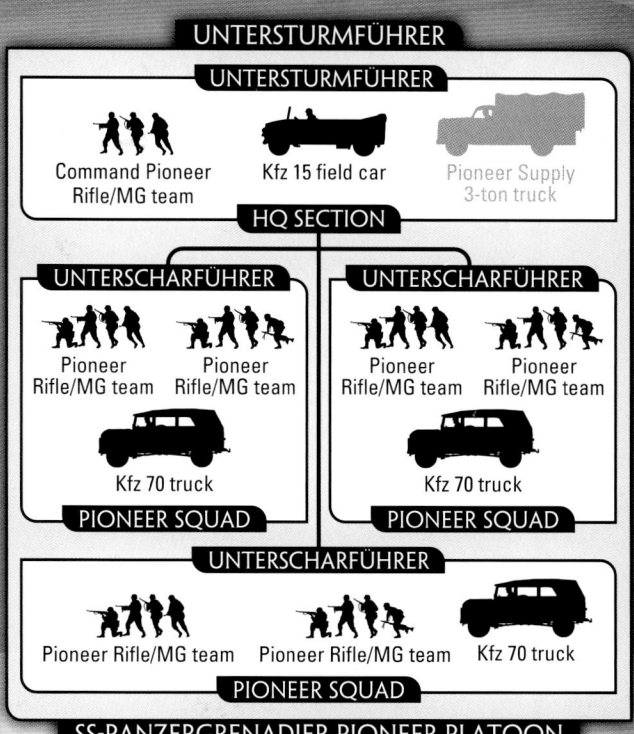

SS-PANZERGRENADIER PIONEER PLATOON

GEPANZERTE SS-PANZERPIONIER PLATOON

PLATOON

HQ Section with:

	V	T
3 Pioneer Squads	315 points	240 points
2 Pioneer Squads	230 points	170 points

OPTIONS

- Replace Command Pioneer MG team with a Command Pioneer Panzerfaust SMG team for +10 points.

- Add an additional Sd Kfz 251/7 (Pioneer) half-track to each squad for +10 points per half-track.

- Replace any or all Sd Kfz 251/7 half-tracks with Sd Kfz 251/1 (Stuka) half-tracks for +40 points per half-track.

- Add a Pioneer Supply Maultier half-track for +30 points.

- Add a Goliath demolition carrier to one Pioneer MG team for +30 points.

You may replace up to one Pioneer MG team per squad with a Flame-thrower team at the start of the game before deployment.

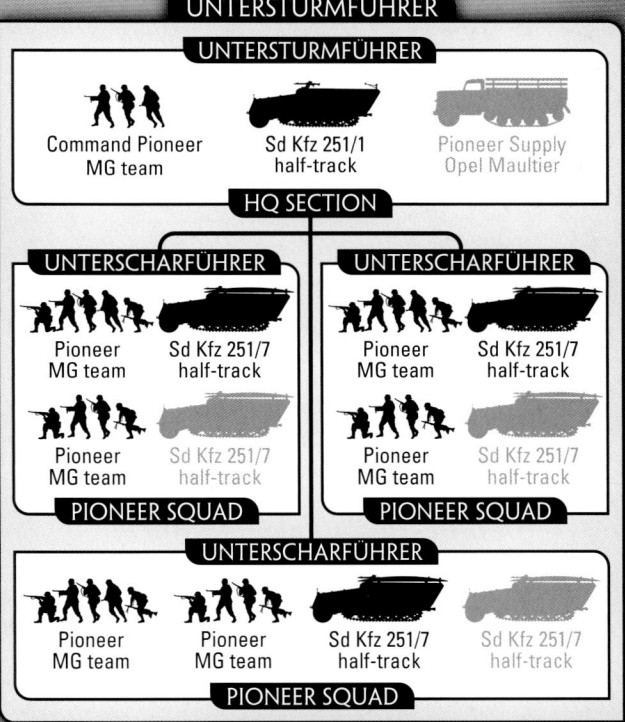

GEPANZERTE SS-PANZERPIONIER PLATOON

Consider yourself fortunate to have a platoon of *SS-Panzerpionier* under your command. Their ability to close quickly and assault enemy infantry in combination with their indifference to enemy armour gives you a striking force unmatched by the enemy.

Adding additional equipment to your *Panzerpionier* platoon provides even greater capability to your pioneer unit. If needed for defence, adding *Panzerfaust* anti-tank rockets, a supply vehicle, or the Goliath can enhance the overall effectiveness of this unit.

You can add an effective artillery capability with the *Stuka* half-track option, or increase the unit's machine-gun support with additional Sd-Kfz 251/7 half-tracks.

MOTORISED SS-ARTILLERY BATTERY

PLATOON

HQ Section with:

	V	T
4 10.5cm leFH18	240 points	185 points
2 10.5cm leFH18	130 points	100 points

OPTION

- Add Kfz 15 field car, Kfz 68 radio truck, and 3-ton trucks for +5 points for the battery.

You must purchase all of the guns from one Gun Section before adding any extra teams from the second Gun Section.

The role of artillery is to pound the enemy into submission. Firing bombardment after bombardment, they will eventually destroy any target. Their bombardments cripple enemy attacks as losses mount and troops are pinned to the ground unable to advance.

OBERSTURMFÜHRER

OBERSTURMFÜHRER

Command SMG team | Kfz 15 field car | Staff team | Kfz 68 radio truck

HQ SECTION

UNTERSCHARFÜHRER

Observer Rifle team | Kübelwagen

10.5cm leFH18 howitzer

3-ton truck

10.5cm leFH18 howitzer

3-ton truck

GUN SECTION

UNTERSCHARFÜHRER

Observer Rifle team | Kübelwagen

10.5cm leFH18 howitzer

3-ton truck

10.5cm leFH18 howitzer

3-ton truck

GUN SECTION

MOTORISED SS-ARTILLERY BATTERY

MOTORISED HEAVY SS-ARTILLERY BATTERY

PLATOON

HQ Section with:

	V	T
4 15cm sFH18	360 points	275 points
2 15cm sFH18	185 points	140 points
4 s10cm K18	445 points	340 points
2 s10cm K18	230 points	175 points

OPTION

- Add Kfz 15 field car, Kfz 68 radio truck, and Sd Kfz 7 (8t) half-tracks for +5 points for the battery.

You must purchase all of the guns from one Gun Section before adding any extra teams from the second Gun Section.

Motorised Heavy SS-Artillery Batteries may not be placed from Ambush within 16"/40cm of enemy teams.

OBERSTURMFÜHRER

OBERSTURMFÜHRER

Command SMG team | Kfz 15 field car | Staff team | Kfz 68 radio truck

HQ SECTION

UNTERSCHARFÜHRER

Observer Rifle team | Kübelwagen

Howitzer or gun

Sd Kfz 7 (8t) half-track

Howitzer or gun

Sd Kfz 7 (8t) half-track

GUN SECTION

UNTERSCHARFÜHRER

Observer Rifle team | Kübelwagen

Howitzer or gun

Sd Kfz 7 (8t) half-track

Howitzer or gun

Sd Kfz 7 (8t) half-track

GUN SECTION

MOTORISED HEAVY SS-ARTILLERY BATTERY

ARMOURED SS-ARTILLERY BATTERY

PLATOON

HQ Section with:

	V	T
6 Wespe	470 points	360 points
4 Wespe	340 points	260 points
3 Wespe	255 points	195 points
2 Wespe	180 points	140 points

OPTIONS

- Replace Observer Rifle teams and Kübelwagen jeeps with Observer Panzer III OP tanks for +10 points per tank.
- Add an Sd Kfz 9 (18t) recovery half-track for +5 points.

You must purchase all of the guns from one Gun Section before adding any extra teams from the second Gun Section.

Observer Panzer III OP tanks cannot launch assaults.

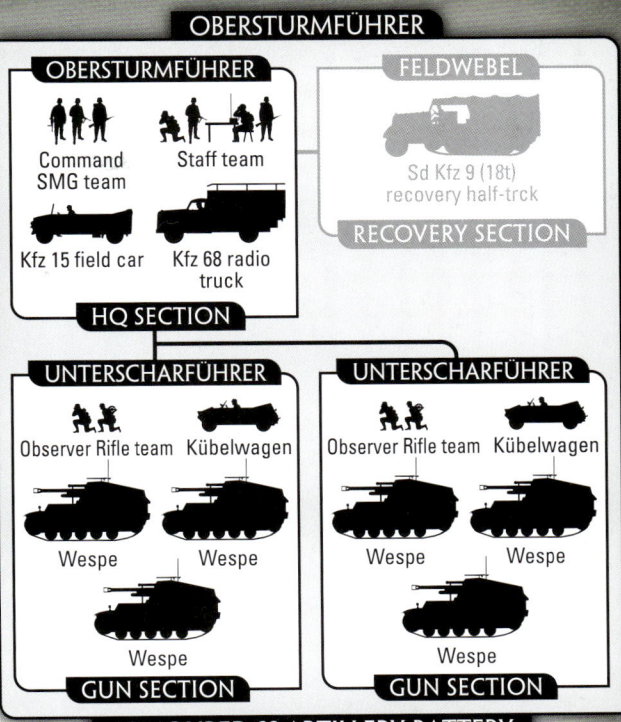

OBERSTURMFÜHRER

OBERSTURMFÜHRER — Command SMG team, Staff team, Kfz 15 field car, Kfz 68 radio truck — HQ SECTION

FELDWEBEL — Sd Kfz 9 (18t) recovery half-trck — RECOVERY SECTION

UNTERSCHARFÜHRER — Observer Rifle team, Kübelwagen, Wespe, Wespe, Wespe — GUN SECTION

UNTERSCHARFÜHRER — Observer Rifle team, Kübelwagen, Wespe, Wespe, Wespe — GUN SECTION

ARMOURED SS-ARTILLERY BATTERY

ARMOURED HEAVY SS-ARTILLERY BATTERY

PLATOON

HQ Section with:

	V	T
6 Hummel	615 points	470 points
4 Hummel	440 points	340 points
3 Hummel	330 points	255 points
2 Hummel	230 points	175 points

OPTIONS

- Replace Observer Rifle teams and Kübelwagen jeeps with Observer Panzer III OP tanks for +10 points per tank.
- Add an Sd Kfz 9 (18t) recovery half-track for +5 points.

You must purchase all of the guns from one Gun Section before adding any extra teams from the second Gun Section.

Observer Panzer III OP tanks cannot launch assaults.

OBERSTURMFÜHRER

OBERSTURMFÜHRER — Command SMG team, Staff team, Kfz 15 field car, Kfz 68 radio truck — HQ SECTION

FELDWEBEL — Sd Kfz 9 (18t) recovery half-trck — RECOVERY SECTION

UNTERSCHARFÜHRER — Observer Rifle team, Kübelwagen, Hummel, Hummel, Hummel — GUN SECTION

UNTERSCHARFÜHRER — Observer Rifle team, Kübelwagen, Hummel, Hummel, Hummel — GUN SECTION

ARMOURED HEAVY SS-ARTILLERY BATTERY

179

SS-ROCKET LAUNCHER BATTERY

PLATOON

HQ Section with:

6 15cm NW41 rocket launchers	180 points
4 15cm NW41 rocket launchers	125 points
3 15cm NW41 rocket launchers	90 points
2 15cm NW41 rocket launchers	65 points

OPTIONS

- Add Kfz 15 field car and Sd Kfz 11 half-tracks for +5 points for the battery.
- Add an Anti-tank Section for +45 points.

The 504. SS-Werfer-Abteilung supported the IV. SS-Panzerkorps having been recently reconstituted as a new unit. As such it is rated **Fearless Trained.**

FEARLESS | TRAINED

You must purchase all of the guns from one Launcher Section before adding any extra teams from the second Launcher Section.

The *Nebelwerfer* is a very cost-effective weapon capable of providing devastating rocket artillery for your defensive positions. Placing them behind terrain away from the prying eyes of enemy reconnaissance will keep them available for the ensuing battle.

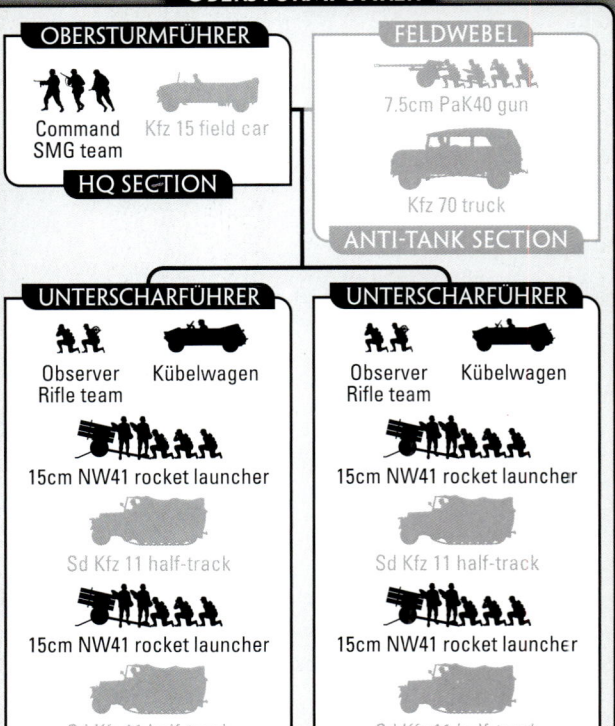

HEAVY SS-ANTI-AIRCRAFT GUN PLATOON

PLATOON

HQ Section with:

	V	T
2 8.8cm FlaK36	190 points	140 points
1 8.8cm FlaK36	100 points	75 points

OPTION

- Model 8.8cm Flak36 anti-aircraft guns with 8 or more crew and increase their ROF to 3 for +10 points per gun.

Good tactical placement of the 8.8cm Flak 36 heavy anti-aircraft gun goes a long way in channelling the enemy towards your desired killing zones. Their long range and high rate of fire can have a decisive result on the battle's outcome.

They can provide the ambush support needed to break the back of enemy armoured spearheads. Placing them in cover helps conceal them from enemy reconnaissance and air attack. Though, if needed, they can add their impressive firepower in clearing the skies of unwanted enemy aircraft.

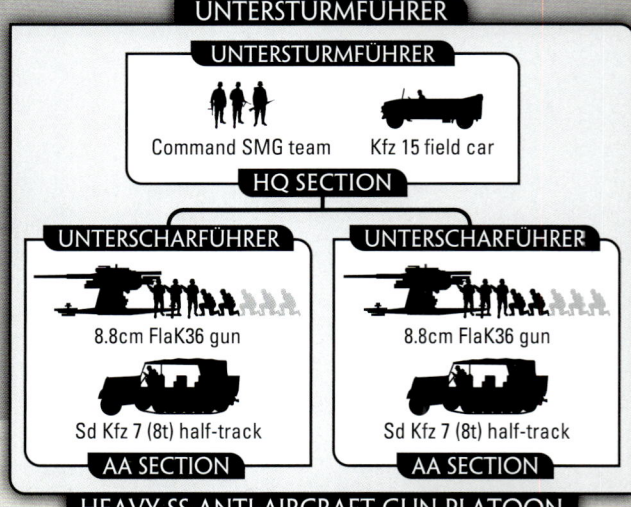

SS-VIELFACHWERFER BATTERY

PLATOON

HQ Section with:

8 Vielfachwerfer auf Maultier	250	points
6 Vielfachwerfer auf Maultier	195	points
4 Vielfachwerfer auf Maultier	130	points
3 Vielfachwerfer auf Maultier	100	points
2 Vielfachwerfer auf Maultier	75	points

OPTIONS

- Model Vielfachwerfer auf Maultier rocket launchers with 5 or more crew and count each rocket launcher as two weapons when firing a bombardment for +5 points per launcher.
- Add Anti-tank Section for +55 points.

DEVASTATING BOMBARDMENT

A *Vielfachwerfer auf Maultier* rocket launcher signals the beginning of your counterattack or offensive. They are very effective against Infantry in the open.

If the Bombarding platoon has nine to thirteen weapons firing, use a Devastating Bombardment Template to determine which teams are hit.

If the Bombarding platoon has fourteen or more weapons, use a Devastating Bombardment Template to determine which teams are hit and re-rolls failed To Hit rolls.

SS-Vielfachwerfer Batteries have supported Waffen-SS since 1943. As such they are rated **Fearless Veteran**.  **FEARLESS** | **VETERAN**

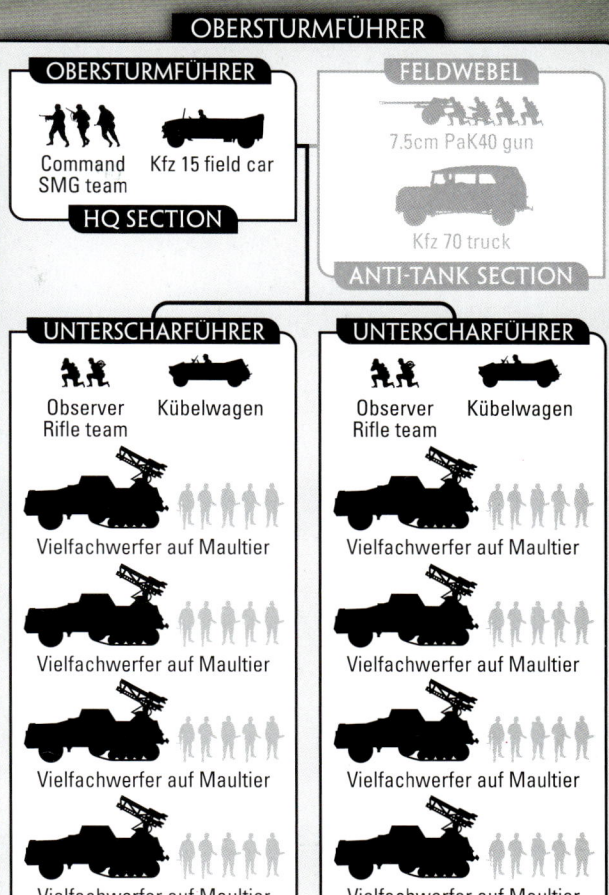

You must purchase all of the guns from one Launcher Section before adding any extra teams from the second Launcher Section.

SS-Vielfachwerfer Batteries use the Armoured Rocket Launcher special rule on page 186.

SELF-PROPELLED SS-ANTI-AIRCRAFT GUN PLATOON

PLATOON

	V	T
3 Sd Kfz 10/5 (2cm)	95 points	75 points
2 Sd Kfz 10/5 (2cm)	65 points	50 points
3 Sd Kfz 7/1 (Quad 2cm)	135 points	105 points
2 Sd Kfz 7/1 (Quad 2cm)	90 points	70 points

With the skies filling rapidly with enemy tank-killing aircraft, the need for effective and efficient anti-aircraft support is critical. Protecting your armour, artillery and infantry as they move into position is essential for victory.

For minimal effort you can insure your forces are well-protected from harassing enemy fighters. Keep them to the rear of your formations away from enemy armour and infantry. This will insure they are available when needed.

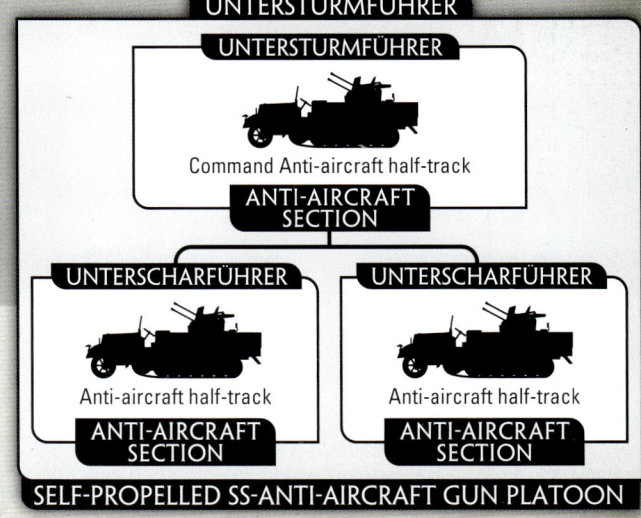

STURMTIGERS IN WARSAW

The *Sturmtiger* made its combat debut during the Warsaw uprising. Two crews of factory engineers manned the *Sturmtiger* assault howitzers giving a hands-on field display of this new weapons' capabilities. Staying well behind the lines, these two behemoths devastated city blocks with their 38cm RW61 shells from up to 4600 metres away.

Built on the chassis of a Tiger heavy tank, the *Sturmtiger* carried 14 rounds of the 600 lb/270 kg shells. Each shell contained 125 kilograms of explosives. It had one hull-mounted 7.62 MG34 machine gun. The massive 15"/38cm rocket-assisted howitzer possessed unprecedented firepower when targeting large targets such as buildings or bridges. Its targets, when hit, were obliterated.

Though not assigned to any particular unit, the first *Sturmtiger* platoon supported the *Sicherungs* (Security) forces of the 3. *Totenkopf SS-Panzerdivision* while they suppressed the Warsaw uprising. The *Sturmtiger* performed its task well, levelling whole areas of buildings in what was once the city of Warsaw. Once they completed their job in Warsaw they were returned to the factory.

Production Quantity: 18	**Combat Weight:** 65,000 kg
Crew: 5	**Speed:** 40 km/h
Length: 6.28m	**Horsepower:** 700
Armour	**Vehicle Range:** 120 km
Front: 100mm at 65 degrees	**Armament:**
Side: 60 mm at 80 degrees	One 380mm StuM RW61 L/5.4
Rear: 80 mm at 81 degrees	One 7.92mm MG34
Top: 25 mm at 0 degrees	**Ammunition Carried:** 14 rounds

HEAVY ASSAULT HOWITZER PLATOON

PLATOON

2 Gun Sections	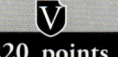 220 points	-

ROCKET ASSAULT HOWITZER

A Sturmtiger may fire as an Artillery Battery.

- *Each Sturmtiger fires as a separate Artillery Battery.*
- *Each Sturmtiger only makes one attempt to Range In on its target.*
- *Once Ranged In, roll one die To Hit for each target under the template, as normal for an Artillery Bombardment.*
- *The score needed To Hit is always 5+ with no modifiers for number of guns firing and no re-rolls.*
- *Armoured vehicles make an Armour Save based on their Top armour.*
- *All other teams are Destroyed on successful hits.*

UNTERSTURMFÜHRER

UNTERSTURMFÜHRER	UNTERSCHARFÜHRER
Command Sturmtiger	Sturmtiger
HQ SECTION	GUN SECTION

HEAVY ASSAULT HOWITZER PLATOON

A Heavy Assault Howitzer platoon is manned by factory engineers, not SS. They are Allies to the SS and follow the Allies rules in the rulebook.
They are rated as
Reluctant Conscript.

BUNKERFEUER

If a Sturmtiger shoots at a Bunker or Building, it Ranges In on a roll of 5+. If it successfully Ranges In on a Bunker, the Bunker is automatically Destroyed. If it successfully Ranges In on a Building, every team in the Building is automatically Destroyed.

GERMAN SPECIAL RULES

These special rules reflect the doctrine and training that give German soldiers their edge in battle. These rules only apply to German teams and platoons.

DOCTRINE

STORMTROOPERS

German soldiers have inherited the Stormtrooper ethos from their fathers in the First World War. While other armies wait for orders, the Stormtroopers race ahead to take their objective.

Any German platoon with a Platoon Command team may attempt a Stormtroopers move in its Assault Step.

Roll a Skill Test for each platoon.

- *If the test is passed, the platoon treats the Assault Step as a Movement Step in which it can move up to another 4"/10cm, regardless of its normal movement distance.*
- *If the test is failed, the platoon cannot move any further this turn.*

Either way, a platoon that attempts to make a Stormtroopers move cannot take any part in an assault in the same turn.

Although it is not the Movement Step, teams making a Stormtroopers move may Mount and Dismount as if it was the Movement Step.

Platoons cannot make Stormtroopers moves if they moved At the Double. Bogged Down or Bailed Out vehicles cannot make Stormtroopers moves.

Only Armoured vehicles can make Stormtroopers moves if they are Pinned Down. Other types of team cannot make Stormtroopers moves if they are Pinned Down.

A team that used Eyes and Ears to reveal Gone to Ground enemy troops cannot make a Stormtroopers move.

Immobile Gun teams and Gun teams that shot earlier in the turn cannot make a Stormtrooper move. Teams of any type that have fired an Artillery Bombardment cannot make a Stormtroopers move. Trains cannot make a Stormtroopers move.

KAMPFGRUPPE

The *Kampfgruppe*, or battle group, is an important part of German military operations. Leaders at every level form mission-specific task forces from any troops available.

At the start of the game before any platoons are Deployed and before any Combat Attachments are made, any company with a 2iC Command team may take up to half of the teams (counting the Platoon Command team) from any Combat or Weapons platoons in the company and place them in a special Kampfgruppe Platoon. However, you must leave at least two Tank teams or three other teams in each of the Combat or Weapons platoons you take teams from. Teams placed in the Kampfgruppe Platoon are no longer part of their original platoons.

If a platoon attaching troops to the Kampfgruppe has Transport teams, you must take the Transport team that normally carries a team into the Kampfgruppe with it. If that Transport team normally carries other teams as well, they must join the Kampfgruppe as well.

The Kampfgruppe Platoon may include any HQ Support Weapons (including Infantry teams). Any remaining HQ Support Weapons must be attached out as normal.

A Kampfgruppe made from the teams of a Fortified Platoon is not a Fortified Platoon and may not contain any of its component platoon's Fortifications.

The 2iC Command team is the command team for this platoon, ceasing to be an Independent team, but still remains a 2iC Command team and a Warrior. If the 2iC Command team is Destroyed, they may use the Appointing New Commanders rule (see the main rulebook) to appoint a new Platoon Command team immediately before being Destroyed.

The Kampfgruppe Platoon has the lowest Motivation rating and the lowest Skill rating of any of the teams in it.

MISSION TACTICS

Before battle, every soldier in the company is briefed on their mission and how it relates to the overall battle plan. Far from compromising security, this trust allows any soldier to take over when their superior is killed.

If a Platoon Command Infantry team is Destroyed, another team takes over immediately. Remove any other Infantry team in the platoon that is within Command Distance of the Command team and replace it with the original Platoon Command team.

If a Platoon Command Tank team is Destroyed, another team takes over immediately. Nominate any other Tank team in the platoon that is within Command Distance of the Destroyed Command team to be the new Platoon Command team.

If there is no team of an appropriate type within Command Distance, then the Platoon Command team is Destroyed and the platoon is left leaderless.

The original Platoon Command team can still use this rule while a Warrior team is leading the platoon and acting as its Command team, but the rule does not apply to the Warrior team.

TANKS

SCHÜRZEN

Schürzen are the thin armoured plates that are welded to the sides of some German tanks to protect them from infantry anti-tank weapons, like anti-tank rifles and bazookas.

> *When a tank that is protected by Schürzen is hit by a weapon with a Firepower of 5+ or 6 on the Side armour by shooting and fails its Armour Save, roll a special 4+ Schürzen save:*
>
> - *If the save is successful the Schürzen protects the tank from the side shot.*
> - *If the save is not successful the shot penetrated the side armour as normal.*

BEGLEIT ESCORTS

While aboard, the *Begleit* troops man the assault gun's hull machine gun, freeing the gun's crew to concentrate on providing cannon support.

> *Assault guns with Tank Escorts use all the normal Tank Escort rules. In addition, an assault gun with Tank Escorts may fire its Hull MG at the same time as its hull-mounted main gun. This does not prevent the Tank Escorts from shooting as well.*

TIGER ACES

Tiger tank crews were highly skilled and led by some of the best tank commanders ever, with the likes of Michael Wittmann and Otto Carius among them. To reflect the abilities and experience of these exceptional soldiers, Flames Of War gives each platoon of Tiger tanks its own Tiger Ace Skill. This skill gives them the ability to perform the incredible feats of the real tankers that manned these formidable machines.

> *At the start of the game before deployment, roll a die for each platoon (including any Kampfgruppe) made entirely of Tiger I E and Königstiger tanks and each Independent Warrior team mounted in a Tiger I E and Königstiger tank, then look up their Tiger Ace Skill on the table below.*
>
> *If the Tiger Ace Skill is specifically for the Platoon Command team or an Independent Warrior team, it only applies to that team unless it is Schnell or For the Fatherland, in which case the whole platoon receives the benefit. If you roll a 6, your platoon is fortunate to be led by an exceptionally talented officer with his own Tiger Ace Skill in addition to that of the platoon as a whole!*

SCHNELL!

This platoon believes in the importance of speed and most of their kills have been racked up through being at the right place at the right time!

The platoon may make Stormtroopers moves on a roll of 2+.

CLEVER HANS!

The drivers of this platoon are masters of their vehicles. No matter how difficult the terrain or how firmly stuck they appear to be, they'll have their tank through the obstacle in no time.

Tanks in this platoon re-roll failed Bogging Checks to cross Rough Terrain and Skill Tests to free a Bogged Down tank. If they have Wide Tracks, they roll to free themselves after re-rolling the Bogging Check.

FOR THE FATHERLAND!

The soldiers of this platoon love their country. They know how important their part in Germany's final victory is and will fight to their utmost to secure it.

The platoon passes Motivation Tests on a roll of 2+.

EVERY SHOT COUNTS!

The outstanding teamwork and excellent gunners of this platoon ensure that almost every shot hits its target.

Tiger tanks in this platoon re-roll any failed roll to hit when they shoot.

RAPID FIRE!

Superb teamwork between the commander, the gunner, and the loader means that this platoon maintains a high rate of fire in combat.

The 8.8cm tank guns of the Tiger tanks in this platoon have ROF 3.

TOP ACE!

The commander of this platoon is a top gun, a real Kanone!

Roll again to determine the platoon's Tiger Ace Skill. If you roll 6 again, you may choose the platoon's skill.

In addition, roll a third time to generate an extra skill for the Platoon Command tank. This gives the platoon commander two Tiger Ace Skills (the one for the platoon as a whole, and their own one specific to them). If you roll a 6 or the same number as the platoon's Tiger Ace Skill, you may choose the platoon commander's extra skill.

ARMOURED INFANTRY

MOUNTED ASSAULT

German armoured infantry units are specifically trained to fight from their half-tracks. Whereas most armoured infantry use their half-tracks as battle taxis to get them close to the enemy then dismount, German Panzergrenadier platoons assault light opposition still mounted in their half-tracks.

Armoured Half-tracked Transport teams in a platoon with the Mounted Assault special rule are Mounted Assault Transports.

A Mounted Assault Transport has a dual nature. While empty, it is just an ordinary Transport team, except that they do not have to be Sent to the Rear when empty—they can remain on the table and continue to fight.

When carrying Passengers, a Mounted Assault Transport is treated as a Tank team for everything except Platoon Morale Checks. If it is carrying two or more Passenger teams, it also has Tank Escorts.

As a Tank team, a Mounted Assault Transport team carrying Passengers may Charge into Contact and fight in assault combat. Since they are Open-topped Tank teams, the platoon is still forced to Fall Back by five hits in Defensive Fire.

If a half-track is Destroyed by Defensive Fire, place any surviving Passengers 2"/5cm away from the teams they were charging. If a half-track is Destroyed during the assault, any Passengers are Destroyed with it.

The Passengers do not need to Dismount to Counterattack. They can stay Mounted or Dismount as they wish. If they Dismount to Counterattack, their Transport team will be Sent to the Rear.

If the platoon voluntarily Breaks Off instead of Counterattacking while still mounted in their half-tracks, they may use the Break Off Through the Enemy rule as if they were Tank teams.

TANK ESCORTS

Escort troops are assigned to protect their assault gun at all costs. They spot threats to the vehicle and see them off before they can cause any harm.

TANK ESCORTS ARE NOT TEAMS

The infantry assigned to escort a tank platoon make that their highest priority. If the escorts on one tank are killed, soldiers from other tanks will take their place, spreading themselves thin, but making sure that every tank is protected.

Tank Escorts are not teams in their own right. They are an addition to a Tank team that gives it extra capabilities. Tank Escorts cannot be Destroyed while their tank survives and cannot survive the loss of their tank.

TANK ESCORTS SHOOTING

Infantry trained as tank escorts can shoot from their perches on the tank, provided it moves slowly. If it speeds up, they are far too busy clinging on to shoot.

A Tank team with Tank Escorts can shoot with its Tank Escorts if it did not move more than 6"/15cm in the Movement Step.

Tank Escorts shoot as if they are separate teams from their Tank teams and may shoot at a different platoon from that which their Tank team is engaging. They shoot as the appropriate type of Infantry team (so German Begleit Assault Rifle Tank Escorts shoot as an Assault Rifle team), but always have a ROF of 1 with no penalty for the tank's movement.

If assaulting Infantry teams Sneak Up on a Tank team that has Tank Escorts, the Tank Escorts can Defensive Fire, despite the Tank team being unable to Defensive Fire with any of its own weapons.

TANK ESCORTS ASSAULTING

Escort infantry don't just protect their tank, they actively hunt out the enemy at close quarters too.

Tank teams with Tank Escorts roll To Hit with their escorts as well as the Tank team itself giving them two rolls to Hit in assaults in most circumstances. Even when the Tank team is Bogged Down or Bailed Out, the Tank Escorts can still Roll to Hit on their own.

A hit from a Tank Escort cannot be allocated to a vehicle. If only vehicles are available to allocate the hit to, then ignore the hit.

TANK ESCORTS IN ROUGH TERRAIN

One of the roles of tank escort infantry is to clear enemy tank-hunters out of buildings and other places inaccessible to the tanks themselves.

Assaulting Tank teams with Tank Escorts can elect not to Charge into Contact or to halt its Charge into Contact before entering, crossing, or moving in Rough Terrain. If they do this, they do not need to take a Bogging Check for assaulting a team in Rough Terrain, but they can only Roll to Hit with their Tank Escorts, not with the Tank teams they are on.

TANK ESCORTS FALL BACK

While tank escorts are supposed to escort their tanks in assaults, they are only flesh and blood. Enough firepower from the defenders will cause them to hang back until the tanks have softened the enemy up.

If the total number of hits from Defensive Fire is five or more, any Tank Escorts on Assaulting Tank teams may not roll To Hit until the platoon Counterattacks.

WEAPONS

ARMOURED ROCKET LAUNCHER

The *15cm NW41 Nebelwerfer* rocket launcher (known as 'Moaning Minnie' or 'Screaming Mimi' to the Allies) was an effective weapon that could deliver the firepower of an entire artillery battery in a single salvo. The problem was that in doing so it gave away its position by the trails of smoke left by the rockets when they fired.

The *15cm Panzerwerfer 42* was invented to solve this problem. It was an armoured half-track carrying a Nebelwerfer rocket launcher. As soon as the rockets fired, it moved to a new location, safe from enemy counterbattery fire.

> An Artillery platoon equipped with Armoured Rocket Launcher teams that fired may take a Skill Test in the Assault Step instead of making a Stormtroopers move or taking part in an Assault:
>
> - If they pass the Skill Test, remove the Smoke Trail markers from all Rocket Launcher teams.
> - Otherwise, they are too slow in getting away leaving them vulnerable to counterbattery artillery fire.

RECOILLESS GUNS

The *Fallschirmjäger* needed artillery light enough to be capable of being dropped by parachute, yet heavy enough to destroy entrenched infantry and marauding tanks. They found the solution in recoilless guns.

These weapons vent propellant gasses from the rear of the gun to counteract the recoil of the shell being fired. Although this creates a huge cloud of dust and flying debris behind the gun, it does allow it to do away with the heavy recoil-absorbing carriage of most artillery.

> Recoilless guns do not count as Concealed if they fired in their last Shooting Step as the dust cloud gives away their position.
>
> Because of the danger from flying debris, you cannot fire a recoilless gun when any part of a friendly team is directly behind the gun and within 2"/5cm. Recoilless guns cannot fire from within buildings.

BREAKTHROUGH GUN

Some weapons are just so powerful that there is no chance of surviving a hit from them. These heavy guns are often mounted in tanks and self-propelled guns designed to break through enemy defensive lines.

> Infantry teams, Gun teams and Unarmoured vehicles automatically fail their saves when hit by a Breakthrough Gun or Bunker Buster. This does not apply to Artillery Bombardments.

STUKA ZU FUSS

Sd Kfz 251 half-tracks in the third platoon of the *Panzerpionierkompanie* are fitted with *Wurfrahmen*, frames for six huge 28cm rockets. These were so devastating that they were called *Stuka zu Fuss*—Stukas on foot.

> Each Stuka zu Fuss rocket launcher can only fire one bombardment in a game, but counts as six weapons firing.
>
> You may take up to six attempts to Range In with a Stuka zu Fuss rocket launcher. Each failed attempt reduces the number of weapons firing by one. If you fail all six attempts to Range In, the rockets have all been fired, but missed, and no bombardment is possible.
>
> If multiple Stuka zu Fuss rocket launchers fire, place a separate template and roll to Range In separately with each as if they were separate artillery batteries.
>
> The half-track is still available to transport the pioneer team after the Stuka zu Fuss fires.

MOTORCYCLE RECONNAISSANCE

Motorcycle and jeep-mounted infantry are fast-moving troops always in the forefront of the advance. Their speed is their greatest asset as they race forward, dismounting to occupy forward positions or to launch assaults on the enemy.

MOTORCYCLE RECONNAISSANCE

Motorcyclists quickly learn to use their speed and small size to survive.

While still on their vehicles, Motorcycle Reconnaissance teams are Recce teams. Once they Dismount, they cease to be Recce teams. If the platoon was Reorganising, it continues to Reorganise after Dismounting.

DISMOUNTING MOTORCYCLES

Motorcycle scouts fight mounted against light opposition, but when things get tough, they dismount and finish the fight on foot, while the drivers take their vehicles to the rear.

A platoon may Dismount all of its Motorcycle Reconnaissance teams and Send their vehicles to the Rear at the start of your Movement Step. If they do so, replace each Motorcycle Reconnaissance team with the equivalent dismounted Infantry team placed anywhere under the area covered by the Motorcycle Reconnaissance team at the start of their movement, e.g. when a Motorcycle Reconnaissance Motorcycle MG team Dismounts it is replaced with a normal MG team. Even if they do not move further, the Dismounted teams are considered to have moved.

When a Bogged Down or Bailed Out Motorcycle Reconnaissance team Dismounts, it immediately ceases being Bogged Down or Bailed Out.

A Motorcycle Reconnaissance platoon may elect either to only Dismount its Man-packed Gun teams, or to Dismount all of its teams. It may not Dismount some of its Infantry teams and leave others mounted.

Motorcycle Reconnaissance teams may only Dismount if you have appropriate Infantry and Man-packed Gun teams for the dismounted soldiers to replace them with, otherwise, they must stay on their vehicles.

STARTING DISMOUNTED

Sometimes motorcyclists are called on to defend their gains, requiring them to start the fight dismounted.

You may start the game with all of the Motorcycle Reconnaissance teams in a platoon Dismounted by replacing each team with the equivalent Infantry team. They may Remount later in the game as if they had sent their vehicles to the rear.

REMOUNTING MOTORCYCLES

Once the enemy have been overcome, you may need to bring your motorcycles forward again to continue the advance.

A platoon that has Dismounted its Motorcycle Reconnaissance teams may bring them forward again at the start of your Movement Step provided that every team in the platoon is not:

- *within 16"/40cm of any enemy team within Line of Sight, unless Concealed by Terrain from it, or*
- *within 4"/10cm of any enemy team, or*
- *within 8"/20cm of any enemy Recce team that is in Line of Sight (apart from Recce teams that are Bogged Down, Bailed Out, or moved at the Double).*

If they do so, replace each Dismounted Motorcycle Reconnaissance team with the equivalent vehicle placed so that it covers the team it is replacing at the start of their movement.

Once Remounted in this way, the Motorcycle Reconnaissance teams can move as normal. If there is insufficient space for all of the Motorcycle Reconnaissance teams' vehicles, take the normal movement each team as it is replaced, using your movement to make enough space.

The Motorcycle Reconnaissance teams are treated as moving, even if they do not move after Remounting, so if a Motorcycle Reconnaissance team is placed in Rough Terrain, it must take a Bogging Check whether it moves or not.

Motorcycle Reconnaissance teams that Remounted may not shoot in the Shooting Step nor assault in the in the Assault Step. They may however move in either step if they have a special rule such as Huszár or Stormtroopers that allows them to do so.

187

SS-SPECIAL RULES

Waffen-SS troops use all of the German special rules on pages 183 to 187 as well as the following division-specific special rules.

TOTENKOPF

VICTORY AT ANY COST

The training, experience, and close comradeship between the *Totenkopf SS-Panzergrenadiere* was legendary. Their long experience of fighting the Red Army on the Eastern Front taught them that defeat was not an option. They expected to win every battle and would sacrifice everything to win.

During an assault, any Combat platoon from the 3. SS-Panzerdivision Totenkopf SS-Panzerkampfgruppe or SS-Panzergrenadierkampfgruppe may Destroy one Infantry or Tank team from that platoon that is currently participating in that assault and either reroll a Motivation Test to Counterattack or reroll a Roll to Hit.

The Destroyed team must be In Command when removed and may not be an Independent team. You may do this multiple times in an assault as long as there are teams available to Destroy.

PANZER ACE

The *3. Totenkopf SS-Panzerdivision* had been fighting on the Eastern Front for over three years. Though their casualties have been extraordinarily high, their remaining veterans and leaders were some of the premier *Panzertruppen* of the *Wehrmacht*.

If your SS-Panzerkampfgruppe is from 3. SS-Panzerdivision Totenkopf, roll a die before the game begins for the Company Command Tank team. Look up the Tiger Ace Skills table on page 184 to find the acquired skill.

If the roll is a 6, simply choose the skill for the commander rather than rolling further. Where the skill refers to Tiger tanks or 8.8cm tank guns, treat it as referring to the commander's own tank and its main gun.

WIKING

DANISH AND FLEMISH REGIMENTS

5. SS-Panzerdivision Wiking contained Danish and Flemish regiments. Each possessed individual national characteristics.

A SS-Panzerkampfgruppe or SS-Panzergrenadierkampfgruppe from the 5. SS-Panzerdivision Wiking must be made up of either Danish (*) or Flemish troops (* *).*

MASTER PLAN

The Danish *Panzergrenadier* Regiment planned their operations well and in close co-operation with the Danish Artillery Regiment. Their meticulous planning ensured accurate artillery fire for the initial attacks.

When firing artillery bombardments using Motorised, Motorised Heavy, Armoured or Armoured Heavy SS-Artillery Batteries in support of the Danish Panzergrenadier Regiment, Roll to Hit as if the artillery are Veterans.

FOLLOWING ORDERS

The Flemish soldiers of the Belgian Regiment were adamant in following their assigned task to the letter. They never varied from their attack plans.

Platoons from the Flemish Regiment may make a Stormtroopers Move while Pinned Down and may move towards the enemy while making this move.

POLITICAL ASSASSINS

Wiking SS-Division officers had a nasty habit of identifying and eliminating the Communist Komissars that they encountered on the Eastern Front.

When Infantry teams from a Gepanzerte SS-Panzergrenadier Platoon or a SS-Panzergrenadier Platoon in a SS-Panzerkampfgruppe or SS-Panzergrenadierkampfgruppe from the 5. SS-Panzerdivision Wiking shoot at a platoon that contains a Soviet Komissar team, you may make a second roll for each hit scored on the platoon.

- *If you roll a 5 or 6, you can choose to mark the Komissar team as a priority target similar to the Gun Tank rule in the rulebook.*
- *Otherwise, the hits are allocated as normal.*

When used, this rule overrides all other rules about hit allocation except those regarding valid targets.

NEDERLAND

Instant Readiness

The training, experience, and close comradeship between the *4. SS-Freiwilligen-Panzergrenadier Brigade Nederland* was legendary. Their long experience fighting the Red Army taught them the importance of instant readiness.

> Any SS-Freiwilligen-Panzergrenadier Platoon, SS-Freiwilligen-Aufklärungs Platoon, or SS-Panzergrenadier Pioneer Platoon from a Nederland SS-Freiwilligen-Panzergrenadierkompanie, that is not Pinned Down before the Assault Step, that becomes Pinned Down during the Assault Step, can take a Motivation Test.
>
> • If the test is passed, remove the Pinned Down Marker.
>
> • Otherwise, the platoon remains Pinned Down.
>
> If the platoon is Pinned Down by Defensive Fire, they still fall back and the assault is over.

Following Orders

The Dutch soldiers of the *4. SS-Freiwilligen-Panzergrenadier Brigade Nederland* were adamant in following their assigned task to the letter. They never varied from their attack plans.

> Platoons from a Nederland SS-Freiwilligen-Panzergrenadier-kompanie may make a Stormtroopers Move while Pinned Down and may move towards the enemy while making this move.

4. SS-Freiwilligen-Panzergrenadier Brigade Nederland collar patches.

NORDLAND

Master Plan

The *11. SS-Freiwilligen-Panzergrenadierdivision Nordland* planned their operations well and in close co-operation with their Artillery Regiments. Their meticulous planning ensured accurate artillery fire for the initial attacks.

> When firing artillery bombardments using Motorised SS-Artillery Batteries or Motorised Heavy SS-Artillery Batteries in support of a Nordland SS-Freiwilligen-Panzergrenadierkompanie, Roll to Hit as if the artillery are Veterans.

To the Bitter End

The men of the *11. SS-Freiwilligen-Panzergrenadierdivision Nordland* quickly earned a reputation for not backing down in close combat. They would often fight on to the bitter end, without concern for their safety and unwilling to give ground.

> When any SS-Freiwilligen-Panzergrenadier Platoon, SS-Freiwilligen-Aufklärungs Platoon, or SS-Panzergrenadier Pioneer Platoon from a Nordland SS-Freiwilligen-Panzergrenadierkompanie fails its Motivation Test to Counterattack during an assault, it has the option to either Break Off as normal or continue the assault.
>
> If the platoon elects to continue the assault, they fight on as if they had passed the Motivation Test. However, if they fail any further Motivation Tests to Counterattack in this assault, then the platoon is immediately Destroyed as if they had failed a Platoon Morale Check.

 # FLORIAN GEYER AND MARIA THERESIA

Local Knowledge

The *8. SS-Kavalleriedivision Florian Geyer* and *22. SS-Freiwilligen-Kavalleriedivision Maria Theresia* are made up of Hungarian, Transylvanian and Serbian Germans, some of who are Budapest natives. In cooperation with the Hungarians of the Budapest defence battalions they make good use the sewer and drainage system to move about the city undetected and to launch surprise attacks.

> A SS-Kavallerieschwadron may deploy a German or Hungarian platoon entirely made up of Infantry or Gun teams, or a mix of both, in Immediate Ambush (see the rulebook) in addition to any Ambush deployment that may be allowed by the Mission played. The Immediate Ambushing platoon must be taken from the platoons that are to be deployed on table during deployment.

TANK TEAMS

Name *Weapon*	Mobility *Range*	Front *ROF*	Armour Side *Anti-tank*	Top *Firepower*	Equipment and Notes
TANKS					
Panzer M14/41	Standard Tank	3	2	1	Co-ax MG, Twin hull MG.
4.7cm KwK 47/32 gun	*24"/60cm*	*2*	*7*	*4+*	
T-70 obr 1943 (captured)	Standard Tank	4	2	1	Co-ax MG, Limited vision, Wide tracks, Unreliable.
4.5cm KwK(r) gun	*24"/60cm*	*1*	*7*	*4+*	
T-34 obr 1942 (captured)	Standard Tank	6	5	1	Co-ax MG, Hull MG, Fast tank, Wide tracks, Limited vision, Unreliable.
7.62cm KwK(r) gun	*32"/80cm*	*2*	*9*	*3+*	
T-34/85 obr 1943 (captured)	Standard Tank	7	5	1	Co-ax MG, Hull MG, Unreliable.
8.5cm KwK(r) gun	*32"/80cm*	*2*	*12*	*3+*	
M4 (M4A2 Sherman) (captured)	Standard Tank	6	4	1	Co-ax MG, Hull MG, Unreliable.
7.5cm KwK(a) gun	*32"/80cm*	*2*	*10*	*3+*	
Panzer IV H	Standard Tank	6	3	1	Co-ax MG, Hull MG, Protected ammo, Schürzen.
7.5cm KwK40 gun	*32"/80cm*	*2*	*11*	*3+*	
Panzer IV/70 (A)	Slow Tank	8	3	1	Hull MG, Overloaded, Schürzen.
7.5cm PaK42 gun	*32"/80cm*	*2*	*14*	*3+*	*Hull mounted.*
Panzer IV/70 (V)	Slow Tank	9	3	1	Hull MG, Overloaded, Schürzen.
7.5cm PaK42 gun	*32"/80cm*	*2*	*14*	*3+*	*Hull mounted.*
Panther D	Standard Tank	10	5	1	Co-ax MG, Hull MG, Wide tracks, Unreliable.
7.5cm KwK42 gun	*32"/80cm*	*2*	*14*	*3+*	
Panther A or G	Standard Tank	10	5	1	Co-ax MG, Hull MG, Wide tracks.
7.5cm KwK42 gun	*32"/80cm*	*2*	*14*	*3+*	
Tiger I E	Slow Tank	9	8	2	Co-ax MG, Hull MG, Protected ammo, Wide tracks.
8.8cm KwK36 gun	*40"/100cm*	*2*	*13*	*3+*	*Slow traverse.*
Königstiger (Henschel)	Slow Tank	15	8	2	Co-ax MG, Hull MG, Overloaded.
8.8cm KwK43 gun	*40"/100cm*	*2*	*16*	*3+*	*Slow traverse.*
TANK-HUNTERS					
PaK40 auf RSO	Slow Tank	0	0	0	
7.5cm PaK40 gun	*32"/80cm*	*2*	*12*	*3+*	*Hull mounted.*
Marder II	Standard Tank	1	0	0	AA MG.
7.5cm PaK40 gun	*32"/80cm*	*2*	*12*	*3+*	*Hull mounted.*
Marder III H	Standard Tank	1	0	0	Hull MG.
7.5cm PaK40 gun	*32"/80cm*	*2*	*12*	*3+*	*Hull mounted.*
Marder III M	Standard Tank	0	0	0	AA MG.
7.5cm PaK40 gun	*32"/80cm*	*2*	*12*	*3+*	*Hull mounted.*
Hetzer	Standard Tank	7	2	1	Hull MG, Overloaded.
7.5cm PaK39 gun	*32"/80cm*	*2*	*11*	*3+*	*Hull mounted.*
Jagdpanzer IV	Standard Tank	7	3	1	Hull MG, Protected ammo, Schürzen.
7.5cm StuK40 gun	*32"/80cm*	*2*	*11*	*3+*	*Hull mounted.*
Hornisse	Standard Tank	1	1	0	AA MG, Protected ammo.
8.8cm PaK43 gun	*40"/100cm*	*2*	*16*	*3+*	*Hull mounted.*
Elefant	Slow Tank	15	8	2	Hull MG, Overloaded, Unreliable.
8.8cm PaK43 gun	*40"/100cm*	*2*	*16*	*3+*	*Hull mounted.*

Name	Mobility	Armour Front	Side	Top	Equipment and Notes
Weapon	*Range*	*ROF*	*Anti-tank*	*Firepower*	

ASSAULT GUNS

StuG M42 75/18	Standard Tank	4	2	1	
7.5cm StuK 75/18 gun	*24"/60cm*	*2*	*9*	*3+*	*Hull mounted.*
StuG G	Standard Tank	7	3	1	Hull MG, Protected ammo, Schürzen.
7.5cm StuK40 gun	*32"/80cm*	*2*	*11*	*3+*	*Hull mounted.*
StuH42	Standard Tank	7	3	1	Hull MG, Protected ammo, Schürzen.
10.5cm StuH42 gun	*32"/80cm*	*2*	*10*	*2+*	*Hull mounted, Breakthrough gun, Smoke.*
StuG IV	Standard Tank	7	3	1	Hull MG, Protected ammo, Schürzen.
7.5cm StuK40 gun	*32"/80cm*	*2*	*11*	*3+*	*Hull mounted.*
Brummbär	Slow Tank	9	5	1	Hull MG, Overloaded, Schürzen.
15cm StuH43 gun	*16"/40cm*	*1*	*13*	*1+*	*Bunker buster, Hull mounted.*
Sturmtiger	Slow Tank	12	8	2	Hull MG, Overloaded.
Firing bombardments	*48"/120cm*	*-*	*6*	*1+*	*Rocket assault howitzer, Bunkerfeuer.*

INFANTRY GUNS (SP)

Sd Kfz 251/2 (8cm)	Half-tracked	1	0	0	AA MG.
8cm GW34 mortar	*24"/60cm*	*2*	*2*	*3+*	*Smoke, Minimum range 8"/20cm.*
Firing Bombardments	*40"/100cm*	*-*	*2*	*6*	*Hull mounted, Portee, Smoke bombardment.*
Sd Kfz 251/9 (7.5cm)	Half-tracked	1	0	0	AA MG.
7.5cm KwK37 gun	*24"/60cm*	*2*	*9*	*3+*	*Hull mounted.*
Grille H	Standard Tank	2	1	0	AA MG.
15cm sIG33 gun	*16"/40cm*	*1*	*13*	*1+*	*Bunker buster, Hull mounted.*
Firing bombardments	*56"/140cm*	*-*	*4*	*2+*	
Grille K	Standard Tank	0	0	0	AA MG.
15cm sIG33 gun	*16"/40cm*	*1*	*13*	*1+*	*Bunker buster, Hull mounted.*
Firing bombardments	*56"/140cm*	*-*	*4*	*2+*	
Sd Kfz 251/16 (Flamm)	Half-tracked	1	0	0	Hull MG.
Two 1.4cm Flammenwerfer	*4"/10cm*	*3 (each)*	*-*	*6*	*Side mounted, Flame-thrower.*
Sd Kfz 233 (7.5cm)	Jeep	2	0	0	Hull MG.
7.5cm KwK37 gun	*24"/60cm*	*2*	*9*	*3+*	*Hull mounted.*

RECONNAISSANCE

Panhard 178 (f)	Wheeled	1	1	0	Co-ax MG, Limited vision.
2.5cm KwK(f) gun	*16"/40cm*	*2*	*6*	*5+*	*No HE.*
BA-10M	Wheeled	1	0	0	Co-ax MG, Hull MG, Limited vision, Unreliable.
4.5cm KwK(r) gun	*24"/60cm*	*2*	*7*	*4+*	
Sd Kfz 221 (MG)	Wheeled	1	0	0	AA MG.
Sd Kfz 221 (2.8cm)	Wheeled	0	0	0	
2.8cm sPzB41 anti-tank rifle	*16"/40cm*	*2*	*7*	*5+*	*Hull mounted, No HE.*
Sd Kfz 222 (2cm)	Wheeled	1	0	0	Co-ax MG.
2cm KwK38 gun	*16"/40cm*	*3*	*5*	*5+*	*Self-defence anti-aircraft.*
Sd Kfz 223 (radio)	Wheeled	1	0	0	AA MG.
Sd Kfz 231 (8-rad)	Jeep	2	0	0	Co-ax MG.
2cm KwK38 gun	*16"/40cm*	*3*	*5*	*5+*	
Sd Kfz 234/2 Puma	Jeep	3	0	0	Co-ax MG
5cm KwK39 gun	*24"/60cm*	*2*	*9*	*4+*	
Sd Kfz 250 (recce)	Half-tracked	1	0	0	Hull MG, AA MG.
Sd Kfz 250/9 (2cm)	Half-tracked	1	0	0	Co-ax MG.
2cm KwK38 gun	*16"/40cm*	*3*	*5*	*5+*	*Self-defence anti-aircraft.*
Panzer II L Luchs	Light Tank	3	1	1	Co-ax MG.
2cm KwK38 gun	*16"/40cm*	*3*	*5*	*5+*	
Motorcycle MG team	Jeep	-	-	-	Motorcycle reconnaissance, Dismount as MG team.
MG	*16"/40cm*	*3*	*2*	*6*	*Hull-mounted, Vehicle MG.*
Motorcycle Panzerfaust SMG team	Jeep	-	-	-	Motorcycle reconnaissance, Dismount as Panzerfaust SMG team.
When firing as SMG	*4"/10cm*	*3*	*1*	*6*	*Hull-mounted, Vehicle MG.*
When firing as Panzerfaust	*4"/10cm*	*1*	*12*	*5+*	*Awkward layout.*

| Name | Mobility | Armour Front | Side | Top | Equipment and Notes |
Weapon	*Range*	*ROF*	*Anti-tank*	*Firepower*	
ARTILLERY (SP)					
Wespe	Standard Tank	1	1	0	AA MG, Protected ammo.
10.5cm leFH18M howitzer	*24"/60cm*	*1*	*10*	*2+*	*Hull mounted, Breakthrough gun, Smoke.*
Firing bombardments	*72"/180cm*	*-*	*4*	*4+*	*Smoke bombardment.*
Hummel	Standard Tank	1	1	0	AA MG, Protected ammo.
15cm sFH18 howitzer	*24"/60cm*	*1*	*13*	*1+*	*Bunker buster, Hull mounted, Smoke.*
Firing bombardments	*80"/200cm*	*-*	*5*	*2+*	*Smoke bombardment.*
Vielfachwerfer auf Maultier	Half-tracked	1	0	0	AA MG, Armoured rocket launcher.
8cm rocket launcher	*56"/140cm*	*-*	*2*	*6*	*Rocket launcher.*
Panzerwerfer 42 (Maultier)	Half-tracked	0	0	0	AA MG, Armoured rocket launcher.
15cm RW42 rocket launcher	*64"/160cm*	*-*	*3*	*4+*	*Rocket launcher, Smoke bombardment.*
Panzer III OP	Standard Tank	3	1	1	Co-ax MG, Protected ammo.
2cm Kwk38 gun	*16"/40cm*	*3*	*5*	*5+*	
Panzer III OP	Standard Tank	5	3	1	Hull MG.
ANTI-AIRCRAFT (SP)					
Sd Kfz 10/5 (2cm)	Half-tracked	-	-	-	
2cm FlaK38 gun	*16"/40cm*	*4*	*5*	*5+*	*Anti-aircraft.*
Armoured Sd Kfz 10/5 (2cm)	Half-tracked	0	0	0	
2cm FlaK38 gun	*16"/40cm*	*4*	*5*	*5+*	*Anti-aircraft.*
Sd Kfz 7/1 (Quad 2cm)	Half-tracked	-	-	-	
2cm FlaK38 (V) gun	*16"/40cm*	*6*	*5*	*5+*	*Anti-aircraft.*
Armoured Sd Kfz 7/1 (Quad 2cm)	Half-tracked	0	0	0	
2cm FlaK38 (V) gun	*16"/40cm*	*6*	*5*	*5+*	*Anti-aircraft.*
Sd Kfz 7/2 (3.7cm)	Half-tracked	-	-	-	
3.7cm FlaK43 gun	*24"/60cm*	*4*	*6*	*4+*	*Anti-aircraft.*
Armoured Sd Kfz 7/2 (3.7cm)	Half-tracked	0	0	0	
3.7cm FlaK43 gun	*24"/60cm*	*4*	*6*	*4+*	*Anti-aircraft.*
Flakpanzer 38(t) (2cm)	Standard Tank	0	0	0	
2cm FlaK38 gun	*16"/40cm*	*4*	*5*	*5+*	*Anti-aircraft.*
Möbelwagen (3.7cm)	Standard Tank	0	0	0	
3.7cm FlaK43 gun	*24"/60cm*	*4*	*6*	*4+*	*Anti-aircraft.*
Wirbelwind (Quad 2cm)	Standard Tank	3	1	0	Hull MG.
2cm FlaK38 (V) gun	*16"/40cm*	*6*	*5*	*5+*	*Anti-aircraft.*
ARMOURED TRAINS					
Locomotive	Train	3	3	2	Locomotive.
Artillery Car	Train	3	3	2	Artillery Car, Protected ammo, Four passenger-fired side train MG (two each side).
10.5cm leFH18/40 howitzer	*24"/60cm*	*1*	*10*	*2+*	*Deck turret, Breakthrough gun, Smoke.*
Firing bombardments	*72"/180cm*	*-*	*4*	*4+*	*Smoke bombardment.*
Anti-aircraft Car	Train	3	3	2	Artillery Car, Protected ammo, Four passenger-fired side train MG (two each side).
10.5cm leFH18/40 howitzer	*24"/60cm*	*1*	*10*	*2+*	*Deck turret, Breakthrough gun, Smoke.*
Firing bombardments	*72"/180cm*	*-*	*4*	*4+*	*Smoke bombardment.*
with 2cm FlaK38 (V) gun	*16"/40cm*	*6*	*5*	*5+*	*Turntable, Anti-aircraft.*
with 3.7cm FlaK43 gun	*24"/60cm*	*4*	*6*	*4+*	*Turntable, Anti-aircraft.*
Tank-hunter Car	Train	6	3	1	Artillery Car, Co-ax MG, Protected ammo, Schürzen.
7.5cm KwK40 gun	*32"/80cm*	*2*	*11*	*3+*	
Infantry Car	Train	3	3	2	Infantry Car, Four passenger-fired train MG (2 each side).
Staff Car	Train	3	3	2	Infantry Car, Four passenger-fired train MG (2 each side).
Light Artillery Car	Train	3	3	2	Artillery Car, Six train MG, AA MG turret.
Two 7.5cm FK02/26 (p)	*24"/60cm*	*2*	*8*	*3+*	*Deck turrets.*
Firing bombardments	*72"/180cm*	*-*	*3*	*6*	
with 2cm FlaK38 (V) gun	*16"/40cm*	*6*	*5*	*5+*	*Turntable, Anti-aircraft.*

Heavy Artillery Car	Train	3	3	2	Artillery Car, Six train MG, AA MG turret.
7.5cm FK02/26 (p) gun	24"/60cm	2	8	3+	Deck turret.
Firing bombardments	72"/180cm	-	4	5+	
10cm leFH14/19 (p) howitzer	24"/60cm	1	8	2+	Deck turret, Breakthrough gun.
Firing bombardments	72"/180cm	-	4	5+	Smoke bombardment.
Assault Car	Train	3	3	2	Infantry Car, Four train MG, Passengers.
Panzer 38(t)	Standard Tank	3	1	1	Co-ax MG, Protected Ammo.
3.7cm KwK38(t) gun	24"/60cm	2	6	4+	

VEHICLE MACHINE-GUNS

Vehicle MG	16"/40cm	3	2	6	ROF 1 if other weapons fire.
Train MG	16"40cm	2	2	6	Retain ROF 2 if other weapons fire.
MG turret	16"40cm	2	2	6	Self-defence Anti-aircraft.

GUN TEAMS

Weapon	Mobility	Range	ROF	Anti-tank	Firepower	Notes
MG42 HMG	Man-packed	24"/60cm	6	2	6	
7.5cm LG40 recoilless gun	Man-packed	16"/40cm	2	9	3+	Recoilless.
8.8cm RW43 (Püppchen) launcher	Man-packed	16"/40cm	1	11	5+	
8cm GW42 (Stummelwerfer) mortar	Man-packed	24"/60cm	2	2	3+	Smoke, Minimum range 8"/20cm.
Firing bombardments		32"/80cm	-	2	6	Smoke bombardment.
8cm GW34 mortar	Man-packed	24"/60cm	2	2	3+	Smoke, Minimum range 8"/20cm.
Firing bombardments		40"/100cm	-	2	6	Smoke bombardment.
10.5cm NbW35 mortar	Man-packed	40"/100cm	-	3	4+	Smoke bombardment.
12cm sGW43 mortar	Light	56"/140cm	-	3	3+	
7.5cm leIG18 gun	Light	16"/40cm	2	9	3+	Gun shield, Smoke.
Firing bombardments		48"/120cm	-	3	6	
15cm sIG33 gun	Heavy	16"/40cm	1	13	1+	Bunker buster, Gun shield.
Firing bombardments		56"/140cm	-	4	2+	
2cm FlaK38 gun	Light	16"/40cm	4	5	5+	Anti-aircraft, Gun shield, Turntable.
2cm Flakvierling 38 gun	Immobile	16"/40cm	6	5	5+	Anti-aircraft, Gun shield, Turntable.
3.7cm FlaK43 gun	Immobile	24"/60cm	4	6	4+	Anti-aircraft, Gun shield, Turntable.
3.7cm PaK36 gun	Light	24"/60cm	3	6	4+	Gun shield.
Firing Stielgranate		8"/20cm	1	12	5+	
5cm PaK38 gun	Medium	24"/60cm	3	9	4+	Gun shield.
7.5cm PaK97/38 gun	Medium	24"/60cm	2	10	3+	Gun shield.
7.5cm PaK40 gun	Medium	32"/80cm	2	12	3+	Gun shield.
7.62cm PaK36(r) gun	Heavy	32"/80cm	2	11	3+	Gun shield.
8.8cm FlaK36 gun	Immobile	40"/100cm	2	13	3+	Gun shield, Heavy anti-aircraft, Turntable.
8.8cm PaK43/41 gun	Immobile	40"/100cm	2	16	3+	Gun shield.
8.8cm PaK43 gun	Immobile	40"/100cm	2	16	3+	Gun shield, Turntable.
7.5cm GebG36 gun	Heavy	16"/40cm	2	9	3+	Gun shield, Smoke.
Firing bombardments		72"/180cm	-	3	6	Smoke bombardment.
7.5cm FK18 gun	Heavy	24"/60cm	2	9	3+	Gun shield, Smoke.
Firing bombardments		64"/160cm	-	3	6	Smoke bombardment.
10.5cm LG40 recoilless gun	Light	16"/40cm	1	10	2+	Gun shield, Recoilless, Breakthrough gun, Smoke.
Firing bombardments		64"/160cm	-	4	4+	
10.5cm leFH18/40 howitzer	Heavy	24"/60cm	1	10	2+	Gun shield, Breakthrough gun, Smoke.
Firing bombardments		72"/180cm	-	4	4+	Smoke bombardment.
10.5cm leFH18 howitzer	Immobile	24"/60cm	1	10	2+	Gun shield, Breakthrough gun, Smoke.
Firing bombardments		72"/180cm	-	4	4+	Smoke bombardment.
s10cm K18 gun	Immobile	32"/80cm	1	15	2+	
Firing bombardments		96"/240cm	-	4	4+	
15cm sFH18 howitzer	Immobile	24"/60cm	1	13	1+	Bunker buster, Smoke.
Firing bombardments		80"/200cm	-	5	2+	Smoke bombardment.
12.8cm K81 gun	Immobile	48"/120cm	1	17	2+	Breakthrough gun.
Firing bombardments		104"/260cm	-	4	3+	
15cm NW41 rocket launcher	Light	64"/160cm	-	3	4+	Rocket launcher, Smoke bombardment.
21cm NW42 rocket launcher	Light	72"/180cm	-	3	3+	Rocket launcher.

INFANTRY TEAMS

Team	Range	ROF	Anti-tank	Firepower	Notes
Rifle team	16"/40cm	1	2	6	
Rifle/MG team	16"/40cm	2	2	6	
MG team	16"/40cm	3	2	6	
SMG team	4"/10cm	3	1	6	Full ROF when moving.
Feldgendarm team	4"/10cm	3	1	6	Full ROF when moving.
Assault Rifle team	8"/20cm	3	1	6	Full ROF when moving.
Panzerschreck team	8"/20cm	2	11	5+	Tank Assault 5.
Flame-thrower team	4"/10cm	2	-	6	Flame-thrower.
Staff team	cannot shoot				Moves as a Heavy Gun team.

Additional Training and Equipment

	Range	ROF	Anti-tank	Firepower	Notes
Panzerfaust	4"/10cm	1	12	5+	Tank Assault 6, Cannot shoot in the Shooting Step if moved in the Movement Step.

Panzerknacker teams are rated as Tank Assault 5. Pioneer teams are rated as Tank Assault 4.

FORTIFICATIONS

Bunkers and Pillboxes

Weapon	Mobility	Range	ROF	Anti-tank	Firepower	Notes
7.5cm PaK40 Pillbox	Immobile	32"/80cm	2	12	3+	
5cm PaK38 Pillbox	Immobile	24"/60cm	3	9	4+	
2cm FlaK Nest	Immobile	16"/40cm	4	5	5+	Anti-aircraft.
Quad 2cm FlaK Nest	Immobile	16"/40cm	6	5	5+	Anti-aircraft.
HMG Pillbox	Immobile	24"/60cm	6	2	6	
HMG Nest	Immobile	24"/60cm	6	2	6	

Turrets

Turret Weapon	Mobility Range	Front ROF	Armour Side Anti-tank	Top Firepower	Equipment and Notes
Panzer I Turret	Immobile	1	1	1	Twin MG.
Panzer II Turret	Immobile	3	1	1	Turret Bunker MG.
2cm KwK38 gun	16"/40cm	3	5	5+	
T-70 Turret	Immobile	4	2	1	Turret Bunker MG.
4.5cm KwK38(r) gun	24"/60cm	1	7	4+	
T-34/76 Turret	Immobile	6	5	1	Turret Bunker MG.
7.62cm KwK34(r) gun	32"/80cm	2	9	3+	
Panther Turret	Immobile	10	5	1	Turret Bunker MG.
7.5cm KwK42 gun	32"/80cm	2	14	3+	

Turret Bunker Machine-guns

Turret Bunker MG	16"/40cm	4	2	6	Cannot shoot if main gun fires.

TRANSPORT TEAMS

Vehicle *Weapon*	Mobility *Range*	Front *ROF*	Side *Anti-tank*	Armour Top *Firepower*	Equipment and Notes
TRUCKS					
BMW motorcycle & sidecar or Kübelwagen jeep	Jeep	-	-	-	Optional Passenger-fired hull MG.
Schwimmwagen	Jeep	-	-	-	Amphibious, Passenger-fired hull MG.
Kettenkrad half-track or Horch Kfz 15 car	Jeep	-	-	-	
Horch, Krupp, or Steyr Kfz 70 truck	Wheeled	-	-	-	
Opel Blitz 3-ton truck	Wheeled	-	-	-	
Opel Maultier	Half-tracked	-	-	-	
Opel Kfz 68 radio truck	Wheeled	-	-	-	
RSO	Slow Tank	-	-	-	
Horse-drawn wagon	Horse-drawn	-	-	-	
TRACTORS					
Sd Kfz 10 (1t), Sd Kfz 11 (3t), or Sd Kfz 7 (8t) half-track	Half-tracked	-	-	-	
Horse-drawn limber	Horse-drawn	-	-	-	
ARMOURED PERSONNEL CARRIERS					
Sd Kfz 250 half-track	Half-tracked	1	0	0	Hull MG, Passenger-fired AA MG.
Sd Kfz 250/10 (3.7cm) *3.7cm PaK36*	Half-tracked *16"/40cm*	1 2	0 6	0 4+	Passenger-fired AA MG. *Hull mounted.*
Sd Kfz 250/11 (2.8cm) half-track *2.8cm sPzB41*	Half-tracked *16"/40cm*	1 2	0 7	0 5+	Passenger-fired AA MG. *Hull mounted, No HE.*
Sd Kfz 251/1 half-track	Half-tracked	1	0	0	Hull MG, Passenger-fired AA MG.
Sd Kfz 251/1 (HMG) half-track	Half-tracked	1	0	0	Hull MG, HMG Carrier, Passenger-fired AA MG.
Sd Kfz 251/7 (Pioneer) half-track	Half-tracked	1	0	0	Hull MG, Passenger-fired AA MG, Assault bridge.
Sd Kfz 251/10 (3.7cm) half-track *3.7cm PaK36*	Half-tracked *16"/40cm*	1 2	0 6	0 4+	Passenger-fired AA MG. *Hull mounted.*
Sd Kfz 251/11 (2.8cm) half-track *2.8cm sPzB41*	Half-tracked *16"/40cm*	1 2	0 7	0 5+	Passenger-fired AA MG. *Hull mounted, No HE.*
Sd Kfz 251/1 (Stuka) half-track *28cm sW40 Rocket Launcher*	Half-tracked *40"/100cm*	1 -	0 3	0 1+	Hull MG, Passenger-fired AA MG. *Hull mounted, Stuka zu Fuss.*
RECOVERY VEHICLES					
Sd Kfz 9 (18t) half-track	Half-tracked	-	-	-	Recovery vehicle.
Bergepanzer III recovery vehicle	Standard Tank	5	3	0	AA MG, Recovery vehicle.
Bergepanther recovery vehicle	Standard Tank	10	5	0	AA MG, Wide tracks, Recovery vehicle.

AIRCRAFT

Aircraft	Weapon	To Hit	Anti-tank	Firepower	Notes
Ju 87D Stuka	Bombs	4+	5	1+	
Ju 87G Stuka	Cannon	3+	11	4+	
Hs 129B	Cannon	2+	9	4+	Flying Tank.
Hs 129B3	Cannon	4+	15	3+	Flying Tank, No HE.
	MG	3+	6	5+	

Hungarian Turán II tanks and riflemen clash with Romanians on their border.

Soviet T-34/85 tanks supporting Romanians bust through Hungarian StuG G assault guns.

Zrínyi assault howitzers, a Turán II tank, and riflemen assault a Romanian artillery battery.

The Hungarian 10th Assault Gun Battalion's Zrínyi assault howitzers destroy another T-34/85.

HUNGARIANS ON THE EASTERN FRONT

The war on the eastern front began on 22 June 1941 with the German invasion of the Soviet Union codenamed Operation Barbarossa. As German forces swept the Soviets before them, they were joined by their ally, Hungary.

In the push and shove of territorial claims during 1938 to 1941 the Hungarians did well. They regained much traditional territory through diplomacy and a little flexing of military muscle. However, the support of Germany and Italy in these endeavours came at a cost, their further commitment to German military campaigns. When Germany invaded Yugoslavia, the Hungarians joined the campaign to regain territory lost in the First World War.

OPERATION BARBAROSSA

Having regained their territory, Hungary was initially reluctant to commit to the German invasion of the Soviet Union. While anti-communist feeling was high in Hungary, the Hungarians had no axe to grind with the Soviet people or any territorial claims to push against them. It was not until 27 June, five days after the German invasion, that they joined the Germans after bombing raids on the Hungarian cities of Kassa and Munkács. There is still much debate all these years later who the raids were conducted by, but regardless of who bombed them, the Hungarian Carpathian Army Group joined the German Seventeenth Army. They pushed the Red Army out of the Carpathian Mountains in to the Ukraine all the way to the Dnestr River.

MOBILE CORPS

The slow moving infantry elements of the Army Group were soon left behind on occupation duties, while the Hungarian mobile troops were formed into a Mobile Corps. They acted as one pincer of the envelopment of Soviet forces at Uman in July, where 100,000 Soviet troops were captured, then continued to advance along the Bug River, assaulting the town of Nikolayev in August.

The Mobile Corps next took part in the battles to encircle Kiev. By the end of 1941 the Mobile Corps had covered over 1000 miles, but was exhausted. It was replaced by the newly raised Hungarian Second Army, which joined German Army Group B on the Don River near Voronezh.

The Soviets attacked the Hungarians as part of their expanding Stalingrad offensive on 16 December 1942. They assaulted the junction of the Hungarian Second Army and Italian Eighth Army to the south. Both armies were overwhelmed and forced to retreat. The Hungarian armoured division, with over 100 tanks and the only major Axis armour in the area, fought hard to hold back the red tide. A new defensive line was finally established further west near Novi Oskol, but, due to losses, the Second Army had ceased to function as a military command.

REBUILDING

In 1943 the remaining Hungarian forces in the Soviet Union were relegated to occupation and security duties, while back in Hungary more forces were mobilised and rebuilt.

The Hungarian Regent and leader Admiral Miklós Horthy was an anti-fascist, making relations between Hungary and Germany difficult. With the war turning against Germany Horthy tried to withdraw Hungary from the war in early 1944, but the Germans met this threat by overrunning Hungary with German ground forces in what amounted to a quick and bloodless invasion.

Now effectively under German command, Hungarian forces were sent east once more in March 1944 as part of German Army Groups North Ukraine and South Ukraine. After the success of the Soviet Operation Bagration in July, more Hungarian troops were committed to shoring up the gaps left by the collapse of German Army Group Centre.

FIGHTING FOR TRANSYLVANIA

By September 1944, Soviet successes led the Red Army to the very doorstep of Hungary. The frontlines were drawn along the Carpathian Mountains in northeastern Hungary and on the Hungarian-Romanian border in Transylvania. Striking before the Red Army could cross Romania, Hungarians took the opportunity to attack the newly Soviet allied Romanians on the weakly-held Transylvanian border. Their aim was to take the passes over the Transylvanian Alps before the Soviets could cross them.

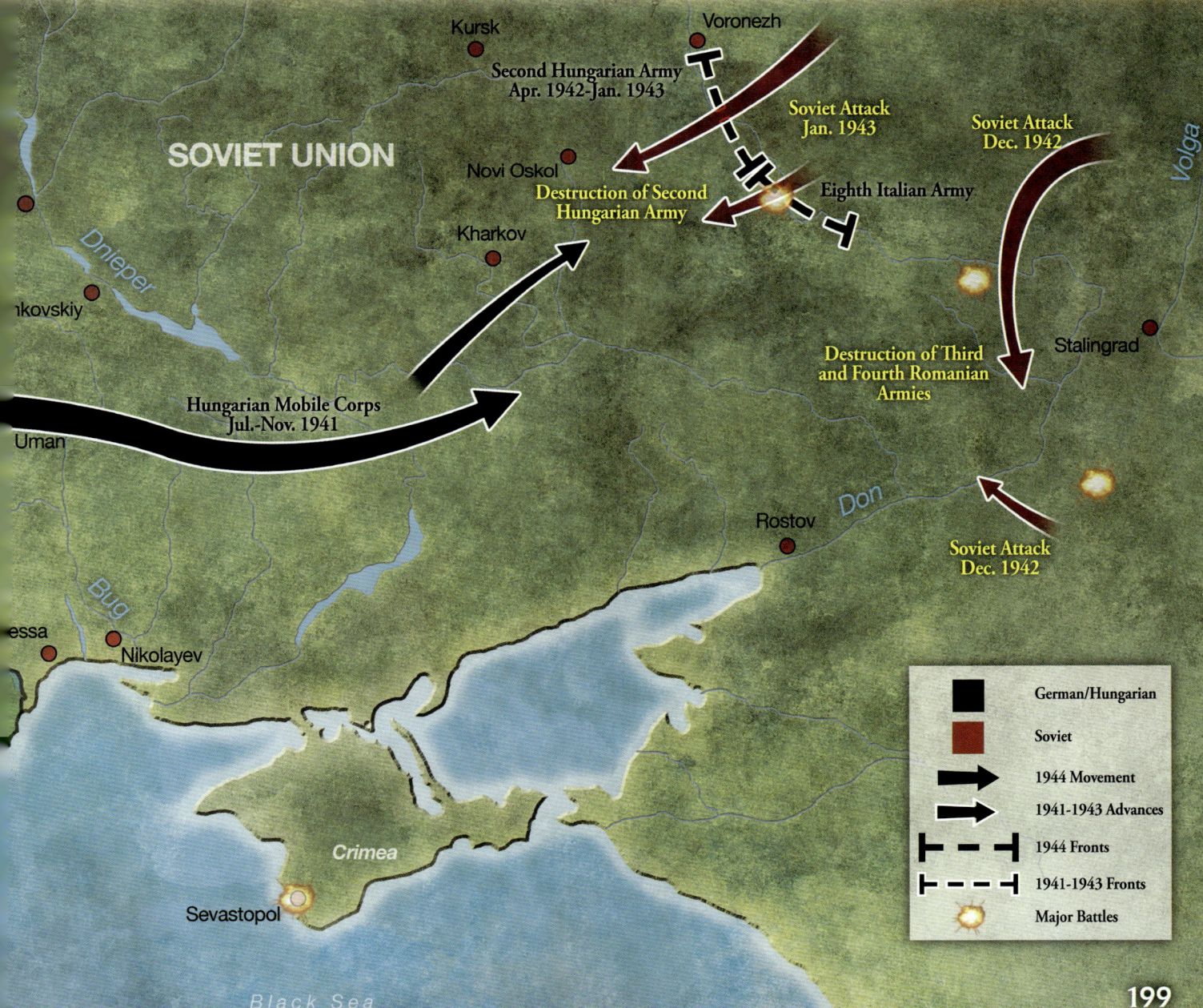

In late August 1944 the Hungarian command decided to make a push towards the Southern Carpathian Mountains to seize the passes before the Red Army could move through them. They faced the Romanian First Army, which was made up of a mix of units still rebuilding after the campaign in the Crimea and training units. The Romanians soon added the reforming Fourth Army to the troops gathering in Transylvania. The campaign was to be entirely conducted by Hungarian troops with the Germans still withdrawing from Romania or committed to fighting further north.

The Hungarian Second Army, under General Lajos Veress, was mobilised for the task with the IX and II Corps. While the IX Corps was to hold Northern Transylvania, it was up to the II Corps to push south. Initially the II Corps had just the 7th and 9th Field Replacement Divisions, but more experienced troops were transferred from the north to take part and the 2nd Armoured Division and 25th Infantry Division arrived in early September. The attack was finally launched on 5 September 1944. By this stage the Hungarian command had given up on their initial plan to push to the Southern Carpathian passes and instead simply wanted to establish a more defensible line along the Maros River.

Meanwhile Soviet units of the Second Ukrainian Front moved through Vulcan Pass in the Southern Carpathians and captured Brassó (Brașov) and Nagyszeben (Sibiu). The Soviets then intended to capture Kolozsvár (Cluj-Napoca), and expected little resistance. The Romanians had also been building up their force in the area and had assembled some 11 divisions, though not all of them were up to full strength.

THE BATTLE FOR TORDA

By the evening of 5 September the 9th Field Replacement Division had taken Torda. On the 7th Field Replacement Division's line there was little or no resistance and they took Borrév without a fight.

On 6 September the German reconnaissance troops observed Soviet tanks in the area of Nagyszeben (Sibiu). However, General Veress did not halt his troops, he considered a defeat of the Romanians as good for the soldiers' morale. Hungarian troops had advanced as far as the Maros River by 7 September. The appearance of Soviets and the Romanian Armoured Division halted Hungarian attacks on 9 September and the Hungarian troops on the Aranyos - Maros line withdrew back to where defence would be made easier by the terrain,

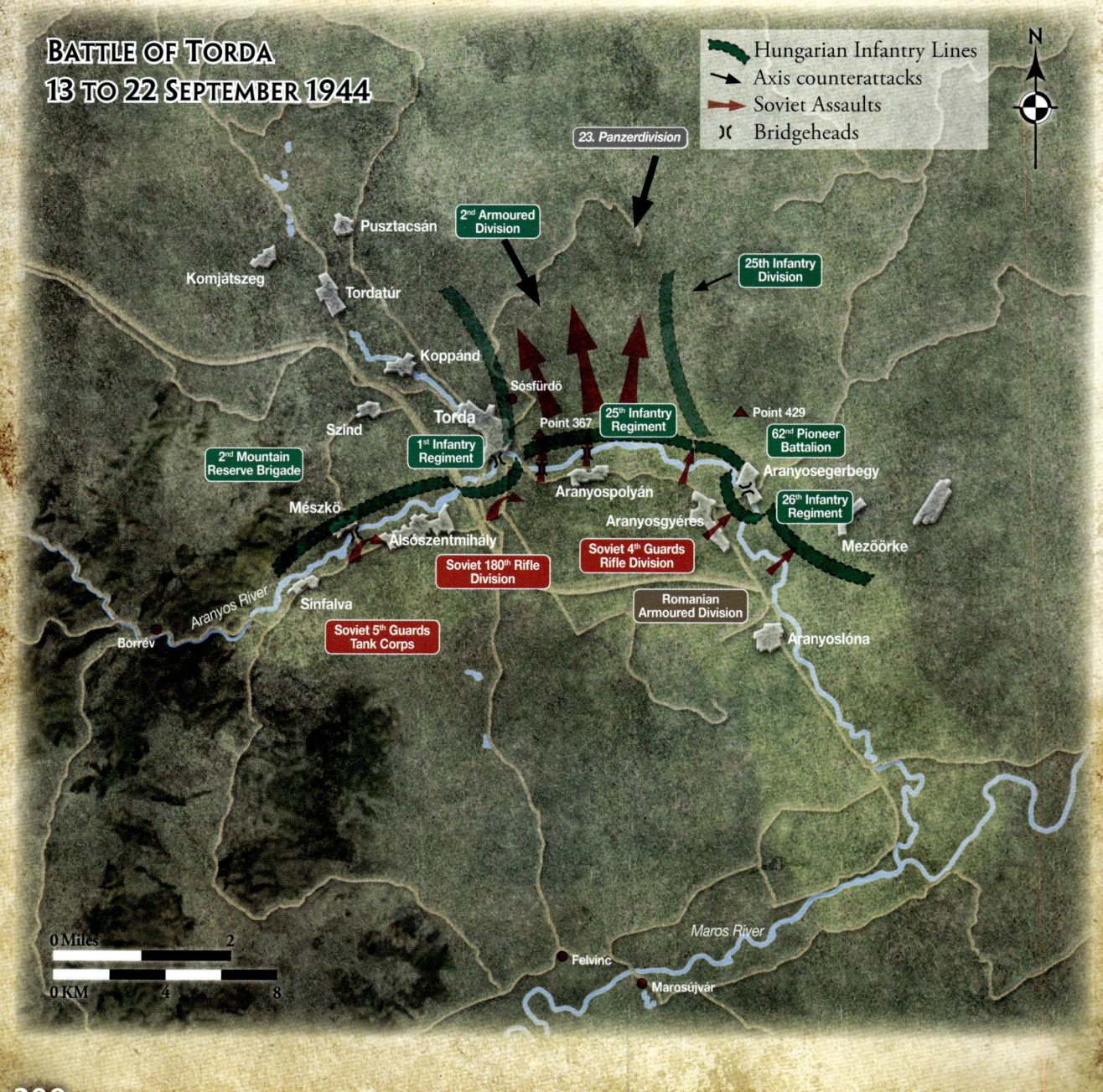

especially at Torda and Aranyosegerbegy (Viişoara). This line, which was already held by three battalions of the veteran 25th Infantry Division, was particularly good for defence with the Maros River running in front of it. It was also protected by the north bank of the river valley, which was flanked by steep hill slopes 60 to 80 metres high.

The Hungarian attack had taken the Soviet command by surprise and they decided to reinforce the area with more troops and attacked towards Torda in cooperation with the Romanian Fourth Army. The Hungarians were able to hold this line for the next few days. The 25th Infantry Division had started taking up positions on 12 September as more battalions arrived, first taking up defensive positions at the river crossing points before Torda and Aranyosegerbegy.

On 13 September the Soviet 5th Guards Tank Corps appeared behind the retreating Hungarian troops west of Torda, causing some panic. An ad hoc defensive battle developed in the evening and into the next day.

In the afternoon the Romanian 20th Infantry Training Division arrived in the area of Aranyospolyán (Poiana) lying east of the city and began to attack in the direction of Sósfürdő (the salt baths). The area was only held by I Battalion/1st Infantry Regiment. After heavy fighting the enemy pushed into the east of Torda. The situation was finally restored with a counterattack by I Battalion/25th Infantry Regiment.

The Soviet 6th Guards Tank Army was not only attacking the left of the city on 13 and 14 September, but was also attempting to out flank to the west of the city. A weak battalion of the 25th Infantry Division was able to halt the Soviet attack launched towards Tordatúr (Tureni). They deployed at the most critical points in small German style blocking groups and the encirclement of the city was stopped.

The motorised 75mm anti-tank guns (PaK40) of the 25th Assault Artillery Battalion's 1st Battery also played an important role in halting the Soviet armour. The previous two days, while under heavy artillery fire, the battery knocked out three tanks, a rocket launcher, and a few other weapons, and seven more Soviets tanks were immobilised.

On 15 September the Soviets launched the first big attack on Torda. This was again in the direction of Sósfürdő, but only after a massive artillery preparation. An attack was launched against the 26th Infantry Regiment's bridgehead at Aranyosegerbegy. The attack proved too powerful for the 26th Infantry Regiment to hold the eastern edge of Torda. However, they managed to slow the attack before withdrawing.

With the situation becoming critical the 2nd Armoured Division counterattacked on 15 and 17 September, but both attacks were halted and Soviets and Romanians were able to hold the plateau stretching to the east of the Torda bridgehead.

Another Hungarian attack began on 19 September after a strong and coordinated artillery preparation. The 25th Infantry Regiment and supporting 10th Assault Artillery Battalion captured the northern river embankment, but the 2nd Armoured Division's 6th Motorised Infantry Battalion was blocked before Point 367. Together they could not eliminate the Soviet bridgehead. Tenacious defence by Soviet Guardsmen halted the Hungarian thrust.

On 22 September, after heavy artillery preparation, the Soviets and Romanians began an attack across the front. The attack first struck along the riverbank northwards, hitting the 2nd Armoured Division, before the full force of the attack hit the Sósfürdő and Szalonnás area against the 25th Infantry Division. The thrust towards Sósfürdő encountered the 10th Assault Artillery Battalion's 1st Battery with six Zrínyi II assault howitzers under the command of Ensign János Bozsoki. Bozsoki's battery knocked out 18 T-34 tanks during the day's fighting preventing the encirclement of Torda.

The Soviets had also attacked the Torda bridgehead from the west. The front was virtually split in two, the gap was only closed by the 3rd Motorized Infantry Regiment's counterattack with supporting armour. In the northeastern sector of the city the defence firmly held, but could not hold the key high ground at Point 429. However, it became clear to the defenders that is was impossible to hold the city on their own against the overwhelming numbers the Soviets and Romanian could throw at them.

On 23 September the German 23. Panzerdivision arrived with two panzergrenadier regiments and about 65 tanks. The German Panzertruppen attacked at dawn. This powerful counterattack was on a wide front, but narrowed when it hit Point 367 where a Soviet bridgehead remained. The situation was more severe in the west, where the Soviet 18th Guard Mechanised Brigade's tanks reached Komjátszeg and the Torda - Kolozsvár road. The situation was stabilised with the deployment of the German troops in the western sector.

The Soviets moved their focus from the west of Torda on 2 October to the southeast in front of Aranyosegerbegy. They attacked the German-Hungarian line, but four Soviet rifle divisions were halted south of Torda.

On 4 October repeated Soviet attacks from the west pushed the front line back east of the Tordá - Kolozsvár road. To the east of Tordá the Soviet-Romanian penetration was further exploited by tank forces, so that by the end of the day Tordá had been almost encircled. To prevent the full encirclement the Hungarian commander, Lieutenant-General Veress, left just a small holding group on the southern edge of Tordá and moved the bulk of his troops to the north of the city, withdrawing the defensive line. With this move Hungarian and German troops had effectively abandoned Tordá.

From 13 September to 4 October the Hungarian Second Army together with German troops had obstructed attempts by 11 Soviet and Rumanian infantry divisions, a Soviet Guards Tank Corps and a Guards Mechanised Corps to breakthrough Tordá towards Kolozsvár.

WITHDRAWAL TO HUNGARY

The withdrawal of German and Hungarian forces into Hungary prompted Hitler to send reinforcements to Hungary and the Hungarians to mobilise a third field army.

The Red Army launched a massive offensive on 6 October. In the opening days Hungarian and German forces were pushed back, despite German counterattacks. The offensive was particularly successful in the south where Romanian and Soviet forces advanced 50 miles (80 km) in just three days. On the opening day the Hungarian First Army was pushed back by the Fourth Ukrainian Front, which was then checked by the German First Panzer Army. In the south the Soviet 53rd Army and the Pliev Cavalry-Mechanised Group (Second Ukrainian Front) broke through the Hungarian Third Army.

THE BATTLE FOR DEBRECEN

On 12 October the loss of Nagyvárad forced the German and Hungarian forces back to the Tisza River. The Soviets launched attacks on 17 October to cut-off the withdraw-ing German Eighth Army and Hungarian First and Second Armies. Fighting centred around Debrecen (Hungary's third largest city) and along the Tisza River.

The Axis troops held on to Debrecen until 20 October. Savage street-fighting by both Hungarian and German infantry and assault guns held off the Soviets until the order to withdraw.

The Soviet thrust past Debrecen by Cavalry-Mechanised forces was cut off by the German *III Panzerkorps* 29 October, and the Soviet thrust was broken up.

Hungarian troops defended the main crossing of the Tisza at Polgár until 31 October. This allowed the German Eighth Army and Hungarian Second Army to withdraw across the river.

The Soviets finally took Debrecen on 20 October and they pushed the Germans and Hungarians back to the Tisza River.

THE SOVIET DRIVE TO BUDAPEST

After a brief pause the Soviets continued their drive on Budapest on 29 October in the south against the Hungarian Third Army. This hastily formed army was made up of reserve and training units and not equipped to deal with the Soviet onslaught. They quickly crumbled before the Soviet 46th Army and retreated.

The Soviet 46th Army captured Kecskemét on 31 October, and by 1 November had advanced up to 30 miles (48 km) all along its front and eventually opened the way towards Budapest. German panzer troops repeatedly attacked the Soviet 46th Army's flanks and halted their advance and they withdrew southeast on 5 November.

In the north, the Soviet 7th Guards Army had pushed across the Tisza and established bridgeheads after three days of fighting. By 4 November they had broken through the

Hungarian 20th Infantry Division, took Cegléd, and were within 10 miles of Budapest. German armoured troops eventually stopped the advance within a few miles of Budapest.

The Tisza line was abandoned in early November and Germans and Hungarians retreated west to a new line between Budapest and Miskolc.

During November the Third Ukrainian Front crossed the west bank of the Danube south of Mohács. The Third Ukrainian Front pushed towards the Lake Balaton-Lake Velence line on 8 December, but were stopped German panzers.

In the centre the Second Ukrainian Front launched a major offensive on 11 November. Resistance was stronger than expected, forcing the Soviets to fight for every inch of ground. After a week of fighting the Soviets had reached Miskolc.

The Second Ukrainian Front continued to advance on the northern flank of Budapest and reach the Börzsönyi Hills on 11 December.

BUDAPEST ENCIRCLED (SEE MAP ON PAGE 135)

On 24 December the 18th Tank Corps of the Third Ukrainian Front and the vanguards of the 6th Guards Tank Army of the Second Ukrainian Front linked up at Esztergom and Budapest was encircled.

Budapest was defended by 33,000 German and 37,000 Hungarian troops. From 4 December these troops came under the German *IX SS-Gebirgskorps*. The city straddles the Danube River, with the hilly suburbs of Buda on the western side and Pest occupying the eastern bank.

RACECOURSE BATTLE

On 6 January the Soviets captured the Pest suburbs Köbánya and Rákosfalva, which allowed them to keep the emergency airfield on the racecourse closed by bombarding it from the north and south.

The following day the Germans and Hungarians started fierce counterattacks to regain the racecourse. The Hungarian 10th Assault Gun Battalion and a mixed German/Hungarian Combat Group were positioned behind the wooden fence of a grandstand. Soon after the position was taken up, and before any attack could be launched, Soviet infantry approached, singing, arm-in-arm and drunk. Hungarian assault gun commander Captain Sándor Hanák ordered his men to fire. They kicked down the fence and fired on the approaching Soviet riflemen. The Soviets broke for the stands to take cover, but the assault guns fired into the rows of seats which took a terrible toll. Reports put the Soviet dead at about 800.

The Germans and Hungarians continued their attack, making it to the far side of the racecourse. A lack of infantry meant the attack could not continue to completely secure the racecourse airfield and it could not be reopened, meaning that supplies to Pest had to be dropped in by parachute. Often the result was the loss of ammunition and food beyond enemy lines due to off-

Through December the Soviets made many attempts to penetrate into the city. They selected sectors held by the Hungarians, but Hungarian armour and German *Feldherrnhalle* and SS troop counterattacks threw back most of the Soviet thrusts.

The Soviets had formed a tight ring around the city by 10 December and began to pound it with heavy artillery. The Soviets tried to batter their way into Budapest on 25 December. The Soviets made gains on 28 December and broke through the eastern bridgehead.

Soviet attacks continued to 30 December. A penetration occurred in the east near Csömör against the Hungarian 10th Infantry Division and they were forced to retreat to a shorter defensive line.

Savage street-fighting continued about the city as the Soviets and Romanians pressed the defences. On the night of 31 December/1 January, the Soviets made a concerted attempt to throw the Axis forces out of Pest with a front-wide assault on the eastern portion of the city. A massive artillery barrage was unleashed all along the front followed by attacks. The defenders were forced to commit their last reserves in an attempt to hold the eastern part of the Hungarian capital.

The Soviets made more concerted attacks between 1 and 15 January, the defenders threw them back, but each time they were forced to shorten their lines. However, Pest's days were numbered. Forces in Pest were finally allowed to withdraw across the Danube to Buda on 16 January. The evacuation lasted until daylight on 18 January. Many Hungarian troops and civilians were trapped when the last bridge was blown.

FALL OF BUDAPEST

On 9 February the German command decided to breakout on the night of 11 February. 16,000 Germans and Hungarians prepared to breakout with what they could carry.

Various groups of Hungarians and Germans made it out in small numbers, but the majority were either killed or taken prisoner. On 13 February the remaining defenders surrendered to the Soviets. After 100 days of battle, Budapest had fallen.

HUNGARIANS FIGHT ON

The bulk of the Hungarian forces were not trapped in Budapest and fought on after the fall of their capital. During December Hungarian units, including the ad hoc Kesseő Group and the Huszár Division and various remnant groups, fought alongside the Germans around Lake Valence to stop the encirclement closing.

In the North the 2nd Armoured Division fought around Sahy near the Slovak border. The Szent László Division (the elite infantry division) also joined them. Both divisions fought delaying actions through to the end of December 1944.

By February the Hungarian First Army was in Slovakia and only the Third Army was still on Hungarian soil, though they were now operating entirely under German command.

FŐHADNAGY
ERVIN TARCZAY

Ervin Tarczay (pronounced ehrr-veen tarch-oi) was born in Pécs in 1919. Initially he served with the Border Guards before retraining in 1942 and joining the 3rd Armoured Regiment in January 1943 as a *Főhadnagy* (Lieutenant).

From April 1944, Tarczay served with the 2nd Company, 1st Battalion of the 3rd Armoured Regiment in Galicia as part of the 2nd Armoured Division. Tarczay earned much experience fighting in a Turán II tank and then, after the division was partially re-equipped with German tanks in May, commanded a company of Tiger I E heavy tanks.

The Soviets launched an offensive in July that pushed back the Hungarians to the frontier in the Carpathian Mountains. The 2nd Armoured Division withdrew and received another batch of German tanks, including five Panther A tanks for Tarczay's company.

In September 1944, Tarczay and the 2nd Armoured Division were moved to the Romanian border with the aim of securing Transylvania from the Romanians.

On 5 September, Tarczay's company quickly overcame token Romanian resistance on their way towards Torda. Ten days later Tarczay's Panthers led the assault east of Torda. Despite becoming separated from the rest of the battalion they fought on through artillery fire without infantry support and destroyed two enemy infantry companies and three anti-tank guns.

The next day the company was back in action outflanking advancing Soviet tanks. The Soviets were forced to withdraw. In the afternoon Tarczay took his own Panther forward to secure the left flank of the battalion's position. He knocked out two tanks and an anti-tank gun.

Tarczay continued to hold off Soviet probes, taking a toll on the enemy until 6 October. During the fighting around Torda,

Tarczay and his Panthers knocked out 11 tanks, 17 anti-tank guns, 20 machine-gun positions, and a *Katyusha*.

Despite this, the Allied offensive could not be held and the 2nd Armoured Division was withdrawn back to the Tisza River on the Hungarian Plain. On 25 October, during an action near Polgár, Tarczay's company overran a Soviet anti-tank battery, destroying three guns and capturing another three. During the action, two T-34 tanks approached from the flank. Tarczay let them come within short range before knocking them out.

The Panthers destroyed 16 enemy tanks and assault guns during the fighting between 6 and 25 October. Tarczay directed his tanks from the open hatch of his Panther, with little regard for his own safety.

On 30 October, Tarczay's company was surrounded. He led the breakout, smashing his way through a Soviet rifle battalion. His Panther took a hit, but the thick armour deflected the round. Tarczay crushed the offending anti-tank gun under his Panther's tracks. They next ran over an artillery battery, quickly destroying it.

By late November Tarczay and his company were withdrawn north of Budapest to halt the Soviet advance. The division fought a withdrawal back to lines northwest of Budapest.

In January 1945 Tarczay was promoted to *Százados* (Captain). He took delivery of 27 Panzer IV and two Panther tanks on 8 January 1945. Until the end of February they fought around Székefehérvár between lakes Balaton and Velence.

Tarczay was Knighted (and gained the title '*vitéz*' meaning valiant) on 15 March and he went on leave to marry. He rejoined the remnants of the 2nd Armoured Division on 17 March and was engaged in the battles against the Soviet drive on Vienna until his death, around 19 March, from wounds.

HUNGARIAN PANTHERS AT TORDA

In the summer of 1944 the Germans agreed to supply the Hungarians with German armoured vehicles. Five Panthers and twenty Panzer IV H tanks were transferred to the 2nd Armoured Division in the Carpathian Mountains. Ervin Tarczay's 2nd Company received the five Panthers.

After Romania switched sides Hungarian forces were ordered to take the passes of the Southern Carpathians. Tarczay's Company was rushed to Transylvania.

The border town of Torda became the focus of the Hungarian 2nd Armoured Division's efforts during September. The extremely able Lieutenant Ervin Tarczay, commanding Hungary's only Panther company, led their attacks.

By 15 September his Panthers were leading an assault east of Torda. His company was separated from the rest of the battalion, had no supporting infantry and was receiving heavy artillery and anti-tank fire. However, they kept pressing on and destroyed two enemy infantry companies and three anti-tank guns before withdrawing. The next day Tarczay's Panthers outflanked advancing Soviet tanks and forced them to withdraw after Tarczay's Panther had knocked out one tank. During the afternoon, showing great initiative, Tarczay's Panthers joined a motorised infantry battalion during an assault, to attack a Soviet position without orders. Tarczay took his Panthers forward and knocked out two tanks and an anti-tank gun to secure the left flank of the battalion.

On 22 September, east of Torda, the two currently operational Panthers outflanked a Soviet thrust, halting and destroying an entire Red Army infantry battalion. In the afternoon Tarczay repeated this flanking manoeuvre and he knocked out two T-34s and an anti-tank gun. During the battle Tarczay's Panther was disabled, but he immediately mounted another tank.

On 24 September the 1st and 2nd Battalion, Hungarian 3rd Tank Regiment had 17 battle-worthy tanks: 2 Panthers, 6 Panzer IV and 9 Turán tanks. The tanks were withdrawn to Hill 348 near Nagy Ördöngös to act as reserve. The next day, after repairs, the number of Panthers went up to five and three Tigers arrived.

The number of active Panthers went down to two again on 26 September after the 2nd Armoured Division launched a counterattack to push the assaulting Soviet forces out of the Hungarian lines. However, Soviet pressure finally told and the 2nd Armoured Division was forced to withdraw in to Hungary in October.

Between 15 September and 6 October Tarczay's Panther company knocked out 11 tanks, 17 anti-tank guns, 20 machine-gun positions and a *Katyusha* rocket launcher during the fighting around Torda.

CHARACTERISTICS

Főhadnagy Ervin Tarczay is a Warrior and is rated **Fearless Veteran**.

Tarczay may join any Harckocsizó Század and replaces the Company Command Turan II or Panzer IV H tank for +40 points, or Tiger I E or Panther A tank for +75 points.

AROUND THE FLANKS

Tarczay likes to get around the flanks of his enemy to catch them off guard. Movement is the key to this and Tarczay always keeps his platoons moving.

> *Tarczay and any platoon he has joined may choose to make a Stormtroopers move (page 183) instead of a normal Huszár move (page 227).*
>
> *This means you can shoot in the Shooting Step as normal and choose to make a Stormtroopers move instead of assaulting in the Assault Step.*

AS STRAIGHT AS AN ARROW

Tarczay always fought from his open turret hatch and was able to direct the fire of his gunners with skill and precision because of the excellent visibility this provided.

> *Tarczay re-rolls misses with his tank's main gun.*

DESTROY THOSE ANTI-TANK GUNS!

Tarczay seemed to be particularly good at taking out guns with his tank tracks. Often an attack on a gun position would end with Tarczay's tank nested in the middle of the twisted wreckage of a Soviet gun line.

> *Tarczay may re-roll any failed saves when hit by a Gun team in Defensive Fire during an Assault.*

HARCKOCSIZÓ SZÁZAD
TANK COMPANY

(TANK COMPANY)

HEADQUARTERS

Harckocsizó Század HQ · 207

You must field one platoon from each box shaded black and may field one platoon from each box shaded grey.

DIVISIONAL SUPPORT PLATOONS

COMBAT PLATOONS

ARMOUR
Harckocsizó Platoon · 207

ARMOUR
Harckocsizó Platoon · 207

ARMOUR
Harckocsizó Platoon · 207

WEAPONS PLATOONS

ARMOUR
Harckocsizó Platoon · 207

ARMOUR
Light Harckocsizó Platoon · 208

INFANTRY
Motorised Pioneer Platoon · 208

REGIMENTAL SUPPORT PLATOONS

ANTI-AIRCRAFT
Self-propelled Anti-aircraft Platoon · 209

ANTI-AIRCRAFT
Self-propelled Anti-aircraft Platoon · 209

ALLIED PLATOONS

Heer Platoons in your force are Allies and follow the Allies rules in the rulebook.

ARMOUR
Light Harckocsizó Platoon · 208
Rohamágyús Platoon · 215
German Panzer Platoon · 73
German Feldhermhalle Panzer Platoon · 137

INFANTRY
Gépkocsizó Lövész Platoon · 211
Puskás Platoon · 217
Önkéntes Puskás Platoon · 221
Assault Pioneer Platoon · 224

INFANTRY
Gépkocsizó Lövész Platoon · 211
Motorcycle Scout Platoon · 209

ARTILLERY
Motorised Artillery Battery · 213
German Motorised Artillery Battery · 167

ARTILLERY
Rocket Launcher Battery · 225

AIRCRAFT
Air Support · 224
German Air Support · 172

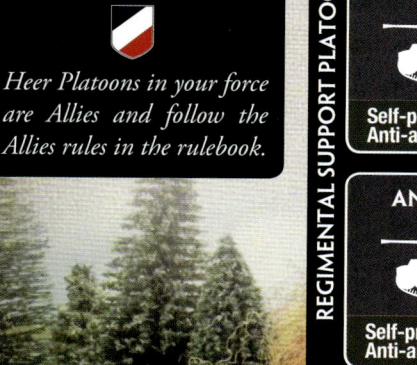

206

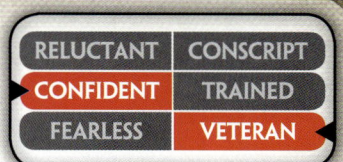

MOTIVATION AND SKILL

The tanks of the Hungarian 1st and 2nd Armoured Division are commanded by veterans of the battles on the Don in 1942 to 1943. During 1943, new recruits trained with these veterans and the Germans in both German and Hungarian tanks and tactics. They are some of the best troops in the Hungarian armies. A Harckocsizó Század is rated **Confident Veteran**.

RELUCTANT	CONSCRIPT
CONFIDENT	TRAINED
FEARLESS	VETERAN

HEADQUARTERS

HARCKOCSIZÓ SZÁZAD HQ

HEADQUARTERS

2 Turán I	115 points
2 Panzer IV H	180 points
2 Turán II	115 points
1 Tiger I E	195 points
1 Panther A	185 points

OPTION

- Add Famo recovery vehicle for +5 points.

SZÁZADOS

SZÁZADOS — Company Command Tank

ÖRMESTER — Famo Recovery vehicle

RECOVERY SECTION

2iC Command Tank

COMPANY HQ

HARCKOCSIZÓ SZÁZAD HQ

A Harckocsizó Század HQ is a Huszár Platoon.

Hungarian Tiger I E tanks do not get German Tiger Ace Skills.

You must field at least one Harckocsizó Platoon entirely equipped with the same model of tank as the Company HQ.

In 1944 the Hungarians fielded two armoured divisions and both were heavily engaged in the battles to defend Hungary against the red tide.

The 1st Armoured Division was mainly armed with Hungarian-manufactured Turán and Toldi tanks. The 2nd Armoured Division had a mix of Hungarian and German tanks. Initially they took the field with Turán tanks, but received German tanks and assault guns as replacements.

COMBAT PLATOONS

HARCKOCSIZÓ PLATOON

PLATOON

5 Turán I	290 points
4 Turán I	230 points
3 Turán I	170 points
4 Panzer IV H	360 points
3 Panzer IV H	270 points
3 Turán II	170 points
3 Tiger I E	585 points
2 Tiger I E	390 points
3 Panther A	550 points
2 Panther A	365 points

FŐHADNAGY

FŐHADNAGY — Command Tank

HQ TANK

ÖRMESTER — Tank / Tank

HARCKOCSIZÓ SECTION

ÖRMESTER — Tank / Tank

HARCKOCSIZÓ SECTION

HARCKOCSIZÓ PLATOON

You may only field up to two Harckocsizó Platoons equipped with Tiger I E tanks in your force.

You may only field up to two Harckocsizó Platoons equipped with Panther A tanks in your force.

Harckocsizó Platoons are Huszár Platoons.

The 1st Armoured Division's Turán I and II tanks first entered combat on the Transylvanian frontier in September 1944. The 2nd Armoured Division received German tanks, including ten Tiger I E heavy tanks, to make up for losses of Turán tanks in May 1944 while fighting in Galicia. In August the 2nd Armoured division received five Panther A tanks which came under the command of *Főhadnagy* Ervin Tarczay.

WEAPONS PLATOONS

LIGHT HARCKOCSIZÓ PLATOON

PLATOON

5 Toldi II	195 points
4 Toldi II	155 points
3 Toldi II	115 points

OPTION

- Replace Toldi II tanks with Toldi IIa tanks for +5 per tank.

Light Harckocsizó Platoons are Reconnaissance and Huszár Platoons.

The Toldi II light tank is armed with a 20mm anti-tank rifle, but in 1944 some units began to receive new 40mm armed Toldi IIa and III tanks with more hitting power.

MOTORISED PIONEER PLATOON

PLATOON

HQ Section with:

3 Pioneer Squads	270 points
2 Pioneer Squads	190 points

OPTIONS

- Replace the Command Pioneer Rifle/MG team with a Command Panzerfaust Pioneer SMG team for +10 points.
- Add 3-ton trucks for +5 points for the platoon.
- Add Pioneer Supply 3-ton truck for +25 points.

Motorised Pioneer Platoons are Huszár Platoons.

Each tank battalion had its own Pioneer Platoon for clearing mines and obstacles, assaulting minor fortified positions, repairing bridges and various other engineering tasks. In the desperate fighting of 1944 many of these pioneers fought alongside the tanks and motorised infantry of the 1st and 2nd Armoured Divisions.

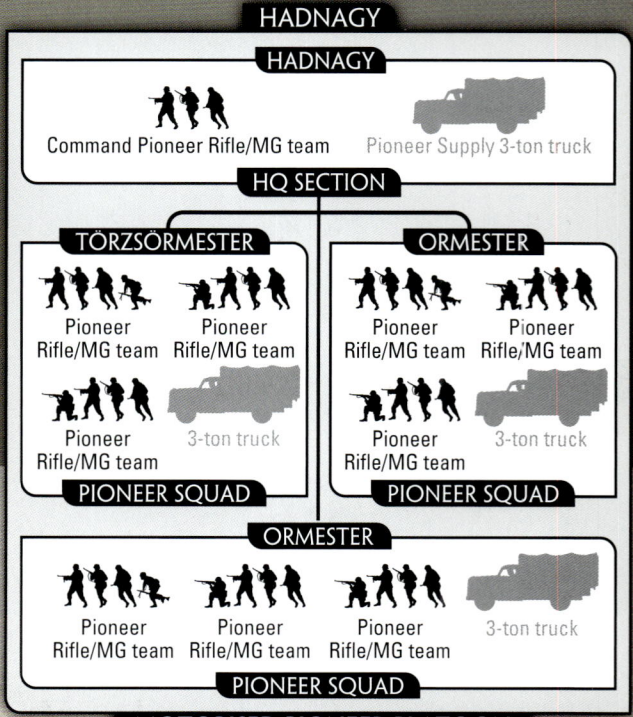

REGIMENTAL SUPPORT PLATOONS

SELF-PROPELLED ANTI-AIRCRAFT PLATOON

PLATOON

2 Nimrods	115 points

HADNAGY

HADNAGY	TIZEDES
Command Nimrod	Nimrod
HQ SECTION	ANTI-AIRCRAFT SECTION

SELF-PROPELLED ANTI-AIRCRAFT PLATOON

Self-propelled Anti-aircraft Platoons are Huszár Platoons.

The Swedish designed self-propelled anti-aircraft guns were made under license in Hungary as the Nimrod.

DIVISIONAL SUPPORT PLATOONS

MOTORCYCLE SCOUT PLATOON

PLATOON

HQ Section with:

3 Scout Squads	220 points
2 Scout Squads	155 points

OPTION

• Replace the Command Motorcycle Rifle/MG team with a Command Panzerfaust Motorcycle SMG team for +10 points.

Motorcycle Scout Platoons use the Motorcycle Reconnaissance rules on page 187 and are Reconnaissance Platoons while mounted.

The armoured reconnaissance battalion of an armoured division contains a variety of specialist scouting units, including two companies of motorcycle mounted scouts.

These highly mobile troops fight alongside the heavy, medium and light tanks by moving ahead of the division to probe the enemy's positions, discover ambushes, and locate enemy strongpoints.

HADNAGY

HADNAGY
Command Motorcycle Rifle/MG team
HQ SECTION

TÖRZSÖRMESTER	ÖRMESTER
Motorcycle Rifle/MG team	Motorcycle Rifle/MG team
Motorcycle Rifle/MG team	Motorcycle Rifle/MG team
SCOUT SQUAD	SCOUT SQUAD

ÖRMESTER

Motorcycle Rifle/MG team	Motorcycle Rifle/MG team
SCOUT SQUAD	

MOTORCYCLE SCOUT PLATOON

A Motorcycle Scout Platoon is a Huszár Platoon.

GÉPKOCSIZÓ LÖVÉSZ SZÁZAD
MOTORISED INFANTRY COMPANY

(INFANTRY COMPANY)

HEADQUARTERS

Gépkocsizó Lövész Század HQ *211*

You must field one platoon from each box shaded black and may field one platoon from each box shaded grey.

DIVISIONAL SUPPORT PLATOONS

COMBAT PLATOONS

INFANTRY
Gépkocsizó Lövész Platoon *211*

INFANTRY
Gépkocsizó Lövész Platoon *211*

INFANTRY
Gépkocsizó Lövész Platoon *211*

MACHINE-GUNS
Gépkocsizó Lövész Machine-gun Platoon *212*

MACHINE-GUNS
Gépkocsizó Lövész Machine-gun Platoon *212*

WEAPONS PLATOONS

MACHINE-GUNS
Gépkocsizó Lövész Machine-gun Platoon *212*

ARTILLERY
Gépkocsizó Lövész Mortar Platoon *212*

ANTI-TANK
Motorised Anti-tank Platoon *213*

INFANTRY
Motorised Pioneer Platoon *208*

RECONNAISSANCE
Motorcycle Scout Platoon *209*

ARMOUR
Harckocsizó Platoon *207*
Light Harckocsizó Platoon *208*
Rohamágyús Platoon *215*
German Panzer Platoon *73*
German Feldherrnhalle Panzer Platoon *137*

ARMOUR
Harckocsizó Platoon *207*
Light Harckocsizó Platoon *208*

INFANTRY
Puskás Platoon *217*
Önkéntes Puskás Platoon *221*
Assault Pioneer Platoon *224*

ARTILLERY
Motorised Artillery Battery *213*
German Motorised Artillery Battery *167*

ARTILLERY
Rocket Launcher Battery *225*

ANTI-AIRCRAFT
Self-propelled Anti-aircraft Platoon *209*
Anti-aircraft Platoon *226*
Heavy Anti-aircraft Platoon *226*

AIRCRAFT
Air Support *224*
German Air Support *172*

ALLIED PLATOONS

Heer Platoons in your force are Allies and follow the Allies rules in the rulebook.

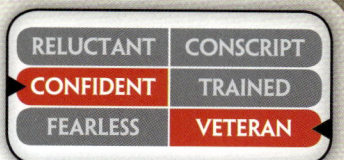

MOTIVATION AND SKILL

The troops of the motorised infantry of the 1st and 2nd Armoured Divisions are commanded by veterans of the battles on the Don in 1942 to 1943. They are some of the best troops in the Hungarian armies. A *Gépkocsizó Lövész Század* is rated **Confident Veteran**.

RELUCTANT	CONSCRIPT
CONFIDENT	TRAINED
FEARLESS	**VETERAN**

HEADQUARTERS

GÉPKOCSIZÓ LÖVÉSZ SZÁZAD HQ

HEADQUARTERS

Company HQ	30 points

OPTIONS

- Replace the Command Rifle teams with Command Páncélvadész SMG teams for +5 points per team or with a Command Panzerfaust SMG teams for +15 points per team.
- Add Panzerschreck teams for +20 points per team.

A Gépkocsizó Lövész Század HQ is a Huszár Platoon.

GÉPKOCSIZÓ LÖVÉSZ SZÁZAD HQ

The Gépkocsizó Lövész Század (Motorised Infantry Company) provides the infantry of the Hungarian armoured divisions. They are some of the most experienced troops available to the Hungarians with many of them seeing service on the eastern front in the Soviet Union between 1941 and 1944.

COMBAT PLATOONS

GÉPKOCSIZÓ LÖVÉSZ PLATOON

PLATOON

HQ Section with:

4 Rifle Squads	185 points
3 Rifle Squads	145 points
2 Rifle Squads	105 points

OPTIONS

- Replace the Command Rifle/MG team with a Command Páncélvadász SMG team for +5 points or Command Panzerfaust SMG team for +10 points.
- Add Botond trucks for +5 points for the platoon.

Motorised Rifle Platoons are Huszár Platoons.

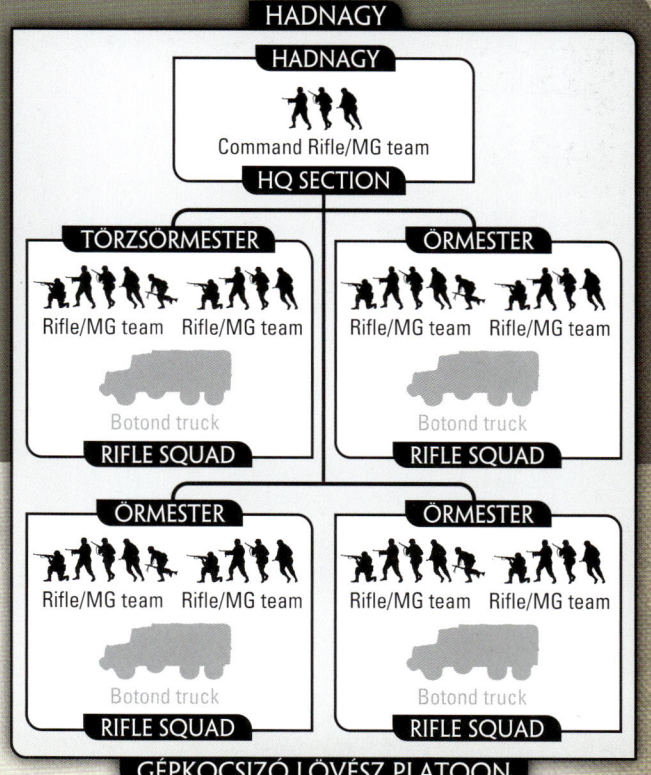

GÉPKOCSIZÓ LÖVÉSZ PLATOON

The *Gépkocsizó Lövész* (Motorised Infantry) platoons of the Motorised Infantry Regiments provide each division with three battalions. These skilled riflemen keep pace with the rapid movements of the tanks to provide support when it is needed, ready to take captured positions and hold them against counterattack.

Some platoons have been issued German *Panzerfaust* anti-tank weapons, while others make-do with improvised satchel charges and anti-tank mines as *Páncélvadász* (tank hunter) teams.

Gépkocsizó Lövész Machine-gun Platoon

Platoon

HQ Section with:

3 Machine-gun Sections	105 points
2 Machine-gun Sections	70 points

A Gépkocsizó Lövész Machine-gun Platoon may make Combat Attachments to Gépkocsizó Lövész Platoons.

The old 7/31M Schwarzlose 8mm water-cooled machine-gun provided sustained fire support to the motorised infantrymen. With a rate-of-fire of 350 rounds per minute it can pour out enough fire to halt any enemy infantry attack.

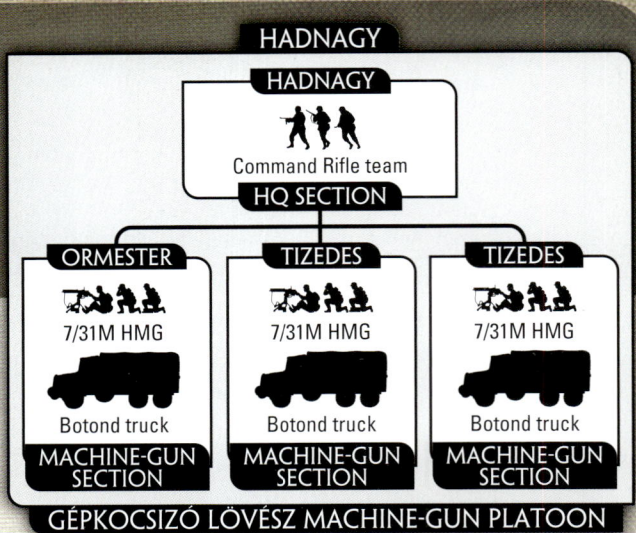

WEAPONS PLATOONS

Motorised Anti-Tank Platoon

Platoon

HQ Section with:

3 75mm 40M gun	155 points
2 75mm 40M gun	105 points

Option

• Add Botond trucks for +5 points for the platoon.

With the ever increasing armour of the Red Army's tanks the Hungarians made sure they secured numbers of the German 7.5cm PaK40 anti-tank gun for use by their forces. These were designated the 75mm 40M gun. This excellent weapon proved just as capable in the hands of the Hungarians as it had with the Germans, knocking out anything but the heaviest of Soviet tanks.

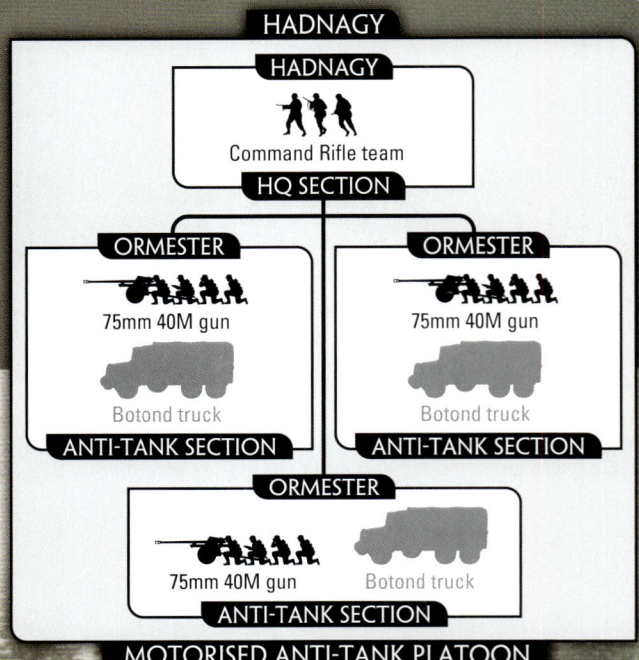

GÉPKOCSIZÓ LÖVÉSZ MORTAR PLATOON

PLATOON

HQ Section with:

2 Mortar Sections	115 points
1 Mortar Section	65 points

Another important weapon in the motorised infantrymen's arsenal was the medium mortar. The Hungarians use the 36/39M 81mm mortar produced by the Hungarian State Weapons Factory at Diósgyőr. It could provide smoke as well as high-explosive bombardments on enemy positions.

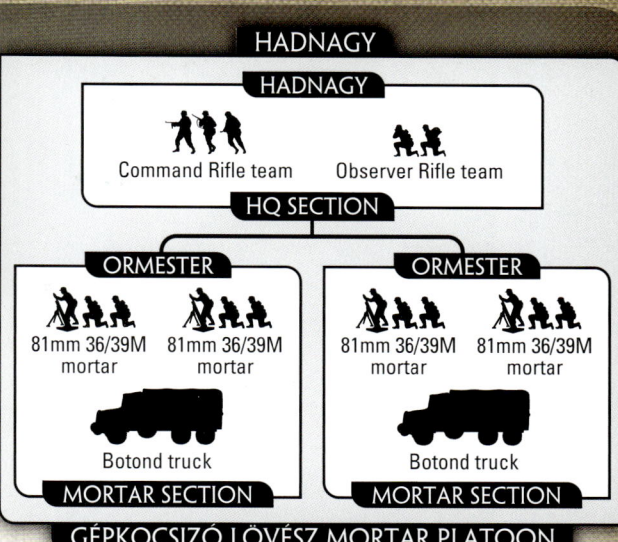

MOTORISED ARTILLERY BATTERY

PLATOON

HQ Section with:

4 105mm 37M howitzers	195 points
2 105mm 37M howitzers	110 points
4 149mm 14/31M howitzers	225 points
2 149mm 14/31M howitzers	125 points

OPTION
• Add 3-ton trucks for +5 points for the platoon.

A Motorised Artillery Battery is a Huszár Platoon.

Motorised Artillery Batteries equipped with 149mm 14/31M howitzers may not be placed from Ambush within 16"/40cm of enemy teams.

ROHAMÁGYÚS ÜTEG
ASSAULT GUN BATTERY

(TANK COMPANY)

HEADQUARTERS

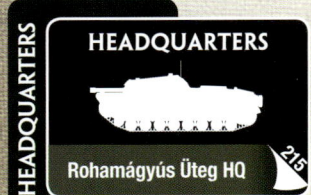

Rohamágyús Üteg HQ | 215

You must field one platoon from each box shaded black and may field one platoon from each box shaded grey.

DIVISIONAL SUPPORT PLATOONS

COMBAT PLATOONS

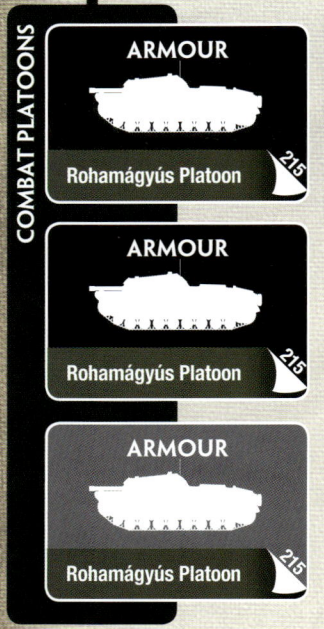

ARMOUR
Rohamágyús Platoon | 215

ARMOUR
Rohamágyús Platoon | 215

ARMOUR
Rohamágyús Platoon | 215

ALLIED PLATOONS

Heer Platoons in your force are Allies and follow the Allies rules in the rulebook.

ARMOUR

Harckosizó Platoon | 207
German Panzer Platoon | 73
German Feldherrnhalle Panzer Platoon | 137

ANTI-TANK

Assault Anti-tank Platoon | 222

INFANTRY

Puskás Platoon | 217
Önkéntes Puskás Platoon | 221
Motorcycle Scout Platoon | 209
German Panzergrenadier Platoon | 81
German Feldherrnhalle Panzergrenadier Platoon | 141
German SS-Kavallerie Platoon | 147

INFANTRY

Puskás Platoon | 217
Önkéntes Puskás Platoon | 221
Pioneer Platoon | 224
Assault Pioneer Platoon | 224

RECONNAISSANCE

Armoured Car Platoon | 223

ARTILLERY

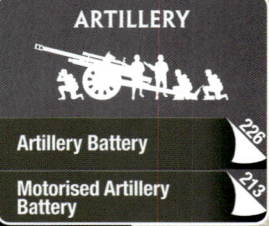

Artillery Battery | 226
Motorised Artillery Battery | 213

ARTILLERY

Rocket Launcher Battery | 225

ANTI-AIRCRAFT

Anti-aircraft Platoon | 226

AIRCRAFT

Air Support | 224
German Air Support | 172

It was planned to equip all the assault gun forces with the *Zrínyi*. However, slow production meant German-made StuG G and Hetzer assault guns were also used.

Hungarian Assault Artillery Badge.

MOTIVATION AND SKILL

Hungarian assault artillery troops are the elite of the artillery branch of service. They undergo extensive training with artillery and tank veterans, as well as additional instruction from the Germans. Once training is complete they are issued with the latest Hungarian Zrínyi or German StuG G or Hetzer assault guns. They fight hard to support their foot-slogging brothers of the infantry divisions. A Rohamágyús Üteg is rated **Confident Veteran.**

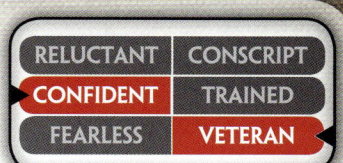

RELUCTANT	CONSCRIPT
CONFIDENT	TRAINED
FEARLESS	**VETERAN**

HEADQUARTERS

ROHAMÁGYÚS ÜTEG HQ

HEADQUARTERS

Zrínyi II	75 points
StuG G	95 points
Hetzer	85 points

OPTION

• Add Schürzen sideskirts to Zrínyi II assault gun for +5 points.

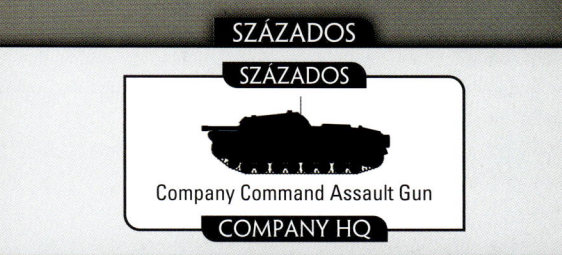

SZÁZADOS

SZÁZADOS

Company Command Assault Gun

COMPANY HQ

ROHAMÁGYÚS ÜTEG HQ

You must field at least one Rohamágyús Platoon entirely equipped with the same model of assault gun as the Company HQ.

A Rohamágyús Üteg HQ is a Huszár Platoon.

Each Assault Gun Battalion had three batteries of ten to sixteen vehicles. The Assault Gun Batteries (*Rohamágyús Üteg*) were often used to support operations as whole batteries, rather than as individual platoons. This lesson was learnt from the hard fighting on the Don where Hungarian tank units were used piecemeal all along the Hungarian lines.

COMBAT PLATOONS

ROHAMÁGYÚS PLATOON

PLATOON

3 Zrínyi II	225 points
3 StuG G	285 points
5 Hetzer	425 points
4 Hetzer	340 points
3 Hetzer	255 points

OPTION

• Add Schürzen sideskirts to all Zrínyi II assault guns for +5 points for the platoon.

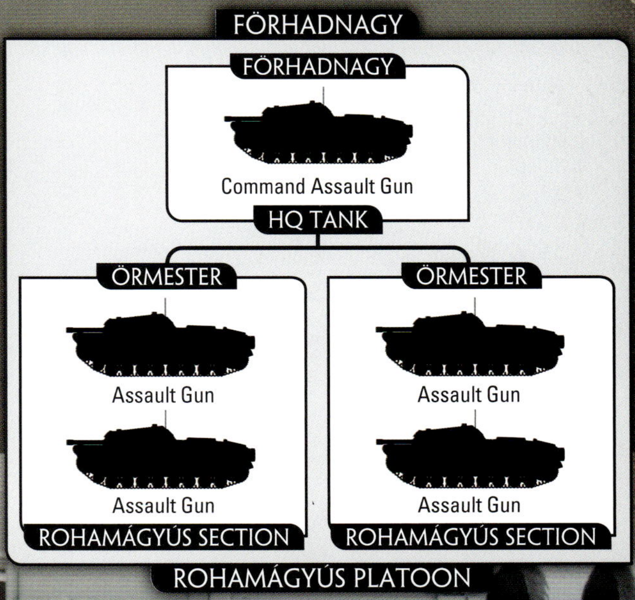

Rohamágyús Platoons are Huszár Platoons.

FÖRHADNAGY

FÖRHADNAGY

Command Assault Gun

HQ TANK

ÖRMESTER

Assault Gun

Assault Gun

ROHAMÁGYÚS SECTION

ÖRMESTER

Assault Gun

Assault Gun

ROHAMÁGYÚS SECTION

ROHAMÁGYÚS PLATOON

PUSKÁS SZÁZAD
RIFLE COMPANY

(INFANTRY COMPANY)

HEADQUARTERS

Puskás Század HQ | 217

You must field one platoon from each box shaded black and may field one platoon from each box shaded grey.

Your Company must be either a Puskás Század (marked 🛡) or a Veteran Puskás Század (marked 🛡). Whichever company you choose, you may only choose platoons and options marked with your company symbol or no symbol.

DIVISIONAL SUPPORT PLATOONS

COMBAT PLATOONS

INFANTRY
Puskás Platoon | 217

INFANTRY
Puskás Platoon | 217

INFANTRY
Puskás Platoon | 217

MACHINE-GUNS
Puskás Machine-gun Platoon | 218

MACHINE-GUNS
Puskás Machine-gun Platoon | 218

ALLIED PLATOONS

Heer Platoons in your force are Allies and follow the Allies rules in the rulebook.

WEAPONS PLATOONS

MACHINE-GUNS
Puskás Machine-gun Platoon | 218

ARTILLERY
Puskás Mortar Platoon | 218

ANTI-TANK
Puskás Anti-tank Platoon | 218

REGIMENTAL SUPPORT PLATOONS

ANTI-TANK
Heavy Anti-tank Platoon | 219

ARTILLERY
Heavy Mortar Platoon | 219

INFANTRY
Huszár Platoon | 223
Scout Platoon | 219

INFANTRY
Pioneer Platoon | 224

ARMOUR
Rohamágyús Platoon | 215

ARMOUR
Rohamágyús Platoon | 215
Harckosizó Platoon | 207

INFANTRY
Border Guard Platoon | 225
Pioneer Platoon | 224

INFANTRY
German Panzergrenadier Platoon | 81
German Grenadier Platoon | 27

RECONNAISSANCE
Armoured Car Platoon | 223

ARTILLERY
German Motorised Artillery Battery | 167
Artillery Battery | 226

ARTILLERY
Artillery Battery | 226
Rocket Launcher Battery | 225

ANTI-AIRCRAFT
Anti-aircraft Platoon | 226
Heavy Anti-aircraft Platoon | 226

AIRCRAFT
Air Support | 224

FORTIFICATIONS
Street Fortifications | 157
Field Fortifications | 155

216

MOTIVATION AND SKILL

 The Hungarians have built up their infantry forces rapidly as the threat of Soviet invasion loomed. Divisions have been raised from reservists and new recruits. Training is adequate, but they have little combat experience. However, they will fight hard to protect their homeland. A *Puskás Század* is rated as **Confident Trained**.

RELUCTANT	CONSCRIPT
CONFIDENT	**TRAINED**
FEARLESS	VETERAN

HONVÉD

RELUCTANT	CONSCRIPT
CONFIDENT	TRAINED
FEARLESS	**VETERAN**

16TH, 24TH AND 25TH INFANTRY DIVISIONS

 The 16th, 24th, and 25th Infantry Divisions all served with the Hungarian First Army during the battles for Galicia in 1944. During this period they gained valuable experience they were able to put to good use during the battles for Transylvania and the Tisza River. A Veteran *Puskás Század* is rated as **Confident Veteran**.

HEADQUARTERS

PUSKÁS SZÁZAD HQ

HEADQUARTERS

Company HQ	25 points	30 points

OPTIONS

- Replace the Command Rifle teams with Command Páncélvadász SMG teams for +10 points per team or Command Panzerfaust SMG teams for +15 points per team.
- Add an Anti-tank section with:

1 Panzerschreck team	+20 points	+25 points
2 Panzerschreck teams	+40 points	+50 points

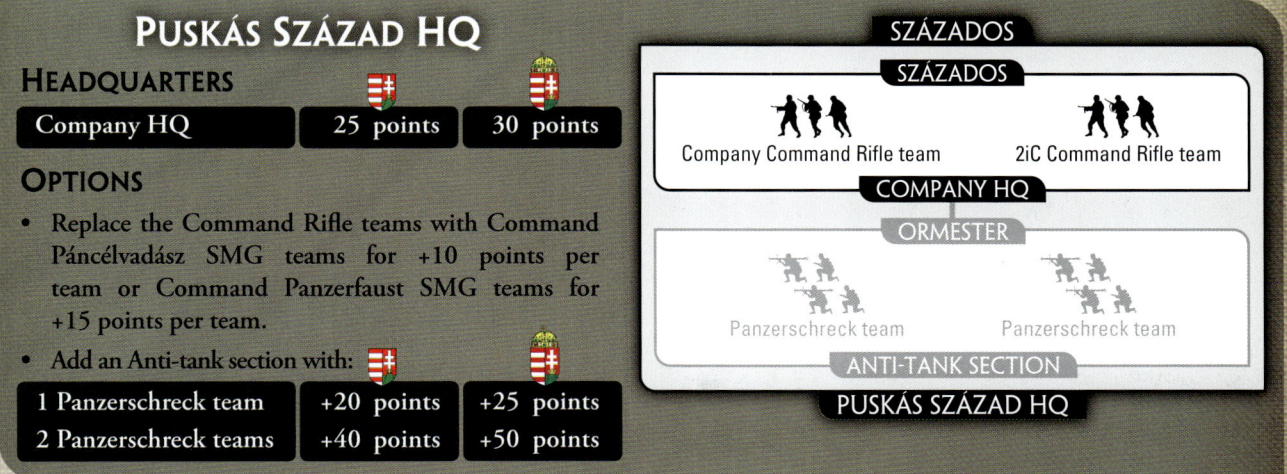

SZÁZADOS

SZÁZADOS

Company Command Rifle team   2iC Command Rifle team

COMPANY HQ

ORMESTER

Panzerschreck team   Panzerschreck team

ANTI-TANK SECTION

PUSKÁS SZÁZAD HQ

The bulk of the Hungarian Army, or *Honvéd* (pronounced hon-veed), was made up of the infantry of the *Puskás Század* (Rifle Company, pronounced poosh-kash sahr-zod). During the mass mobilisation of 1944 the small number of veteran divisions who had been fighting in the Ukraine during the early part of the year were joined by newly mobilised units.

In 1943 and 1944 the Hungarians had also been reforming their organisation and arming their troops with more modern heavy weapons such as German 75mm anti-tank guns, 120mm mortars and German Panzerfaust and Panzershreck anti-tank weapons. They even began manufacturing their own version of the Panzershreck anti-tank rocket launcher.

COMBAT PLATOONS

PUSKÁS PLATOON

PLATOON

HQ Section with:

3 Puskás Squads	160 points	210 points
2 Puskás Squads	110 points	145 points

OPTION

- Replace the Command Rifle/MG team with a Command Páncélvadász SMG team for +5 points or with a Command Panzerfaust SMG team for +10 points.

Hungarian riflemen are either armed with Steyr-Mannlicher M95/31, an unusual bolt action rifle with a unique straight pull action, or the more conventional German K98 Mauser rifle. Their heavy firepower was supplied by the Solothurn M31 light machine-gun or German supplied MG-42 machine-guns.

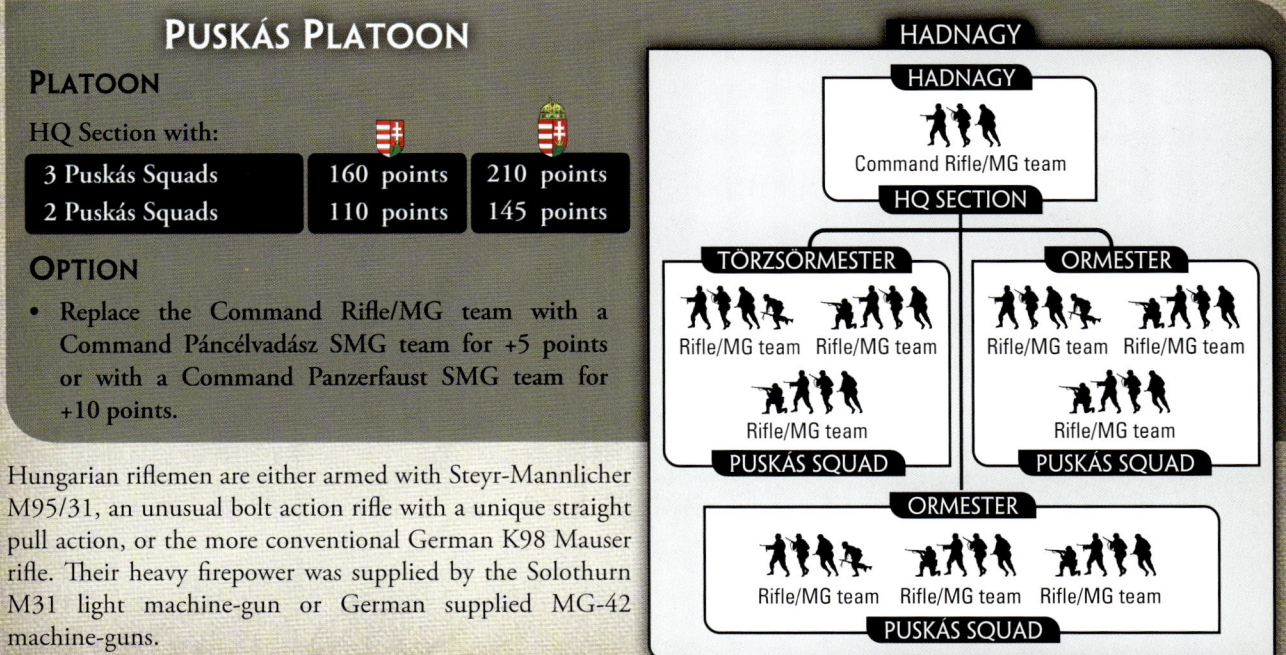

HADNAGY

HADNAGY

Command Rifle/MG team

HQ SECTION

TÖRZSÖRMESTER

Rifle/MG team   Rifle/MG team

Rifle/MG team

PUSKÁS SQUAD

ORMESTER

Rifle/MG team   Rifle/MG team

Rifle/MG team

PUSKÁS SQUAD

ORMESTER

Rifle/MG team   Rifle/MG team   Rifle/MG team

PUSKÁS SQUAD

PUSKÁS PLATOON

Puskás Machine-gun Platoon

Platoon

HQ Section with:

3 Machine-gun Sections	75 points	100 points

A Puskás Machine-gun Platoon may make Combat Attachments to Puskás Platoons.

The Hungarians use the same machine-gun they did during WWI, the Schwarzlose 7/31M.

WEAPONS PLATOONS

Puskás Mortar Platoon

Platoon

HQ Section with:

2 Mortar Sections	85 points	115 points
1 Mortar Section	50 points	65 points

Mortars are ideal support weapons for the infantry. They can be called on to give instant support against assaulting enemy. On attack they can bring fire down on the enemy positions to keep their heads down, while your infantry approach to assault.

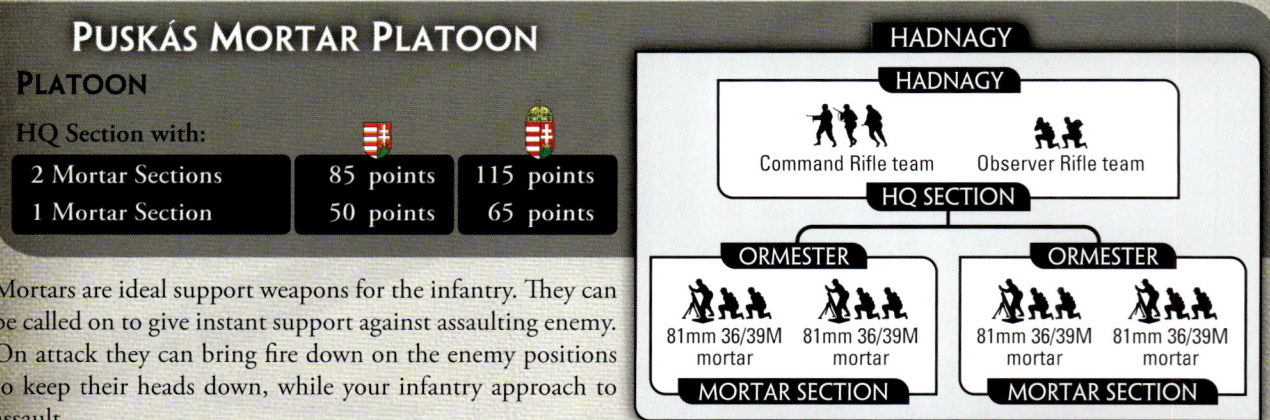

Puskás Anti-tank Platoon

Platoon

HQ Section with:

3 40mm 40M	60 points	80 points

Option

- Add horse-drawn limbers for +5 points for the platoon.

The Hungarians field their own version of the German PaK36 gun that has been modified to take the same 40mm round as their Bofors anti-aircraft gun. This gives the guns slightly better anti-tank performance. They can also be fitted with the Hungarian version of the muzzle fired *Stielgranate* anti-tank grenade.

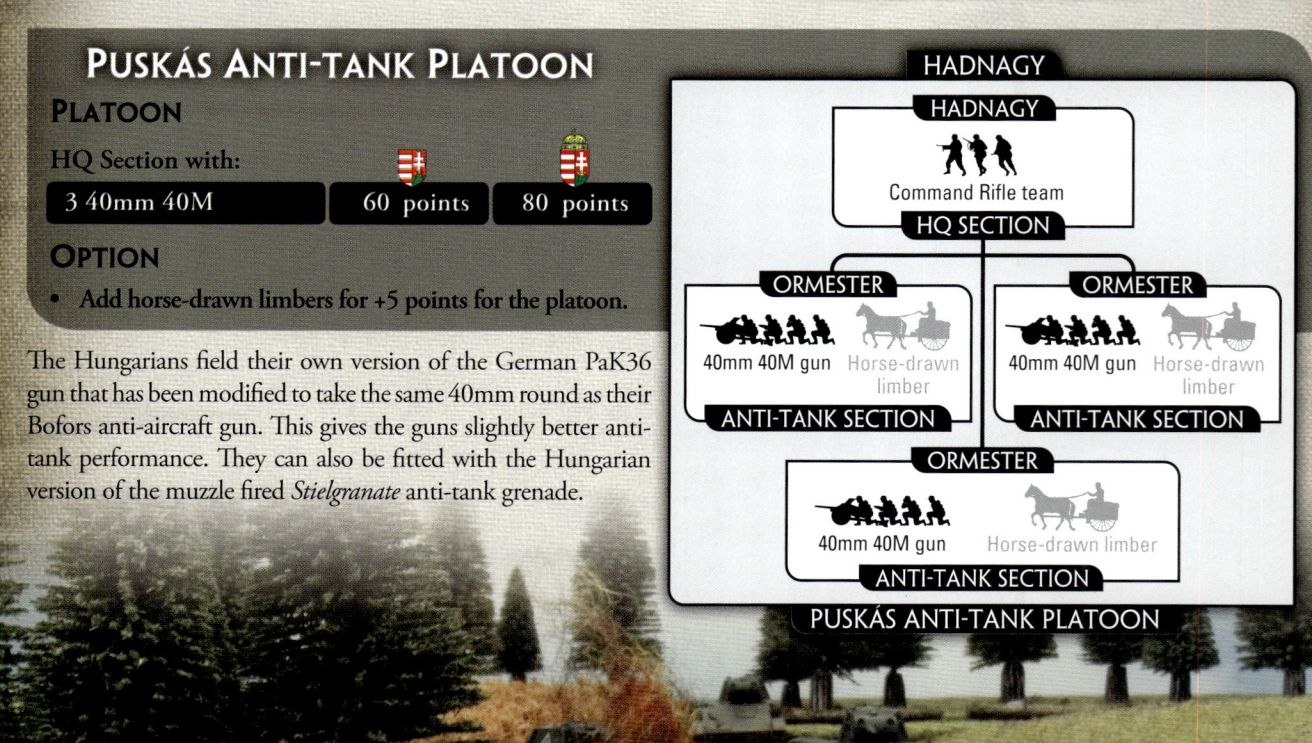

REGIMENTAL SUPPORT PLATOONS

HEAVY ANTI-TANK PLATOON

PLATOON

HQ Section with:

3 75mm 40M	120 points	160 points
2 75mm 40M	80 points	105 points

OPTION

- Add Botond trucks for +5 points for the platoon.

One of the lessons the Hungarians learnt from their fighting on the Don River in 1942 and 1943 was the need to have heavier anti-tank weapons to deal with Soviet tanks like the T-34 and KV-1. They were able to obtain 7.5cm PaK40 guns from the Germans, which they designated the 75mm 40M.

This excellent weapon allowed the Hungarian anti-tank troops to deal with most of the tanks that the Soviets throw at them.

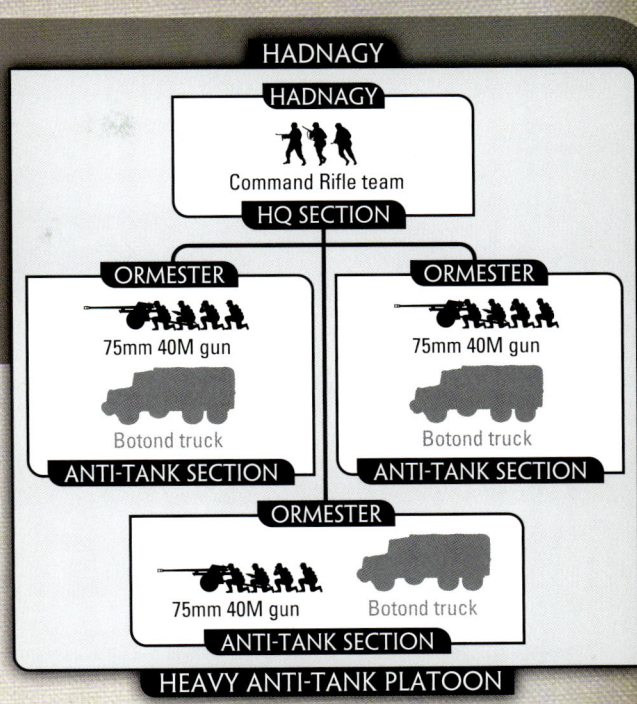

HEAVY MORTAR PLATOON

PLATOON

HQ Section with:

2 Mortar Sections	105 points	140 points
1 Mortar Section	60 points	80 points

OPTION

- Add horse-drawn limbers for +5 points for the Platoon.

During the fighting on the Don in 1942 the Hungarians were able to capture a number of 120mm Soviet mortars. The Hungarians were impressed with this weapon and similar mortars used by the Germans. They soon began to make their own, giving their infantry added heavy support in the front line.

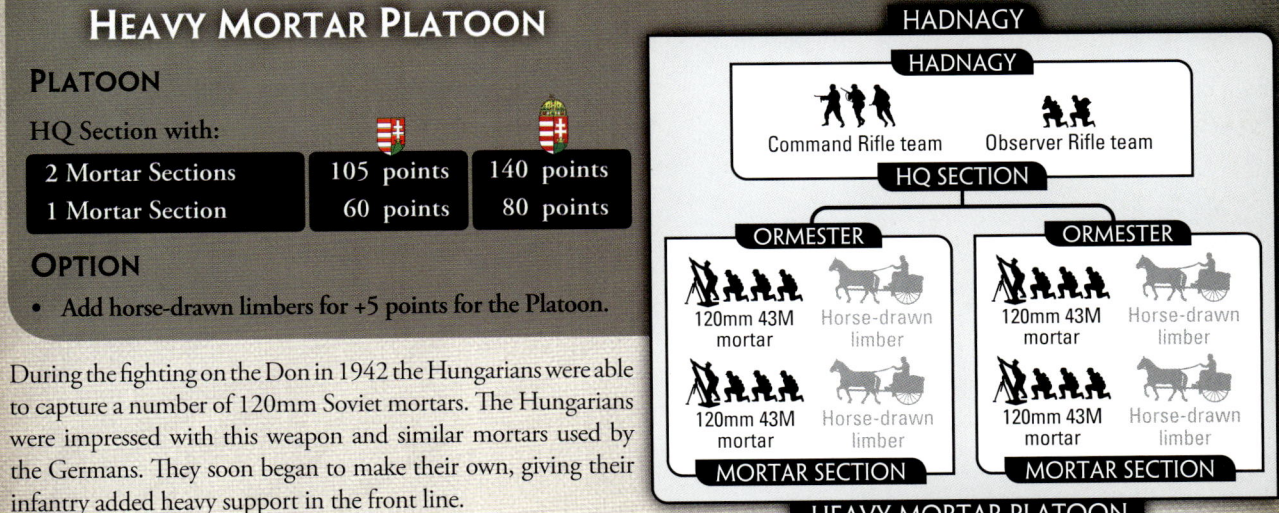

SCOUT PLATOON

PLATOON

HQ Section with:

3 Scout Squads	145 points	190 points
2 Scout Squads	105 points	135 points

OPTION

- Replace the Command Rifle/MG team with a Command Páncélvadász SMG team for +5 points or with a Command Panzerfaust SMG team for +10 points.

A Scout Platoon is a Reconnaissance Platoon.

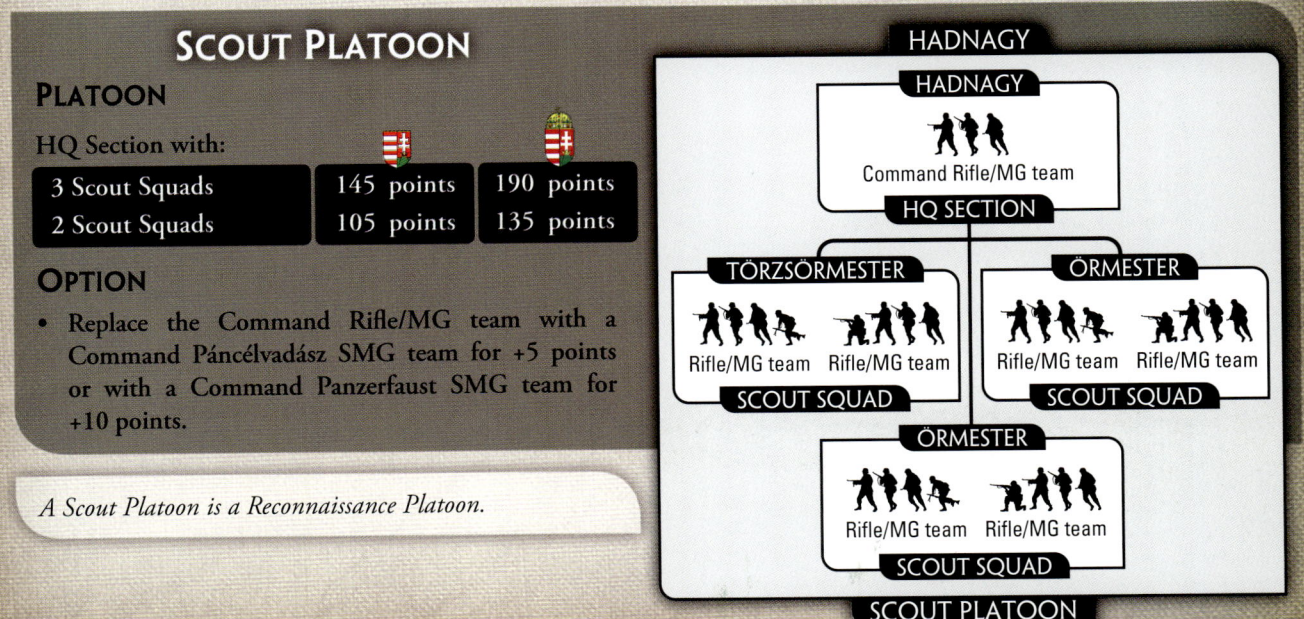

ÖNKÉNTES PUSKÁS SZÁZAD
VOLUNTEER RIFLE COMPANY

(INFANTRY COMPANY)

HEADQUARTERS

HEADQUARTERS

Önkéntes Puskás Század HQ — 221

You must field one platoon from each box shaded black and may field one platoon from each box shaded grey.

COMBAT PLATOONS

INFANTRY

Önkéntes Puskás Platoon — 221

INFANTRY

Önkéntes Puskás Platoon — 221

INFANTRY

Önkéntes Puskás Platoon — 221

MACHINE-GUNS

Önkéntes Puskás Machine-gun Platoon — 222

WEAPONS PLATOONS

MACHINE-GUNS

Önkéntes Puskás Machine-gun Platoon — 222

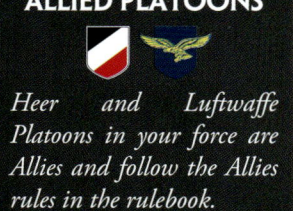

ALLIED PLATOONS

Heer and Luftwaffe Platoons in your force are Allies and follow the Allies rules in the rulebook.

DIVISIONAL SUPPORT PLATOONS

ARMOUR

Rohamágyús Platoon — 215

Armoured Car Platoon — 223

ARMOUR

Rohamágyús Platoon — 215

German Feldherrnhalle Panzer Platoon — 137

ANTI-TANK

Assault Anti-tank Platoon — 222

German Anti-tank Gun Platoon — 165

INFANTRY

Önkéntes Puskás Platoon — 221

German Feldherrnhalle Panzergrenadier Platoon — 141

German SS-Kavallerie Platoon — 147

INFANTRY

Pioneer Platoon — 224

Assault Pioneer Platoon — 224

RECONNAISSANCE

Armoured Car Platoon — 223

ARTILLERY

Puskás Mortar Platoon — 218

German Feldherrnhalle Mortar Platoon — 142

ARTILLERY

Artillery Battery — 226

German Motorised Artillery Battery — 167

ANTI-AIRCRAFT

Anti-aircraft Platoon — 226

Heavy Anti-aircraft Platoon — 226

Luftwaffe Light Anti-aircraft Gun Platoon — 173

German Heavy Anti-aircraft Gun Platoon — 171

AIRCRAFT

Air Support — 224

German Air Support — 172

FORTIFICATIONS

Street Fortifications — 157

The Order of the Valiant (in Hungarian, *Vitézi Rend*) or Knighthood of the Heroes was the first and probably the most important Hungarian order established after the Great War. All the recipients were proven soldiers. The minimal requirements for obtaining the title *vitéz* was linked to the receipt of certain bravery medals. The award also entitled the recipient to a plot of land.

MOTIVATION AND SKILL

A great variety of militia and paramilitary were raised by the Hungarian authorities during the defence of Budapest. Among these were militia, volunteers and police units.

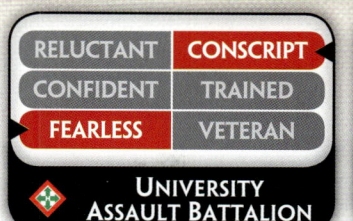
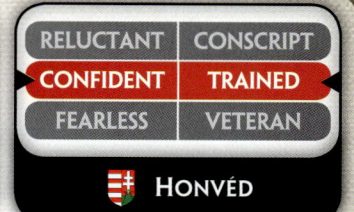

 The volunteers of the University Assault battalion, Vannay Alarm Battalion and Arrow Cross Militia are led by reserve officers. They are brave and know their city well, but many don't have a lot of military training, so are rated **Fearless Conscript.**

 Regular Infantry, Police, and Gendarme battalions were also pressed into action during the siege. These soldiers and police had a higher level of military training and were expected to fight in the front lines. They are rated **Confident Trained.**

HEADQUARTERS

ÖNKÉNTES PUSKÁS SZÁZAD HQ

HEADQUARTERS

| Company HQ | 20 points | 25 points |

OPTIONS

- Replace the Command Rifle teams with Command Páncélvadász SMG teams for +10 points per team or Command Panzerfaust SMG teams for +15 points per team.
- Add Panzerschreck teams for +15 points per team.
- Add Light Mortar teams for +15 points per team.

All the Combat and Weapons platoons of a Önkéntes Puskás Század must be selected from options with the same symbol as the Company HQ.

The Hungarian infantry forces defending Budapest against the Soviet onslaught were truly a mixed bag. Units varied from the volunteer students and workers with limited training, through disciplined Police and *Gendarmerie* (rural police) to regular infantry units with professional officers. However, many of them fought tenaciously to defend their capital city.

COMBAT PLATOONS

ÖNKÉNTES PUSKÁS PLATOON

PLATOON

HQ Section with:		
3 Puskás Squads	110 points	120 points
2 Puskás Squads	75 points	85 points

OPTIONS

- Replace the Command Rifle team with a Command Páncélvadász SMG team for +10 points or Command Panzerfaust SMG team for +15 points.
- Replace all Rifle teams with Rifle/MG teams for +10 points per Puskás Squad.

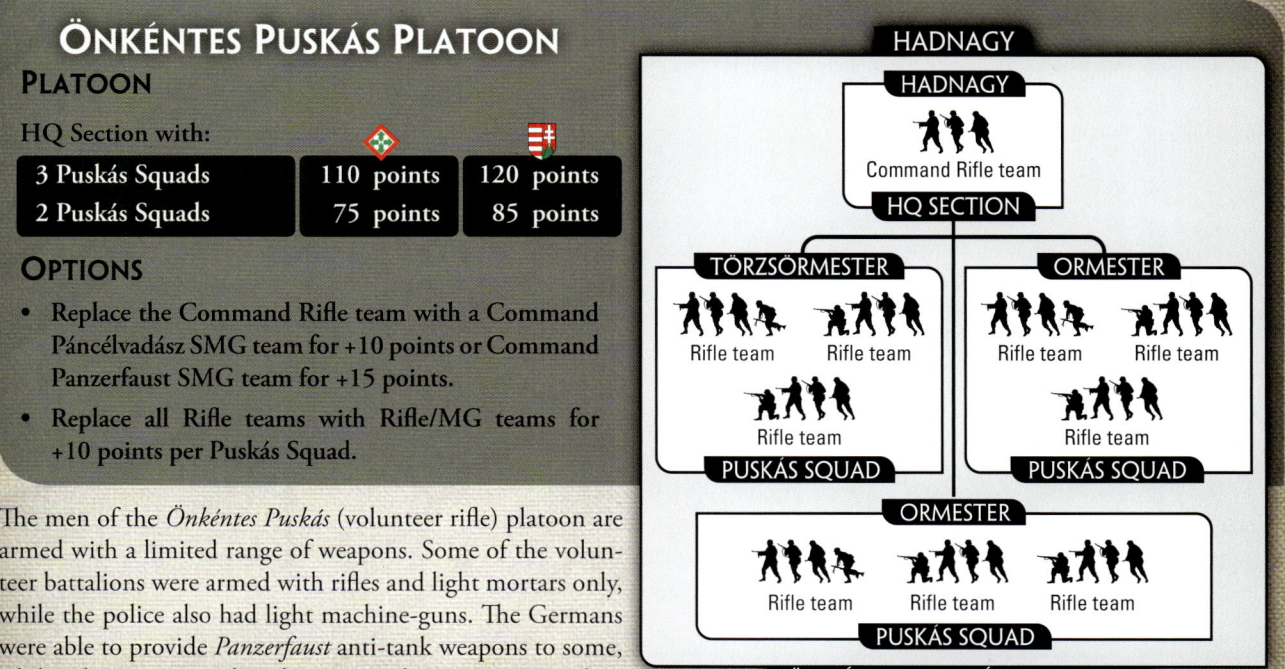

The men of the *Önkéntes Puskás* (volunteer rifle) platoon are armed with a limited range of weapons. Some of the volunteer battalions were armed with rifles and light mortars only, while the police also had light machine-guns. The Germans were able to provide *Panzerfaust* anti-tank weapons to some, while other improvised explosives to take out enemy tanks.

ÖNKÉNTES PUSKÁS MACHINE-GUN PLATOON

PLATOON

HQ Section with:

3 Machine-gun Sections	70 points	75 points

A Önkéntes Puskás Machine-gun Platoon may make Combat Attachments to Önkéntes Puskás Platoons.

Some of the volunteer, police and *Gendarmerie* battalions in Budapest were also issued heavy machine-guns. The regular riflemen of the 10th and 12th Infantry Divisions expect support from machine-guns as part of their training.

The high rate-of-fire provided by these weapons is vital in the tight confines of the street fighting for Budapest. A nest of machine-guns will often stop even the most determined infantry assault.

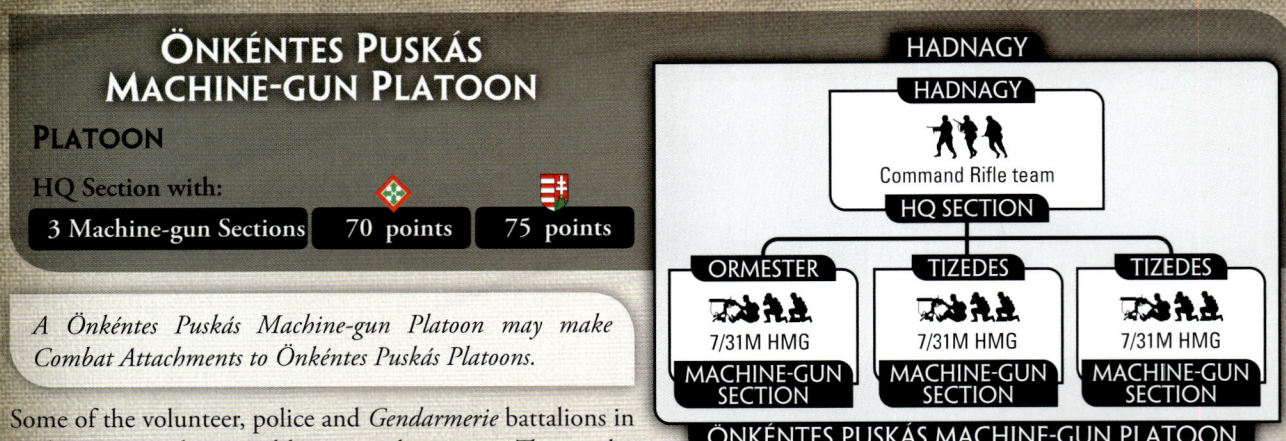

DIVISIONAL SUPPORT PLATOONS

ASSAULT ANTI-TANK PLATOON

PLATOON

HQ Section with:

3 75mm 40M	155 points
2 75mm 40M	105 points

OPTION

• Add Botond trucks for +5 points for the platoon.

The assault artillery are the elite of the artillery arm. An Assault Anti-tank Platoon is rated **Confident Veteran.**

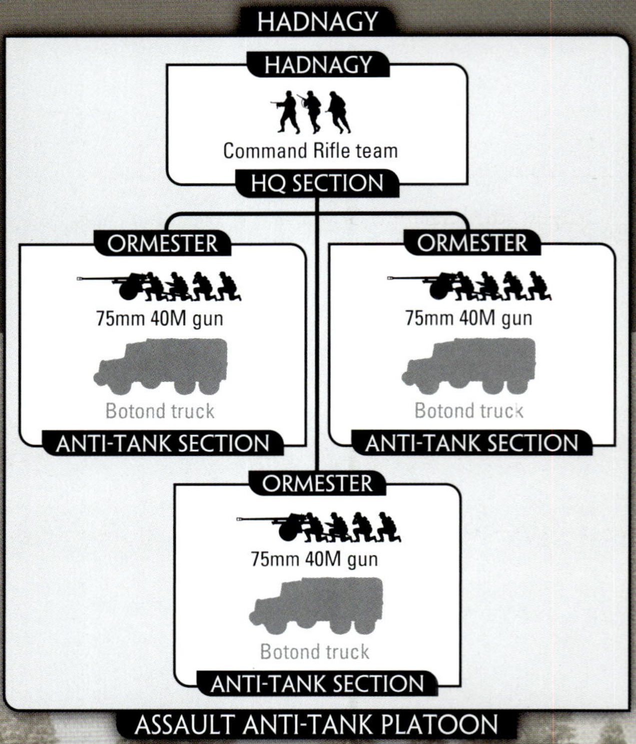

The men of the Hungarian assault artillery were considered elite. While many of these men fought in assault guns like the Zrinyi or Hetzer, some units didn't get vehicles and instead were armed with 75mm 40M (German PaK40) anti-tank guns.

These assault artillery troops were often attached to units that needed a boost to their firepower.

HUNGARIAN DIVISIONAL SUPPORT

MOTIVATION AND SKILL

The Hungarian soldiers could rely on a variety of support from their division. The infantry and armoured troops could call on pioneers, anti-aircraft and artillery. The various police, Gendarmerie, volunteer, and militia battalions fighting in Budapest were also supported by troops from the three regular divisions also trapped inside the city. They are rated **Confident Trained**, except where noted.

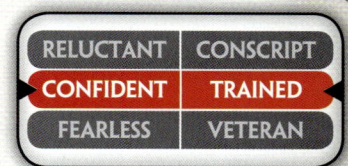

RELUCTANT	CONSCRIPT
CONFIDENT	**TRAINED**
FEARLESS	VETERAN

ARMOURED CAR PLATOON

PLATOON

3 Csaba 39M	85 points
3 Fiat-Ansaldo 35M	60 points
3 Toldi II	95 points

An Armoured Car Platoon is a Reconnaissance and a Huszár Platoon.

In 1944 the infantry were allocated a platoon of Csaba armoured cars to their reconnaissance battalions.

In Budapest the Galánta *Gendarmerie* Battalion operated a company of ten Csaba armoured cars, ten Fiat-Ansaldo tankettes and ten Toldi II tanks. During one assault all the Fiat-Ansaldo 35M tankettes were knocked out bravely at-

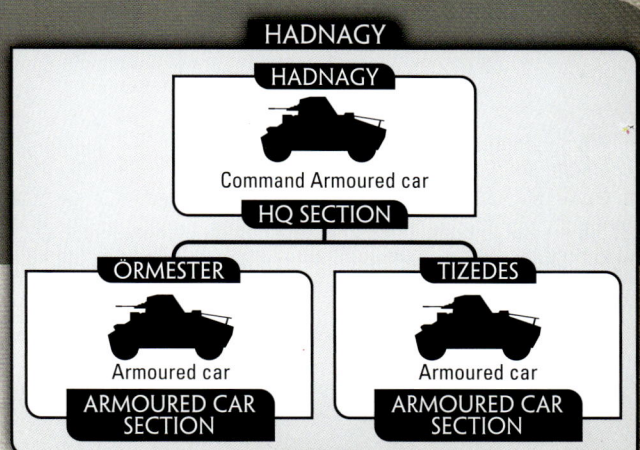

tempting to retake the racecourse in support of police and regular infantry. The Csaba supported the Gendarmes during a counterattack against Vecsés on 1 November 1944.

HUSZÁR PLATOON

PLATOON

HQ Section with:

3 Huszár Squads	220 points
2 Huszár Squads	155 points

OPTIONS

- Replace the Command Rifle/MG team with a Command Páncélvadász SMG team for +5 points or with a Command Panzerfaust SMG team for +10 points.
- Add Light Mortar teams for +30 points per team.

A Huszár Platoon is a Huszár and Reconnaissance Platoon.

The elite Huszár light cavalry of Hungarians has a long and proud history. A Huszár Platoon is rated **Confident Veteran.**

CONFIDENT	**VETERAN**

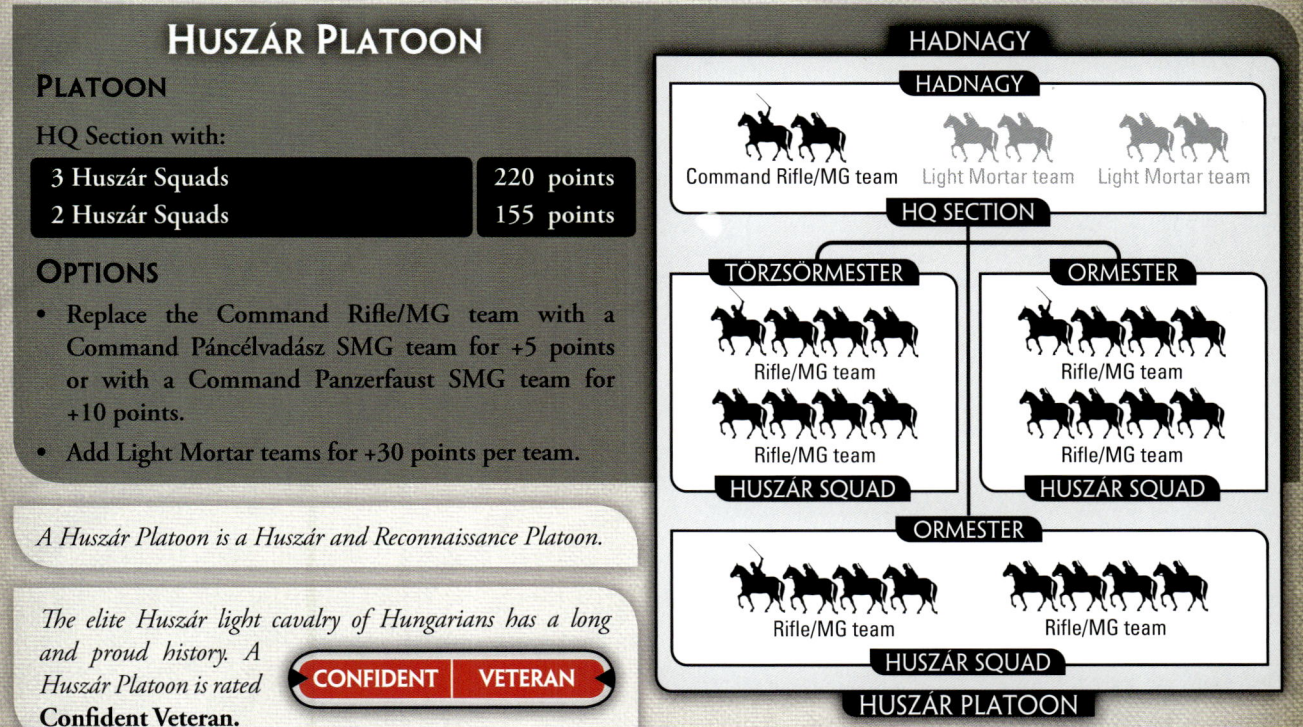

The advantages of the mounted *Huszár* (Hussar, pronounced hoo-sahr) become obvious on the Russian Front. Speed is their major asset, especially on the counterattack, where the enemy is not set for defence and the surprise of a cavalry force appearing on the flank can send even the steadiest company into disarray.

PIONEER PLATOON

PLATOON

HQ Section with:

3 Pioneer Squads	210 points
2 Pioneer Squads	145 points

OPTIONS

- Replace the Command Pioneer Rifle/MG team with a Command Panzerfaust Pioneer SMG team for +10 points.
- Add Pioneer Supply horse-drawn wagon for +20 points.

You may replace up to two Pioneer Rifle/MG teams with Flame-thrower teams at the start of the game before deployment.

Each infantry regiment has an allocation of engineering troops. They provide immediate support to infantry when obstacles, minefields and barricades need to be dealt with. In a pinch, they can also knock out the enemy armour.

ASSAULT PIONEER PLATOON

PLATOON

HQ Section with:

3 Pioneer Squads	270 points
2 Pioneer Squads	190 points

OPTIONS

- Replace the Command Pioneer Rifle/MG team with a Command Panzerfaust Pioneer SMG team for +10 points.
- Add 3-ton trucks for +5 points for the platoon.
- Add Pioneer Supply 3-ton truck for +25 points.

You may replace up to two Pioneer Rifle/MG teams with Flame-thrower teams at the start of the game before deployment.

Assault Pioneers are highly trained professionals. An Assault Pioneer Platoon is rated **Confident Veteran.**

CONFIDENT | VETERAN

AIR SUPPORT

SPORADIC AIR SUPPORT

Ju 87D Stuka	100 points
Ju 87G Stuka	100 points

The Royal Hungarian Air Force used the German-built Ju 87 dive-bomber for ground-attack to support the army in the field. The 102nd Dive Bomber Group operated under the German 4. *Luftwaffe Luftflotte.*

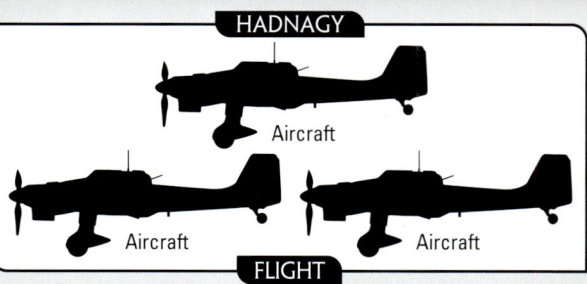

BORDER GUARD PLATOON

PLATOON

HQ Section with:

3 Border Guard Squads	160 points
2 Border Guard Squads	110 points

OPTION

- Replace the Command Rifle/MG team with a Command Páncélvadász SMG team for +5 points.

You may replace up to one Rifle/MG team with a Flame-thrower team at the start of the game before deployment.

The *Határvadász* (border guards, pronounced ho-tar vo-dahs) were equipped much like any other Hungarian rifleman. However, the fortress companies were often issued with flame-throwers to help in the defence of border passes and valleys that they defended.

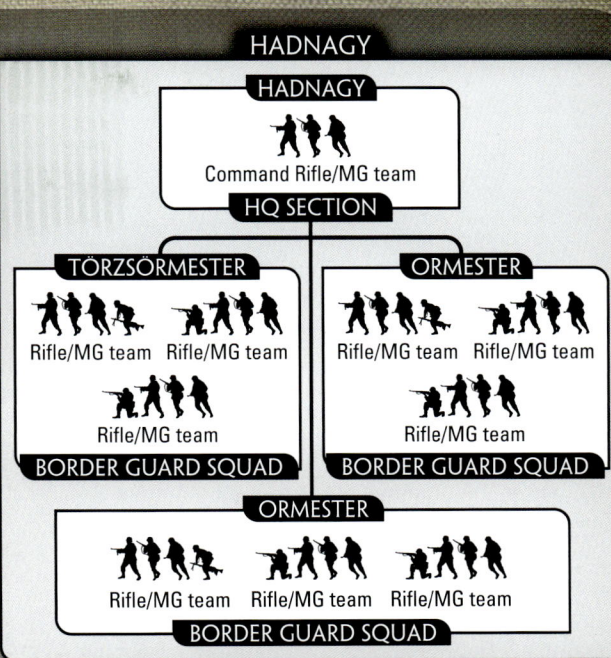

ROCKET LAUNCHER BATTERY

PLATOON

HQ Section with:

6 150mm 43M rocket launchers	150 points
4 150mm 43M rocket launchers	110 points
3 150mm 43M rocket launchers	85 points
2 150mm 43M rocket launchers	60 points

OPTION

- Add a Field Car and Botond trucks to the battery for +5 points.

In 1943 the Hungarians bought the license to produce the German 15cm Nebelwerfer rocket launcher. It was manufactured by the Manfréd-Weisz Company and designated the *43M sorozatvető* (multiple rocket launcher). In 1944 the Hungarian army fielded four battalions (151, 152, 153, 154) with 18 43M *sorozatvető* per battalion.

The rocket launcher battalions fought in Galicia and Hungary during 1944.

ARTILLERY BATTERY

PLATOON

HQ Section with:

4 100mm 14M	140 points
2 100mm 14M	80 points
4 149mm 14/31M	175 points
2 149mm 14/31M	100 points

OPTIONS

- Add horse-drawn wagon and limbers for +5 points for the platoon.
- Replace horse-drawn wagon and limbers with 3-ton trucks for 149mm 14/31M howitzers at no cost.

Artillery Batteries equipped with 149mm 14/31M howitzers may not be placed from Ambush within 16"/40cm of enemy teams.

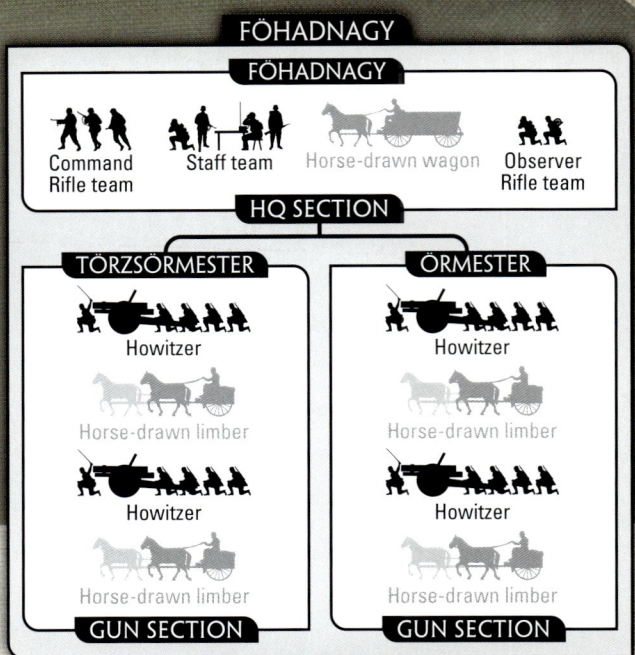

The artillery of the infantry divisions has undergone changes since 1943. The infantry's artillery fights with 100mm 14M Skoda howitzers as the standard weapons instead of the now retired 80mm guns. The medium batteries are now armed with 149mm 14/31M howitzers.

ANTI-AIRCRAFT PLATOON

PLATOON

HQ Section with:

2 40mm 36M	55 points

OPTION

- Add 3-ton trucks for +5 points for the platoon.

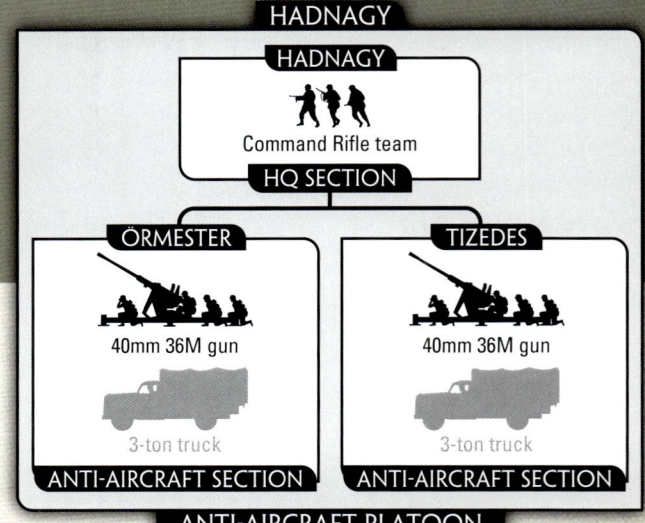

The standard anti-aircraft weapon of all Hungarian divisions is the Swedish designed 40mm Bofors gun. This modern and very effective anti-aircraft weapon is ideal for protecting your troops from roaming Allied ground-attack aircraft.

The Bofors gun's excellent rate-of-fire also makes it an effective weapon against ground targets.

HEAVY ANTI-AIRCRAFT PLATOON

PLATOON

HQ Section with:

2 80mm 29/38M	75 points
2 8.8cm FlaK37	125 points

OPTIONS

- Model 80mm 29/38M gun or 8.8cm FlaK37 gun with eight or more crew and increase their ROF to 3 for +10 points per gun.
- Add 3-ton trucks for +5 points for the platoon.

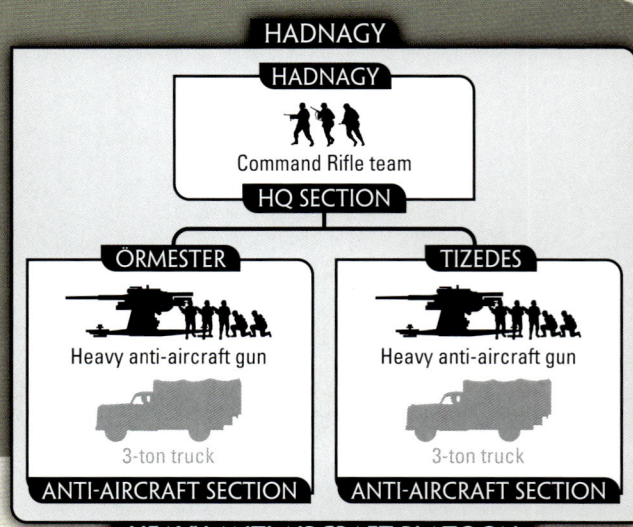

Hungarian heavy anti-aircraft fire was provided by the Bofors 80mm 29/38M guns. Some units got the infamous German '88' (8.8cm FlaK37 gun) in 1944.

HUNGARIAN SPECIAL RULES

HUSZÁR

Hungary has a strong cavalry tradition descended from Magyar horsemen. Hungarian knights often stood alone against the Ottoman Turks as the defenders of Europe and during the 17th to 19th Centuries they supplied the Habsburg Empire's elite light cavalry. Hungarian mobile troops are famed for their aggression and wide sweeping movements.

Hungarian platoons that are Huszár Platoons are so indicated under those platoons.

Any Huszár Platoon with a Command team may attempt a Huszár move at the start of the Shooting Step instead of shooting. If a platoon attempts to make a Huszár move, it may not shoot even if it fails to make a Huszár move.

Roll a Skill test for each platoon:

- *If the test is successful, the platoon may move another 4"/10cm,*
- *Otherwise the platoon cannot move this step.*

All normal rules apply for this movement. Platoons cannot make Huszár moves if they are Pinned Down or have moved At the Double. Bogged Down or Bailed vehicles cannot make Huszár moves.

BREAKTHROUGH GUN

Some weapons are just so powerful that there is no chance of surviving a hit from them. These heavy guns are often mounted in tanks and self-propelled guns designed to break through enemy defensive lines.

Infantry teams, Gun teams and Unarmoured vehicles automatically fail their saves when hit by a Breakthrough Gun or Bunker Buster. This does not apply to Artillery Bombardments.

LOCAL KNOWLEDGE

The Budapest defence battalions were made up of local police, students and workers who knew their city well. The Vannay Alarm Battalion, made up of municipal workers, would use the sewer and drainage system to move about the city undetected and launch surprise attacks.

A Önkéntes Puskás Század may deploy a Hungarian platoon entirely made up of Infantry or Gun teams, or a mix of both, in Immediate Ambush (see the rulebook) in addition to any Ambush deployment that may be allowed by the Mission played. The Immediate Ambushing platoon must be taken from the platoons that are to be deployed on table during deployment.

PREPARING FOR THE COMING STORM

The Hungarians learned harsh lessons on the Don front during the winter of 1942/43, so now when they prepare for defence, they prepare well.

Hungarian platoons may re-roll failed Skill Tests to Dig In.

SCHÜRZEN

Schürzen (German), Köténypáncélzattal (Hungarian) or side-skirt armour are the thin armoured plates that are welded to the sides of some tanks to protect them from infantry anti-tank weapons, like anti-tank rifles and bazookas.

When a tank that is protected by Schürzen is hit by a weapon with a Firepower of 5+ or 6 on the Side armour by shooting and fails its Armour Save, roll a special 4+ Schürzen save:

- *If the save is successful the Schürzen protects the tank from the side shot.*
- *If the save is not successful the shot penetrated the side armour as normal.*

227

HUNGARIAN ARSENAL

TANK TEAMS

Name *Weapon*	Mobility *Range*	Armour Front *ROF*	Side *Anti-tank*	Top *Firepower*	Equipment and Notes
LIGHT TANKS					
Toldi II	Standard Tank	2	1	1	Co-ax MG.
20mm 36M anti-tank rifle	*16"/40cm*	*3*	*5*	*5+*	
Toldi IIa	Standard Tank	2	1	1	Co-ax MG.
40mm 41M gun	*24"/60cm*	*3*	*7*	*4+*	
TANKS					
Turán I	Standard Tank	5	3	1	Co-ax MG, Hull MG, Protected ammo.
40mm 41M gun	*24"/60cm*	*3*	*7*	*4+*	
Turán II	Standard Tank	5	3	1	Co-ax MG, Hull MG, Protected ammo.
75mm 41M gun	*24"/60cm*	*2*	*9*	*3+*	
Panzer IV H	Standard Tank	6	3	1	Co-ax MG, Hull MG, Protected ammo, Schürzen.
7.5cm KwK40 gun	*32"/80cm*	*2*	*11*	*3+*	
Panther A	Standard Tank	10	5	1	Co-ax MG, Hull MG, Wide tracks.
7.5cm KwK42 gun	*32"/80cm*	*2*	*14*	*3+*	
HEAVY TANKS					
Tiger I E	Slow Tank	9	8	2	Co-ax MG, Hull MG, Protected ammo, Wide tracks.
8.8cm KwK36 gun	*40"/100cm*	*2*	*13*	*3+*	*Slow traverse.*
ASSAULT GUNS					
Zrínyi II	Standard Tank	7	2	1	AA MG, Protected ammo.
105mm 43M howitzer	*24"/60cm*	*2*	*10*	*2+*	*Hull mounted, Breakthrough gun.*
StuG G	Standard Tank	7	3	1	Hull MG, Protected ammo, Schürzen.
7.5cm StuK40 gun	*32"/80cm*	*2*	*11*	*3+*	*Hull mounted.*
Hetzer	Standard Tank	7	2	1	Hull MG, Overloaded.
7.5cm PaK39 gun	*32"/80cm*	*2*	*11*	*3+*	*Hull mounted.*
RECONNAISSANCE					
Csaba 39M	Wheeled	1	0	0	Co-ax MG.
20mm 36M anti-tank rifle	*16"/40cm*	*3*	*5*	*5+*	
Fiat-Ansaldo 35M	Half-tracked	1	0	1	Twin MG, Unreliable.
Motorcycle Rifle/MG team	Jeep	-	-	-	Motorcycle reconnaissance, Dismount as Rifle/MG team.
Rifle/MG	*16"/40cm*	*2*	*2*	*6*	*Hull mounted, Vehicle MG.*
Motorcycle Panzerfaust SMG team	Jeep	-	-	-	Motorcycle reconnaissance, Dismount as Panzerfaust SMG team.
When firing as SMG	*4"/10cm*	*3*	*1*	*6*	*Hull-mounted, Vehicle MG.*
When firing as Panzerfaust	*4"/10cm*	*1*	*12*	*5+*	*Awkward layout.*
SELF-PROPELLED ANTI-AIRCRAFT					
Nimrod	Standard Tank	2	1	0	
40mm 36M Bofors gun	*24"/60cm*	*4*	*7*	*4+*	*Anti-aircraft.*
VEHICLE MACHINE-GUNS					
Vehicle MG	*16"/40cm*	*3*	*2*	*6*	*ROF 1 if other weapons fire.*

AIRCRAFT

Aircraft	Weapon	To Hit	Anti-tank	Firepower	Notes
Ju 87D Stuka	Bombs	4+	5	1+	
Ju 87G Stuka	Cannon	3+	11	4+	

GUN TEAMS

Weapon	Mobility	Range	ROF	Anti-tank	Firepower	Notes
7/31M HMG	Man-packed	24"/60cm	6	2	6	
81mm 36/39M mortar	Man-packed	24"/60cm	2	2	3+	Smoke, Minimum range 8"/20cm.
Firing bombardments		40"/100cm	-	2	6	Smoke bombardment.
120mm 43M mortar	Light	56"/140cm	-	3	3+	
40mm 40M gun	Light	24"/60cm	3	7	4+	Gun shield.
Firing Stielgranate		8"/20cm	1	12	5+	
75mm 40M gun (7.5cm PaK40)	Medium	32"/80cm	2	12	3+	Gun shield.
40mm 36M Bofors gun	Immobile	24"/60cm	4	7	4+	Anti-aircraft, Turntable.
80mm 29/38M Bofors gun	Immobile	32"/80cm	2	12	3+	Heavy anti-aircraft, Turntable.
8.8cm FlaK37 gun	Immobile	40"/100cm	2	13	3+	Gun shield, Heavy anti-aircraft, Turntable.
100mm 14M howitzer (100/17)	Immobile	24"/60cm	1	9	2+	Gun shield, Breakthrough gun.
Firing bombardments		72"/180cm	-	4	4+	
105mm 37M howitzer (leFH18)	Immobile	24"/60cm	1	10	2+	Gun shield, Breakthrough gun, Smoke.
Firing bombardments		72"/180cm	-	4	4+	Smoke bombardment.
149mm 14/31M howitzer	Immobile	16"/40cm	1	8	1+	Gun shield, Bunker buster.
Firing bombardments		72"/180cm	-	5	2+	
150mm 43M rocket launcher	Light	64"/160cm	-	3	4+	Rocket launcher, Smoke bombardment.

INFANTRY TEAMS

Team	Range	ROF	Anti-tank	Firepower	Notes
Rifle team	16"/40cm	1	2	6	
Rifle/MG team	16"/40cm	2	2	6	
SMG team	4"/10cm	3	1	6	Full ROF when moving.
Light Mortar team	16"40cm	1	1	4+	Can fire over friendly teams.
Panzerschreck team	8"/20cm	2	11	5+	Tank Assault 5.
Flame-thrower team	4"/10cm	2	-	6	Flame-thrower.
Staff team	cannot shoot				Moves as a Heavy Gun team.

ADDITIONAL TRAINING AND EQUIPMENT

Panzerfaust	4"/10cm	1	12	5+	Tank Assault 6, Cannot shoot in the Shooting Step if moved in the Movement Step.

Páncélvadész and Pioneer teams are rated as Tank Assault 4.

TRANSPORT TEAMS

Vehicle	Mobility	Armour			Equipment and Notes
		Front	Side	Top	
TRUCKS					
Motorcycle team or Field Car	Jeep	-	-	-	
Botond truck or 3-ton truck	Wheeled	-	-	-	
Horse-drawn wagon	Horse-drawn	-	-	-	
TRACTORS					
Horse-drawn limber	Horse-drawn	-	-	-	
RECOVERY VEHICLES					
Famo	Half-tracked	-	-	-	Recovery vehicle.

FORTIFICATIONS

Weapon	Mobility	Range	ROF	Anti-tank	Firepower	Notes
40mm Bofors Nest	Immobile	24"/60cm	4	7	4+	Anti-aircraft.
HMG Pillbox	Immobile	24"/60cm	6	2	6	
HMG Nest	Immobile	24"/60cm	6	2	6	

A Hungarian StuG G knocks out a T-34/85 tank from a concealed position.

Budapest

Hungarian troops and Hetzers defend a barricade against a Soviet Shturmovye Group.

THE FINNISH ARMY IN 1944

The Finnish Army became famous for its epic defence against Soviet invasion in the Winter War (*Talvisota*) in 1939-40. In the Continuation War (*Jatkosota*) in 1941-42 it recaptured the lost Finnish territory, settling down to defend it over the next two years. In the summer of 1944 Finland faced another final massive Soviet invasion.

SOVIET ASSAULT AND BREAKTHROUGH

The relief of the siege of Leningrad enabled large Soviet forces to be positioned on the Karelian Isthmus. By June 1944 the Soviets had massed two armies on the Karelian Isthmus under the command of Army General Leonid Govorov, an expert in assault warfare who had commanded Soviet forces during much of the siege of Leningrad. The 21st Army had 15 divisions and the 23rd Army had 9 divisions, with over 600 tanks including IS-2, T-34/85, ISU-152 and British supplied Churchill tanks. There were another sixteen divisions with the 7th and 32nd Armies in Eastern Karelia. In contrast the Finns had allowed their defences to run down or were not completed.

The hammer fell on 9 June in the Valkeasaari sector of Finnish 10th Division. Hundreds of planes and 300 guns per kilometre of front pounded the Finnish lines. Soviet gunners fired over 80,000 rounds on the Karelian Isthmus that day. Badly maintained positions collapsed and troops retreated.

The Finnish High command ordered a fall back to the VT Line.

On 13 June the Soviets began attacking the VT line at Kuuterselkä and Siiranmäki. At Siiranmäki *Jallkaväki* Regiment 7 (JR7) from 2nd Division was attacked by three Soviet divisions, but fought an epic defence. However at Kuuterselkä village Govorov concentrated on the weakest point in the Finnish lines. 10th Division was defeated and Soviet tanks broke through. This tore open the whole Finnish front on the isthmus, from Kuuterselkä to the Gulf of Finland. The *Panssari* Division (Armoured Division) and Finnish Cavalry Brigade tried to counterattack but could only slow the Soviet tank spearhead.

WITHDRAWAL TO VKT LINE AND THE LOSS OF VIIPURI

By 20 June the Soviet spearheads had reached the VKT line. On the western end of the line was Viipuri, Finland's third largest city. Govorov quickly attacked in strength, with infantry supported by KV and lend-lease Churchill heavy tanks. The defending Finnish 20th Brigade withdrew almost without a fight and Viipuri was lost in a single day. Stalin had reason to be pleased. Two of the three Finnish defence lines had been breached, Viipuri would be the last victory for the Soviet's summer offensive. For the Finns help was beginning to arrive. Two more Finnish divisions (6th and 11th) were transferred from east Karelia. German assistance saw the issuing of Panzerfaust (*Panssarinyrkki*) and Panzerschreck (*Panssarikauhu*), and assault guns. They temporarily transferred the German *122. Infanteriedivision, 303. Sturmgeschütz* (assault gun) brigade, and, most importantly, Air Group Kuhlmey to Finnish command.

BATTLE OF TALI - IHANTALA

The Soviets pressed on their attack and by 30 June the Finns had fallen back to the village of Ihantala. The Finnish 18th and *Panssari* divisions launched the first counterattacks with *Sturmi* assault guns supporting the infantry, but became worn down after a week of fighting and fell back. The 6th Division under General Eric Vihma was thrown into the gap, at first falling back, but then holding. By 1 July Soviet forces were strung out

LAKE LADOGA

IV CORPS

III CORPS

23. ARMY

21. ARMY

LENINGRAD FRONT

GULF OF FINLAND

LENINGRAD

be shifted. Over 12 days the Soviets had lost over 300 tanks and at least 22,000 casualties, while the Finns lost 8000 casualties. The VKT line had been bent but not broken.

OUTFLANKING ATTEMPTS AT VIIPURINLAHTI AND VUOSALMI

As the tide at the Tali-Ihantala battle was beginning to turn against the Soviet forces, attempts were made to attack around either flank of the Ihantala position. To the west and north-west of Viipuri lay the bay of Viipuri (Viipurinlahti in Finnish). The fresh Soviet 59th Army attempted to cross the bay by boat in an 'island hopping' strategy on 1 July. Soviet troops were attacked while still in their boats and immediately after landing on shore and decisively defeated, the last troops being withdrawn by 10 July.

in columns along the road between Tali and Ihantala and suffering increasing losses from artillery, flank attacks by Finnish and German StuG assault guns, infantry with the new anti-tank weapons, and Finnish and German aircraft.

Finally 6th Division's *Jallkaväki* Regiment 12 (JR12) halted the Soviet spearhead at Ihantala Hill on 30 June. Still, fierce battles continued as the Soviets tried to force another breakthrough. Finnish intelligence discovered the time and place and pulverised the forming up area with over 250 guns. The final Soviet attacks were stopped dead. After 9 July Soviet attacks petered out and the area settled down to a stalemate. The Finns dug in and with their artillery support could not

To the east of Tali, the Soviets again tried to outflank the Finns holding along the Vuoksi River line. Soviet 98th corps attacked and forced the defenders across the river in heavy battles lasting five days. The Soviets committed a further three divisions across the river to exploit the bridgehead, leading to a bitter melee on the north shore of the Vuoksi. The combined efforts of the 2nd Division, artillery, air force, tanks and *Jääkäri* (light infantry) of the *Panssari* Division managed to contain the assault force, which became trapped on the open terrain on the north bank. In this battle alone the Soviets lost over 15,000 men and the Finns 6000. JR7 of 2nd Division took almost 75% casualties, but did not break.

Battles North of Lake Ladoga

The second part of the Soviet plan had been to trap Finnish forces in Karelia to the east. The Soviet attack began on the 21 June and on the same day managed to cross the Svir River and gain a significant break-in. Finnish troops from II Corps and Group Aunus then fought a fighting withdrawal by establishing one delaying position after another. By 10 July the Finnish troops had manned the so-called U-line located inside the pre-Winter War border. The Soviet 7th Army was unable to break through U-line. The Soviets then switched focus against II Corps north of the U-line. Battles continued until the beginning of August when the Soviets stopped all their attacks.

Ilomantsi – the Final Motti Battle

In the north, a short distance east of Ilomantsi, from 31 July to 13 August, Finnish and Soviet forces were locked in combat for what would turn out to be the last time. By 4 August two Soviet divisions had been surrounded by Group Raappana. On 9 August the main body of the two divisions broke out through the surrounding Finnish troops to the east with the loss of most of their heavy equipment.

The Price of Peace

By August the front had stabilised to trenches and static defence along roughly the 1940 Finnish – Soviet boundary. A truce was negotiated prior to a full peace treaty coming into effect in September.

The conditions of the treaty required Finland to demobilise most of its army, and eject German forces in Lapland. This was a task troops had little enthusiasm for. It led to a six month campaign that ended in stalemate in the barren arctic landscape. Despite a Finnish amphibious landing at Tornio, and outflanking attempts, the last of the German 20th Mountain army crossed into Norway at Kilpisjärvi in April 1945.

The Butchers Bill

The scale of fighting on the Finnish front in the 1944 summer offensive was immense. In the three months from the start of the Soviet offensive on 9 June to the cease fire on 4 September the Finns lost 60,000 casualties, including 15,000 dead, most in the first month of fighting. For the Finns the losses per day were even more than in the Winter War. Over four years from 1941 to 1944 the Continuation War had cost the Finns 250,000 casualties including 60,000 dead. Still by the end of the war they were the only country in Eastern Europe that remained independent.

The Soviet 1944 offensive had been much better planned than in 1939 and their troops far better trained, but Soviet

losses were still heavy. Sources differ on Soviet casualties, but in total they are likely to have exceeded 150,000 men and 600 tanks for the 1944 summer offensive - a ratio of three Soviet casualties for every one Finnish casualty in the 1944 fighting. Given that the Soviets were attacking through rugged woods and lake country, this is not surprising. Indeed it compares favourably with Allied casualties attacking positions such as Cassino in Italy. (In the Winter War the Soviets lost over 400,000 casualties against 60,000 Finnish, a ratio of seven to one.) Total Soviet Continuation War casualty estimates vary from over 400,000 (Krivosheev) to over 600,000 (Manninen) including 300,000 dead.

ARMOUR IN FINNISH SERVICE

1ST TANK BATTALION

1st Tank Company
HQ (2 T-26 tanks)
1st Platoon (5 T-26 tanks)
2nd Platoon (5 T-26 tanks)
3rd Platoon (5 T-26 tanks)

2nd Tank Company
HQ (2 T-26 tanks)
1st Platoon (5 T-26 tanks)
2nd Platoon (5 T-26 tanks)
3rd Platoon (5 T-26 tanks)

3rd Tank Company
HQ (2 T-26 tanks)
1st Platoon (3 T-34 tanks)
2nd Platoon (4 T-28 tanks)
3rd Platoon (5 T-26 tanks)

2ND TANK BATTALION

4th Tank Company
HQ (2 T-26 tanks)
1st Platoon (5 T-26 tanks)
2nd Platoon (5 T-26 tanks)
3rd Platoon (5 T-26 tanks)

5th Tank Company
HQ (2 T-26 tanks)
1st Platoon (5 T-26 tanks)
2nd Platoon (5 T-26 tanks)
3rd Platoon (5 T-26 tanks)

6th Tank Company
HQ (2 T-26 tanks)
1st Platoon (2 KV-1 tanks)
2nd Platoon (4 T-28 tanks)
3rd Platoon (5 T-26 tanks)

T-34/85 tanks were used to replace T-26 tanks as they were captured.

ASSAULT GUN BATTALION

1st Assault Gun Company
HQ (1 Stu 40 G *Sturmi*)
1st Platoon (3 Stu 40 G *Sturmi* assault guns)
2nd Platoon (3 Stu 40 G *Sturmi* assault guns)
3rd Platoon (3 Stu 40 G *Sturmi* assault guns) (not fully equipped until late July 1944)

3rd Assault Gun Company
HQ (1 Stu 40 G *Sturmi*)
1st Platoon (3 Stu 40 G *Sturmi* assault guns)
2nd Platoon (3 Stu 40 G *Sturmi* assault guns)
3rd Platoon (3 Stu 40 G *Sturmi* assault guns) (not fully equipped until late July 1944)

2nd Assault Gun Company
HQ (1 Stu 40 G *Sturmi*)
1st Platoon (3 Stu 40 G *Sturmi* assault guns)
2nd Platoon (3 Stu 40 G *Sturmi* assault guns)
3rd Platoon (3 Stu 40 G *Sturmi* assault guns) (not fully equipped until late July 1944)

TOIVO ILOMÄKI

Toivo Ilomäki was born in 1917 and worked as a storeman before the Winter War. He participated in the Winter War and the early Continuation war as a member of an anti-tank gun crew initially as an ammunition carrier. However, in the middle of a battle in 1941 when the gunner did not seem to be able to hit anything, he went to the platoon commander and said he would be a better shot. With the permission of the platoon commander he then demonstrated this by shooting into a tree that fell on Soviet infantry.

In 1944 he was the gunner of a PaK40 and destroyed a total of 21 tanks. At the Battle of Sammatus alone he destroyed 16 Soviet tanks over three days, many of them at close range in the middle of a Soviet artillery barrage. These included T-34 tanks and IS-2 heavy tanks. He was the only man in the Finnish army to get all four additional stripes to the tank killer badge and received the Mannerheim Cross for his valour.

After the war Ilomäki returned to his job as a storeman and died in 1965, aged just 48. His gun (with 21 kill marks) is preserved in the armour museum in Parola.

CHARACTERISTICS

Toivo Ilomäki is a Warrior who may be added to upgrade one anti-tank gun team of a Heavy Anti-Tank Platoon for +10 points. He is a **Fearless Veteran** team armed with the same anti-tank gun as the rest of the platoon.

NERVES OF STEEL

Ilomäki was a calm, phlegmatic character whose response to difficult situations was to simply keep doing his job.

> *Toivo Ilomäki's Gun team is never Pinned Down, regardless of the platoon's status.*

ONE SHOT, ONE KILL

Ilomäki is a crack shot, who not only hits his targets, but knew the vulnerable spots to aim for on Soviet tanks.

> *Toivo Ilomäki's Gun team may re-roll any failed To Hit rolls.*

WARRIOR GUN TEAM

Ilomäki is a Warrior Gun team.

> *If the enemy Destroys Ilomäki's Gun team by Shooting or assault (but not as a result of a failed Platoon Morale Check), they roll a die.*
>
> *On a roll of 4+ Ilomäki is killed or seriously wounded and the team is Destroyed.*
>
> *Otherwise, the Finnish player replaces any other friendly Gun team equipped with the same weapon that is within Command Distance of Ilomäki with Ilomäki's Gun team. The Gun team Ilomäki takes over is Destroyed.*

LAURI TÖRNI

Lauri Allan Törni was born in Viipuri in 1919 where his family had a large home. He learned from an Olympic boxing champion and already had a reputation as a tough, disciplined fighter and a fine skier before he joined the Finnish Army in 1938. During the Winter War he proved a brilliantly effective soldier in *Sissi* ski-guerilla units. By the end of the Winter War he had earned the rank of Sergeant.

Despite this, his home in Viipuri was part of the Finnish territory lost to the Soviet Union. This only added to Törni's hatred of communism.

After the Winter War, Törni was determined to continue the fight against the Soviet Union, and travelled to Germany to train with the *Waffen-SS*. In Operation Barbarossa, Törni again proved an excellent soldier, and was decorated with the Iron Cross Second Class. After Finland declared war on the Soviet Union again in the Continuation War, he returned home and re-enlisted in the Finnish Army.

During the Continuation War (1941-1944) Törni proved an excellent combat leader, first with armour and then with light infantry, and was promoted to Captain. His long-range missions were so successful that the Red Army put a price of 3 million Finnish Marks on his head. In 1944 he was transferred to the Karelian Isthmus, scene of the final Soviet offensive. He led a *Jääkäri* unit which made a decisive counterattack and helped stabilise the line after the Soviet breakthrough. He was awarded the Mannerheim Cross, Finland's highest medal for bravery.

Torni's life in the military did not end there. After the war he journeyed to the United States, became a citizen and enlisted in the US army as Larry Thorn. He served with US Special Forces and led missions in places ranging from Iran to Vietnam, being decorated several times. He served two tours in Vietnam before being killed in Laos in 1965.

He was the basis of the character played by John Wayne in the movie Green Berets.

CHARACTERISTICS

Captain Lauri Törni is a Warrior and Command team rated as **Fearless Veteran.** Lauri replaces the Command team of a Jääkäri Platoon for a cost of +65 points. Lauri counts as a Recce team.

He is armed with a Suomi SMG and Satchel Charges with the following ratings: Range: 4"/10cm, ROF: 3, Anti-tank: 1, and Firepower: 6. Like an SMG team, Lauri fires at full ROF when moving and with Satchel Charges counts as having Tank Assault 4.

IMPLACABLE

Törni is brave and daring soldier and this inspires the men who fight alongside him.

Lauri and the Jääkäri Platoon he has joined pass all Motivation Tests on a 2+.

SKILLED SOLDIER

Not only is Törni brave but he knows the skills of his trade well. He puts his soldiering knowledge to good use leading his men.

Lauri and the Jääkäri Platoon he has joined may re-roll any failed Skill Tests.

PANSSARIKOMPPANIA
TANK COMPANY

(TANK COMPANY)

HEADQUARTERS

Panssarikomppania HQ — 239

You must field one platoon from each box shaded black and may field one platoon from each box shaded grey.

COMBAT PLATOONS

ARMOUR
Panssari Platoon — 239

ARMOUR
Panssari Platoon — 239

ARMOUR
Panssari Platoon — 239

ALLIED PLATOONS

Heer Platoons in your force are Allies and follow the Allies rules in the rulebook.

DIVISIONAL SUPPORT PLATOONS

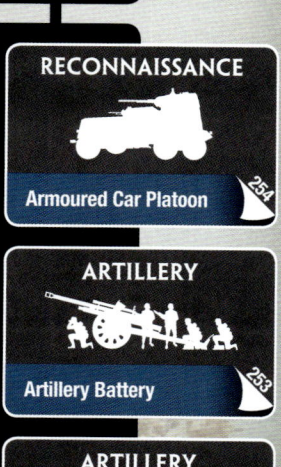

ARMOUR
Captured Panssari Platoon — 245
Assault Gun Platoon — 254
German StuG Platoon — 59

ANTI-TANK
Heavy Anti-tank Platoon — 252

INFANTRY
Jääkäri Platoon — 243
Pioneeri Platoon — 251
German Grenadier Platoon — 27

INFANTRY
Jääkäri Platoon — 243

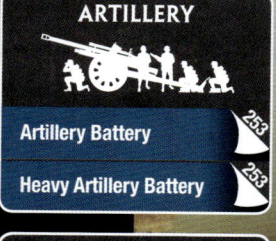

RECONNAISSANCE
Armoured Car Platoon — 254

ARTILLERY
Artillery Battery — 253

ARTILLERY
Artillery Battery — 253
Heavy Artillery Battery — 253

ANTI-AIRCRAFT
Self-propelled Anti-aircraft Platoon — 254

AIR SUPPORT
Air Support — 253
German Air Support — 172

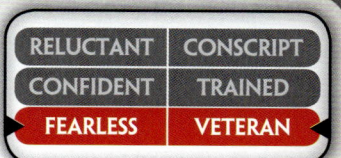

MOTIVATION AND SKILL

Finnish tank units have spent the last two years constantly training. Their commanders are veterans from the earlier fighting. A Panssarikomppania is rated as **Fearless Veteran.**

RELUCTANT	CONSCRIPT
CONFIDENT	TRAINED
FEARLESS	**VETERAN**

HEADQUARTERS

PANSSARIKOMPPANIA HQ

HEADQUARTERS

2 T-26	65 points

OPTION

- Replace up to one T-26 tank with either a T-34/76 tank for +50 points or a T-34/85 tank for +85 points.
- Upgrade T-34/76 tank to have a Cupola for +5 points.
- Add Famo recovery vehicle for +5 points.
- Replace Famo recovery with ISU recovery vehicle for +5 points.

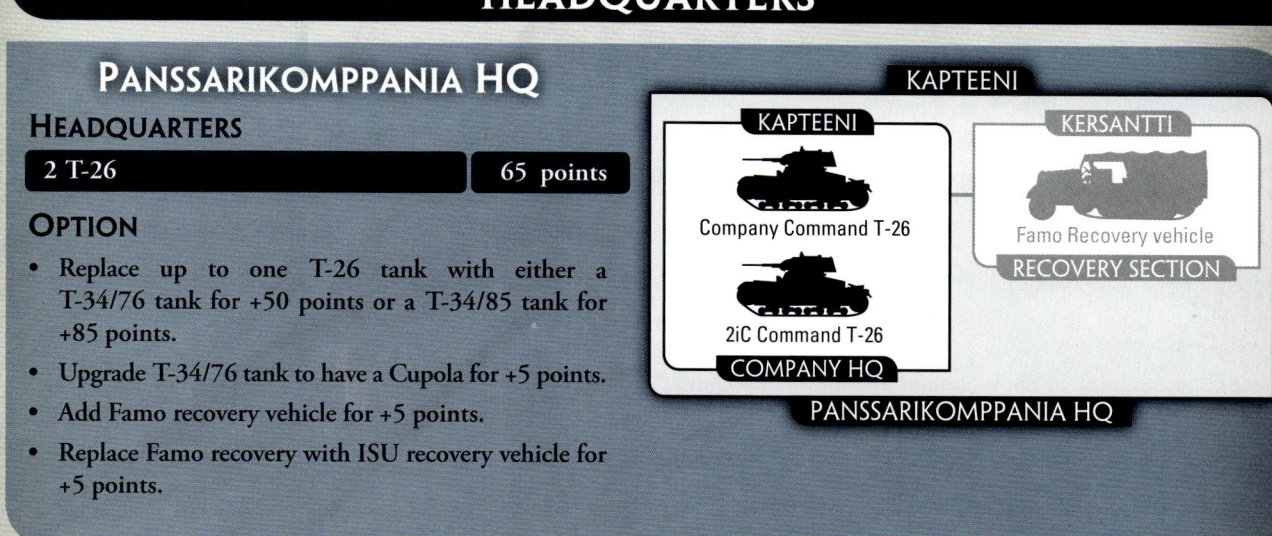

The leaders of the *Panssari* (Tank or Armour — pronounced pans-sa-ree) units are seasoned veterans, who have had up to two years training. They have learnt to survive operating obsolete Soviet light and medium tanks and are now full of confidence in new T-34 tanks. *Panssari* unit commanders are chosen for aggression and initiative, and lead from the front.

COMBAT PLATOONS

PANSSARI PLATOON

PLATOON

5 T-26	160 points
4 T-26	130 points
3 T-26	95 points
4 T-28	180 points
3 T-28	135 points
2 KV-1	215 points
3 T-34/76	245 points

- Upgrade all T-34/76 tanks to have Cupolas for +5 points for the platoon.

3 T-34/85	355 points

Any type of Finnish Company may only field one Panssari Platoon of each of the following tanks:

- *T-34/76*
- *T-34/85*
- *T-28*
- *KV-1*

You may have as many of your Panssari Platoons equipped with T-26 tanks as you like.

Some *Panssari* companies are now a mixture of the captured Soviet medium tanks. However, most brave crews still soldier on in T-26 and T-28 tanks The T-26 tank remains the most numerous tank in the Finnish arsenal.

STURMIKOMPPANIA
ASSAULT GUN COMPANY

(TANK COMPANY)

HEADQUARTERS

HEADQUARTERS

Sturmikomppania HQ | 241

You must field one platoon from each box shaded black and may field one platoon from each box shaded grey.

COMBAT PLATOONS

ARMOUR

Sturmi Platoon | 241

ARMOUR

Sturmi Platoon | 241

ARMOUR

Sturmi Platoon | 241

ALLIED PLATOONS

Heer Platoons in your force are Allies and follow the Allies rules in the rulebook.

DIVISIONAL SUPPORT PLATOONS

ARMOUR

Assault Gun Platoon | 254

German StuG Platoon | 59

ANTI-TANK

Heavy Anti-tank Platoon | 252

INFANTRY

Jalkaväki Platoon | 247

Jääkäri Platoon | 243

Pioneeri Platoon | 251

German Grenadier Platoon | 27

INFANTRY

Jääkäri Platoon | 243

RECONNAISSANCE

Armoured Car Platoon | 254

ARTILLERY

Artillery Battery | 253

ARTILLERY

Artillery Battery | 253

Heavy Artillery Battery | 253

ANTI-AIRCRAFT

Self-propelled Anti-aircraft Platoon | 254

AIR SUPPORT

Air Support | 253

German Air Support | 172

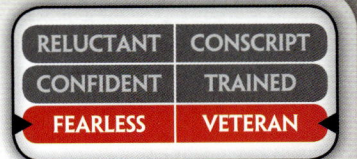

MOTIVATION AND SKILL

Finnish Sturmi (assault guns) units have spent the last two years constantly training, many of the crews have transferred from the tanks to take command of these new fighting vehicles. Their commanders are veterans from the earlier fighting. A Sturmikomppania is rated as **Fearless Veteran.**

RELUCTANT	CONSCRIPT
CONFIDENT	TRAINED
FEARLESS	**VETERAN**

HEADQUARTERS

STURMIKOMPPANIA HQ

HEADQUARTERS

1 Stu 40 G Sturmi	105 points

OPTION

- Add Famo recovery vehicle for +5 points.

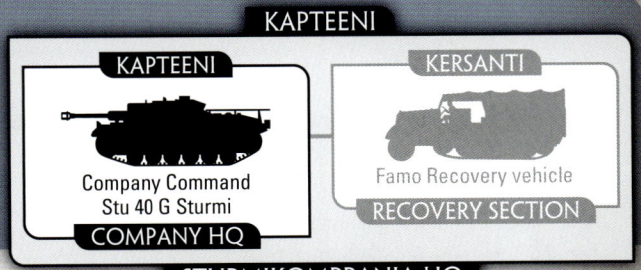

With our new German-supplied *Sturmi* (StuG III G) assault guns, the *Panssari* division now packs quite a punch. The *Sturmi* crews have been training hard for two years, are led by selected officers picked from *Jääkäri* units, and are now ready for battle against the toughest foes.

COMBAT PLATOONS

STURMI PLATOON

PLATOON

3 Stu 40 G Sturmi	315 points

The *Sturmi* assault guns have been placed in a separate battalion. Their 75mm guns, combined with the skill of their Finnish crews, have enabled them to achieve impressive kill ratios, even against Soviet heavy armour.

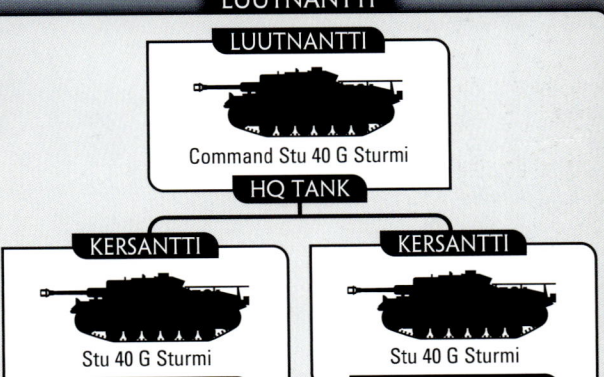

JÄÄKÄRIKOMPPANIA
LIGHT INFANTRY COMPANY

(INFANTRY COMPANY)

HEADQUARTERS

Jääkärikomppania HQ — 243

You must field one platoon from each box shaded black and may field one platoon from each box shaded grey.

COMBAT PLATOONS

INFANTRY

Jääkäri Platoon — 243

INFANTRY

Jääkäri Platoon — 243

INFANTRY

Jääkäri Platoon — 243

ALLIED PLATOONS

Heer Platoons in your force are Allies and follow the Allies rules in the rulebook.

WEAPONS PLATOONS

MACHINE-GUN

Jääkäri Machine-gun Platoon — 244

MACHINE-GUN

Jääkäri Machine-gun Platoon — 244

ARTILLERY

Jääkäri Mortar Platoon — 244

ANTI-TANK

Jääkäri Tank-hunter Platoon — 245

BRIGADE SUPPORT PLATOONS

ANTI-TANK

Jääkäri Heavy Tank-hunter Platoon — 245

ANTI-TANK

Jääkäri Heavy Tank-hunter Platoon — 245

ARMOUR

Captured Panssari Platoon — 245

DIVISIONAL SUPPORT PLATOONS

ARMOUR

Panssari Platoon — 239
Sturmi Platoon — 241
German StuG Platoon — 59

ARMOUR

Panssari Platoon — 239
Assault Gun Platoon — 254
Sturmi Platoon — 241
Heavy Anti-tank Platoon — 252

INFANTRY

Jalkaväki Platoon — 247
Pioneeri Platoon — 251
German Grenadier Platoon — 27

RECONNAISSANCE

Armoured Car Platoon — 254

ARTILLERY

Artillery Battery — 253

ARTILLERY

Artillery Battery — 253
Heavy Artillery Battery — 253

ANTI-AIRCRAFT

Anti-aircraft Platoon — 254

AIR SUPPORT

Air Support — 253
German Air Support — 172

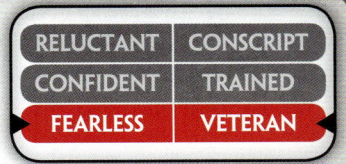

The *Jääkäri* remain the best of the Finnish infantry, and rigorous training and patrolling has kept them so during the quiet of 1943. A *Jääkärikomppania* is rated **Fearless Veteran**.

Any Divisional Support Platoons you take in your force must be marked 🔴 or have no symbol.

RELUCTANT	CONSCRIPT
CONFIDENT	TRAINED
FEARLESS	**VETERAN**

HEADQUARTERS

JÄÄKÄRIKOMPPANIA HQ

HEADQUARTERS

Company HQ	50 points

OPTIONS

- Add an Anti-tank Section with Lahti Anti-tank Rifle teams for +15 points per team.
- Replace Lahti Anti-tank Rifle teams with Close-defence Rifle teams for +5 points per team, Panzerfaust Rifle teams for +10 points per team, or Panzerschreck teams for +15 points per team.
- Add up to three Sniper team for +50 points per team.

KAPTEENI

KAPTENNI

Company Command SMG team   2iC Command SMG team

COMPANY HQ

KERSANTTI

Lahti Anti-tank Rifle   Lahti Anti-tank Rifle

ANTI-TANK SECTION

JÄÄKÄRIKOMPPANIA HQ

A *Jääkärikomppania* (Light Infantry Company — pronounced yay-kar-ree komp-pa-nee-a) is led by a *Kapteeni* (Captain) who has been selected for his skill, courage and initiative. The *Jääkäri* provide much of the cutting edge of the Finnish army. The *Panssari* Division has a brigade of *Jääkäri* that are the finest troops in the army. While much of the army has been used in a static defensive role for the last two years, the *Jääkäri* have been patrolling and engaging the enemy. Their tactics emphasis speed and mobility. A high proportion of their leaders are volunteers or former regular army officers. For them the best form of reconnaissance is attack.

COMBAT PLATOONS

JÄÄKÄRI PLATOON

PLATOON

HQ Section with:

4 Jääkäri Squads	175 points
3 Jääkäri Squads	135 points

OPTIONS

- Replace the Command SMG team with a Command Close-defence SMG team for +5 points or with a Command Panzerfaust SMG team for +10 points.
- Replace all Rifle teams with SMG teams for +15 points per Jääkäri Squad.
- Upgrade a Jääkäri Platoon to be a Reconnaissance Platoon for +60 points for the platoon.

Only one Jääkäri Platoon in a company may be upgraded to a Reconnaissance Platoon.

LUUTNANTTI

LUUTNANTTI

Command SMG team

HQ SECTION

KERSANTTI

Rifle team   Rifle team

JÄÄKÄRI SQUAD

KERSANTTI

Rifle team   Rifle team

JÄÄKÄRI SQUAD

KERSANTTI

Rifle team   Rifle team

JÄÄKÄRI SQUAD

KERSANTTI

Rifle team   Rifle team

JÄÄKÄRI SQUAD

JÄÄKÄRI PLATOON

Jääkäri (pronounced yay-kar-ree) means hunter as well as scout, and these units perform both roles. Whether they are fighting alongside an infantry unit or a *Panssari* unit, they are always at the vanguard. Often composed of men from rural areas, they have excellent fieldcraft skills.

WEAPONS PLATOONS

JÄÄKÄRI MACHINE-GUN PLATOON

PLATOON

HQ Section with:

2 Machine-gun Sections	150 points
1 Machine-gun Section	80 points

Using surprise and quick repositioning of their machine guns, the *Jääkäri* are able to make better use of them in mobile battles.

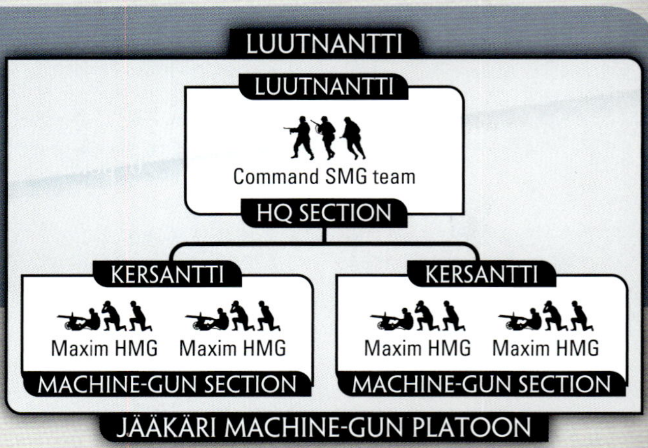

LUUTNANTTI

LUUTNANTTI

Command SMG team

HQ SECTION

KERSANTTI
Maxim HMG   Maxim HMG
MACHINE-GUN SECTION

KERSANTTI
Maxim HMG   Maxim HMG
MACHINE-GUN SECTION

JÄÄKÄRI MACHINE-GUN PLATOON

A Jääkäri Machine-gun Platoon may make Combat Attachments to Jääkäri Platoons.

JÄÄKÄRI MORTAR PLATOON

PLATOON

HQ Section with:

3 Mortar Sections	100 points
2 Mortar Sections	75 points

OPTIONS

• Replace all Light Mortar teams with Tampella 81mm M/35 mortars and add an Observer Rifle team at no cost.

A Jääkäri Mortar Platoon may make Combat Attachments to Jääkäri Platoons.

LUUTNANTTI

LUUTNANTTI

Command SMG team   Observer Rifle team

HQ SECTION

KERSANTTI
Light Mortar team
MORTAR SECTION

KERSANTTI
Light Mortar team
MORTAR SECTION

KERSANTTI
Light Mortar team
MORTAR SECTION

JÄÄKÄRI MORTAR PLATOON

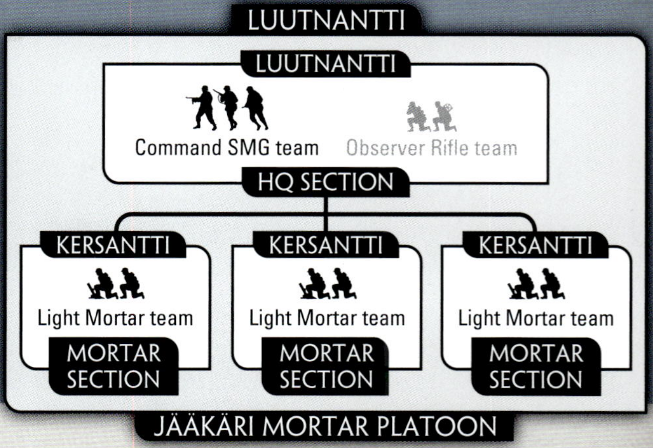

The *Jääkäri* were originally equipped with captured Soviet light mortars for mobile firepower. However with their inclusion in the *Panssari* as an infantry brigade they are now using the same Tampella 81mm mortars as the *Jalkaväki* units.

JÄÄKÄRI TANK-HUNTER PLATOON

PLATOON

HQ Section with:

2 Anti-tank Sections	75 points
1 Anti-tank Section	40 points

OPTIONS

- Replace any Lahti Anti-tank Rifle teams with Close-defence Rifle teams for +10 points per team and Command SMG team with Command Close-defence SMG team for +5 points.
- Replace all Lahti Anti-tank Rifle teams with Panzerfaust Rifle teams for +15 points per team and Command SMG team with Command Panzerfaust SMG team for +10 points.

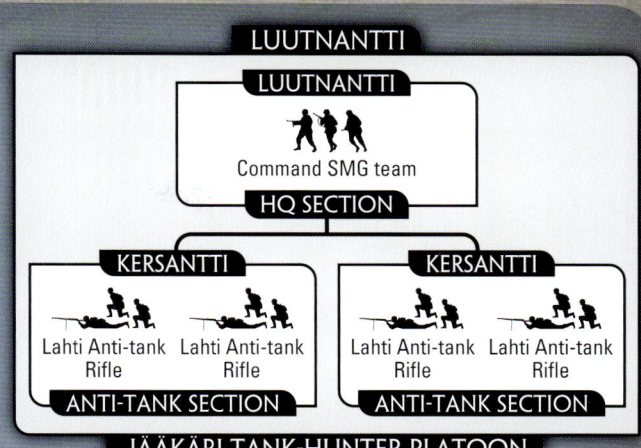

JÄÄKÄRI TANK-HUNTER PLATOON

BRIGADE SUPPORT PLATOONS

JÄÄKÄRI HEAVY TANK-HUNTER PLATOON

PLATOON

HQ Section with:

2 50 PstK/38 guns	70 points
2 75 PstK/40 guns	120 points
1 75 PstK/40 gun	60 points

OPTION

- Add tractors for +5 points for the platoon.

The *Jääkäri* Brigade had first priority in receiving the new German PaK guns and transport to keep mobile.

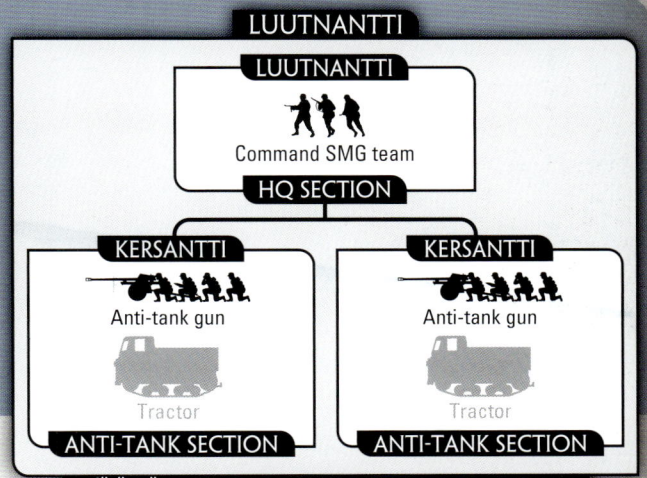

JÄÄKÄRI HEAVY TANK-HUNTER PLATOON

CAPTURED PANSSARI PLATOON

PLATOON

1 ISU-152	105 points

A Captured Panssari Platoon is rated **Fearless Trained**.

FEARLESS | **TRAINED**

CAPTURED PANSSARI PLATOON

During the fighting in June and July 1944 the Finnish troops captured two Soviet ISU-152 assault guns. One was sent to the rear where it was used as a recovery vehicle, while the other was immediately put back into action against its former owners.

JALKAVÄKIKOMPPANIA
INFANTRY COMPANY

(INFANTRY COMPANY)

HEADQUARTERS

Jalkaväkikomppania HQ — 247

You must field one platoon from each box shaded black and may field one platoon from each box shaded grey.

DIVISIONAL SUPPORT PLATOONS

COMBAT PLATOONS

INFANTRY
Jalkaväki Platoon — 247

INFANTRY
Jalkaväki Platoon — 247

INFANTRY
Jalkaväki Platoon — 247

ALLIED PLATOONS

Heer Platoons in your force are Allies and follow the Allies rules in the rulebook.

WEAPONS PLATOONS

MACHINE-GUNS
Jalkaväki Machine-gun Platoon — 248

MACHINE-GUNS
Jalkaväki Machine-gun Platoon — 248

Field Fortifications — 155

ARTILLERY
Jalkaväki Mortar Platoon — 248

REGIMENTAL SUPPORT PLATOONS

ANTI-TANK
Jalkaväki Tank-hunter Platoon — 249

ARTILLERY
Jalkaväki Heavy Mortar Platoon — 249

RECONNAISSANCE
Jalkaväki Scout Platoon — 252

ARMOUR
Panssari Platoon — 239
Assault Gun Platoon — 254
Sturmi Platoon — 241
German StuG Platoon — 59

ANTI-TANK
Heavy Anti-tank Platoon — 252

INFANTRY
Pioneeri Platoon — 251
German Grenadier Platoon — 27

ARTILLERY
Artillery Battery — 253

ARTILLERY
Artillery Battery — 253
Heavy Artillery Battery — 253

ANTI-AIRCRAFT
Anti-aircraft Platoon — 254

AIR SUPPORT
Air Support — 253
German Air Support — 172

246

MOTIVATION AND SKILL

 The Jalkaväki are superbly trained and experienced, although some of the troops grow weary from five years of war. Most troops are rated as **Confident Veteran**.

 Elite *Jalkaväkikompania* units are still as tough as ever. These were often the units thrown in to the line at critical times, and so they played a disproportionate part in the final battles. Units such as Jalkaväki *Regiment 7 (JR7)* of 2nd Division, Jalkaväki *Regiment 12 (JR12)* of 6th Division and 4th Division were noteworthy examples. These veterans are soldiers of great experience and skill and fight with steely determination. These elite platoons are rated **Fearless Veteran**.

Any Divisional Support Platoons you take in your force must have the same unit symbol or no symbol.

HEADQUARTERS

JALKAVÄKIKOMPPANIA HQ

HEADQUARTERS

Company HQ	40 points	50 points

Add Anti-tank Section with:

2 Lahti Anti-tank Rifles	30 points	40 points
1 Lahti Anti-tank Rifle	15 points	20 points

OPTIONS

- Replace Lahti Anti-tank Rifle teams with Close-defence Rifle teams for +5 points per team, Panzerfaust Rifle teams for +10 points per team, or Panzerschreck teams for +15 points per team.
- Add up to three Sniper team for +50 points per team.

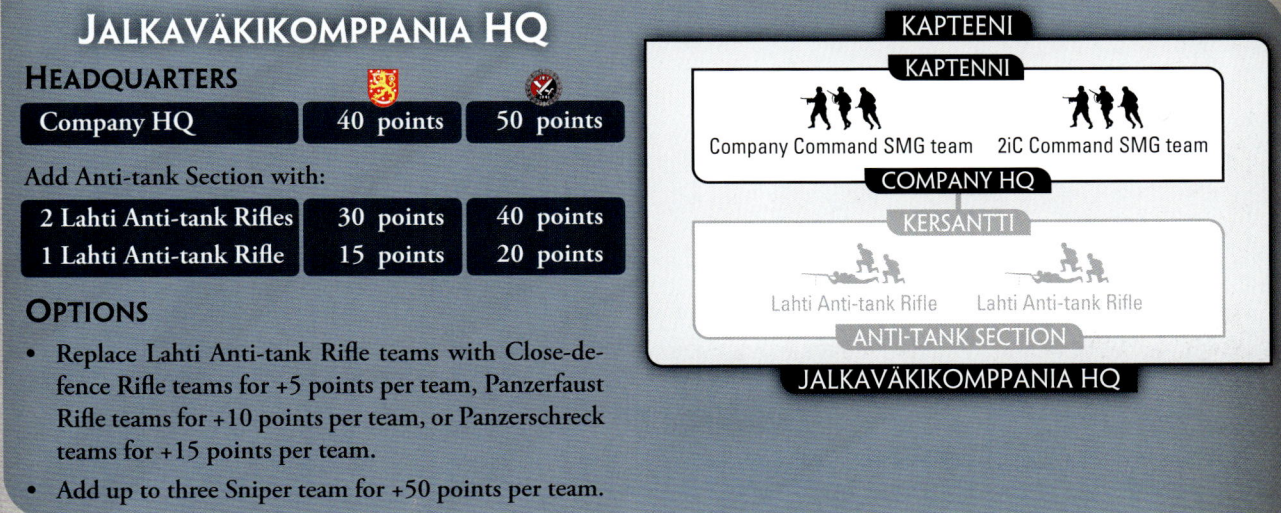

A *Jalkaväkikompania* (Infantry Company — pronounced yal-kar-va-kee komp-pa-nee-a) is led by a *Kapteeni* (Captain — pronounced kap-teen-ee) who is a veteran of years of fighting. Finnish divisions have had to be reduced from three to two regiments each of infantry, but they are still balanced teams with infantry, light infantry (*Jääkäri* scouts), machine guns and mortars. Over the last year enough new guns have been acquired for each regiment to have an anti-tank unit. Divisional support is better than ever, with anti-aircraft, anti-tank and artillery support readily available and even modern assault guns – the *Sturmi*.

COMBAT PLATOONS

JALKAVÄKI PLATOON

PLATOON

HQ Section with:

4 Jalkaväki Squads	185 points	225 points
3 Jalkaväki Squads	145 points	175 points

OPTIONS

- Replace the Command SMG team with a Command Close-defence SMG team for +5 points or with a Command Panzerfaust SMG team for +10 points.
- Replace all Rifle/MG teams with SMG teams at no cost.

Only one Jalkaväki Platoon in a company may replace its Rifle/MG teams with SMG teams.

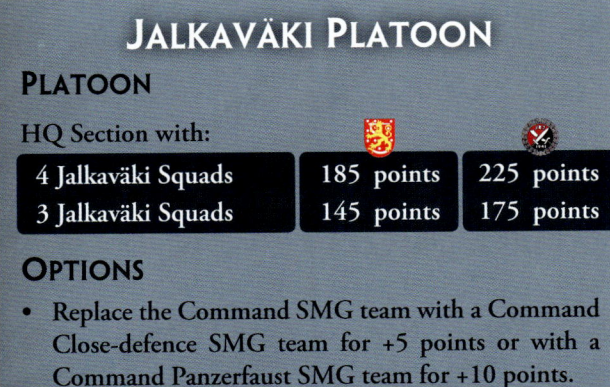
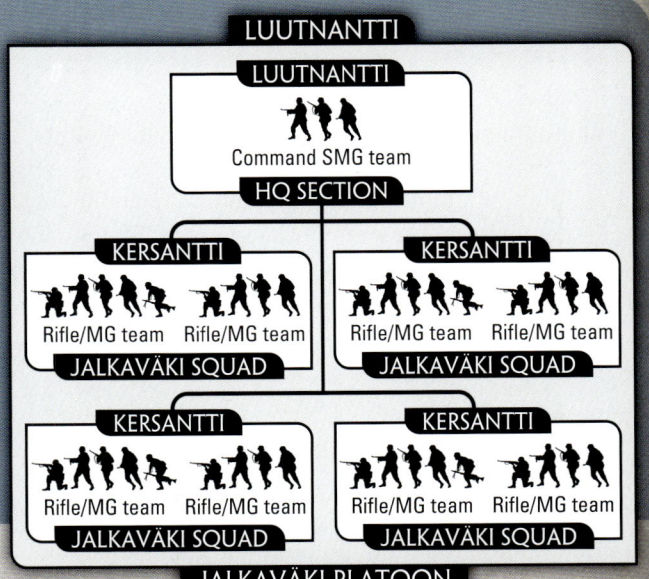

Jalkaväki platoons are the core of the Finnish army. The infantry are well-trained, well-led, combat-seasoned, and comfortable fighting in the difficult terrain and weather.

WEAPONS PLATOONS

JALKAVÄKI MACHINE-GUN PLATOON

PLATOON

HQ Section with:

2 Machine-gun Sections	135 points	150 points
1 Machine-gun Section	70 points	80 points

A Machine-gun Platoon may make Combat Attachments to Jalkaväki Platoon.

Large numbers of Soviet infantry are still the greatest threat to Finnish defences. But the ever reliable Maxim HMG is still at hand to see them off. As long as they keep coming, it will keep firing.

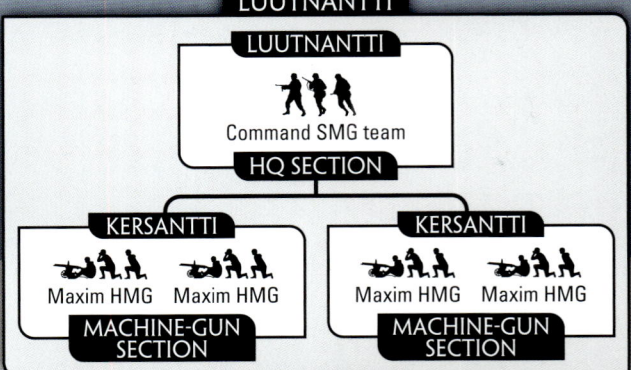

LUUTNANTTI

LUUTNANTTI

Command SMG team

HQ SECTION

KERSANTTI

Maxim HMG   Maxim HMG

MACHINE-GUN SECTION

KERSANTTI

Maxim HMG   Maxim HMG

MACHINE-GUN SECTION

JALKAVÄKI MACHINE-GUN PLATOON

JALKAVÄKI MORTAR PLATOON

PLATOON

HQ Section with:

3 Mortar Sections	90 points	100 points
2 Mortar Sections	65 points	75 points

Our little Tampella M/35 81mm mortars remain a great support weapon for the infantry. Skilled observers can bring their shells down on the Soviets before they have time to react.

LUUTNANTTI

LUUTNANTTI

Command SMG team   Observer Rifle team

HQ SECTION

KERSANTTI

Tampella M/35 81mm mortar

MORTAR SECTION

KERSANTTI

Tampella M/35 81mm mortar

MORTAR SECTION

KERSANTTI

Tampella M/35 81mm mortar

MORTAR SECTION

JALKAVÄKI MORTAR PLATOON

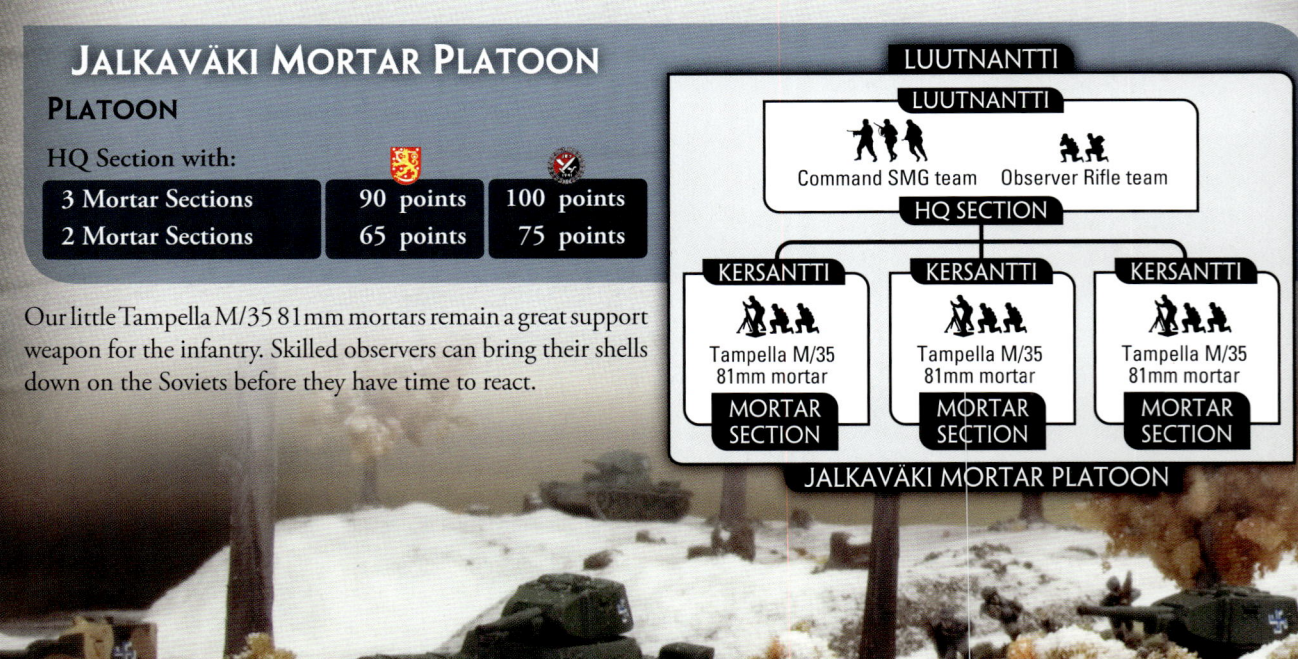

REGIMENTAL SUPPORT PLATOONS

JALKAVÄKI TANK-HUNTER PLATOON

PLATOON

HQ Section with:

3 Anti-tank Sections	155 points	190 points
2 Anti-tank Sections	110 points	135 points
1 Anti-tank Section	65 points	80 points

OPTIONS

- Replace any or all Close-defence Rifle teams with Panzerfaust Rifle teams for +5 points per Anti-tank section and Command SMG team with Command Panzerfaust SMG team for +10 points.

- Replace up to two Close-defence Rifle teams with 76 RK/27 guns each at no cost, 37 PstK/37 guns each for +5 points per gun, or 45 PstK/37 guns each for +5 points per gun.

- Add a Horse-drawn limbers for each gun at no cost or a 3-ton truck or tractor for each gun for +5 points for the platoon.

During the desperate fighting to stem the Soviet invasion the Finns once again formed ad-hoc anti-tank detachments. This time they were armed with *Panssarinyrkki* (Panzerfaust) anti-tank launchers, making them much more deadly than their predecessors armed with Molotov Cocktails.

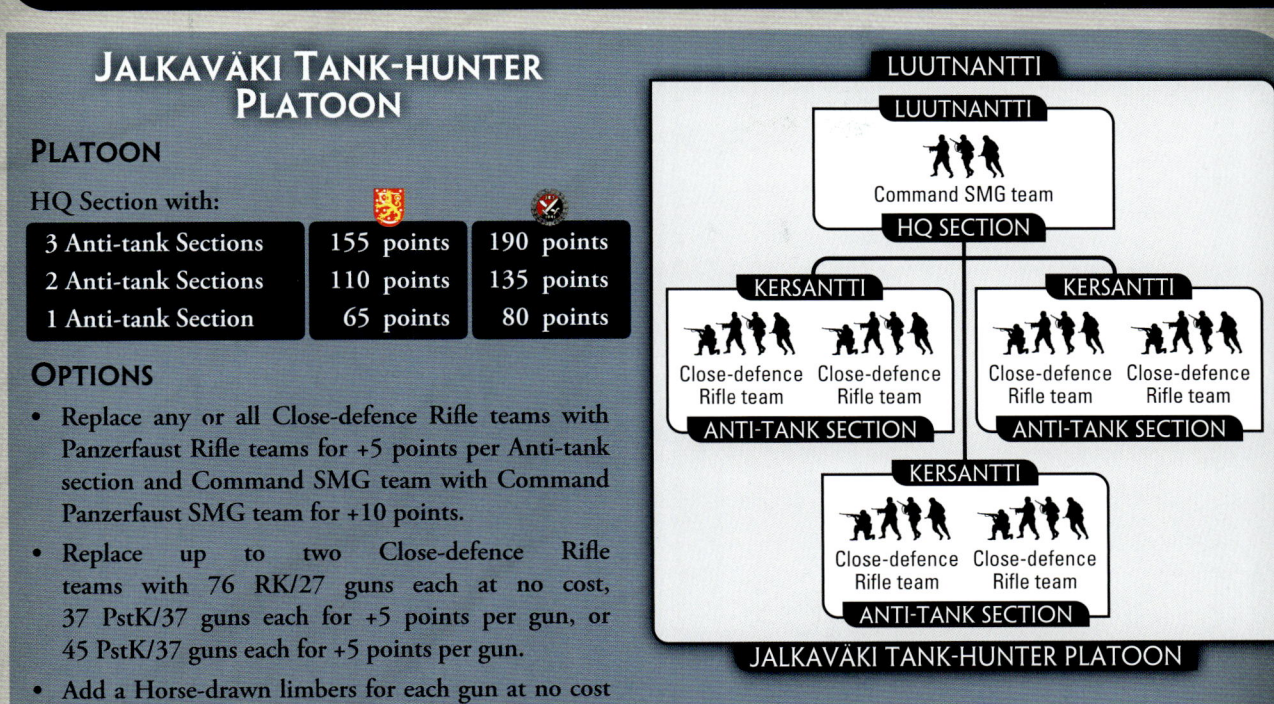

LUUTNANTTI
LUUTNANTTI
Command SMG team
HQ SECTION

KERSANTTI
Close-defence Rifle team — Close-defence Rifle team
ANTI-TANK SECTION

KERSANTTI
Close-defence Rifle team — Close-defence Rifle team
ANTI-TANK SECTION

KERSANTTI
Close-defence Rifle team — Close-defence Rifle team
ANTI-TANK SECTION

JALKAVÄKI TANK-HUNTER PLATOON

JALKAVÄKI HEAVY MORTAR PLATOON

PLATOON

HQ Section with:

3 Mortar Sections	115 points	130 points
2 Mortar Sections	85 points	100 points

OPTIONS

- Add horse-drawn wagons for +5 points for the platoon.
- Replace horse-drawn wagons with 3-ton trucks at no cost.

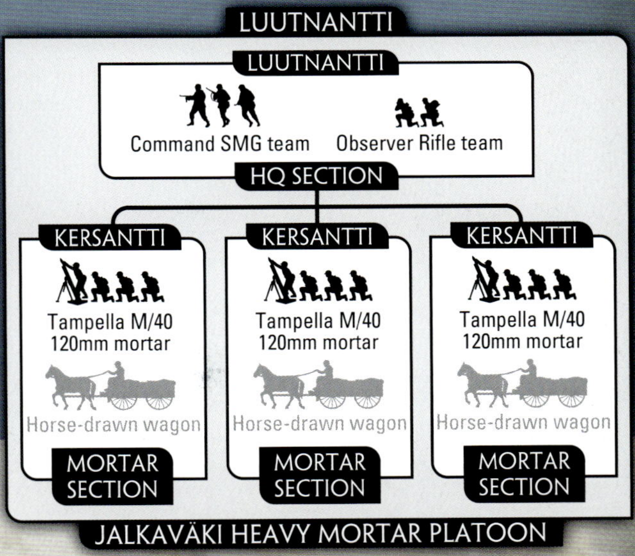

Our own Tampella 120mm mortars along with those we captured from the Soviets provide our infantry with heavy firepower. The heavy mortars' high explosive rounds disrupt Soviet assaults as they advance on our positions.

LUUTNANTTI
LUUTNANTTI
Command SMG team — Observer Rifle team
HQ SECTION

KERSANTTI
Tampella M/40 120mm mortar
Horse-drawn wagon
MORTAR SECTION

KERSANTTI
Tampella M/40 120mm mortar
Horse-drawn wagon
MORTAR SECTION

KERSANTTI
Tampella M/40 120mm mortar
Horse-drawn wagon
MORTAR SECTION

JALKAVÄKI HEAVY MORTAR PLATOON

PIONEERIKOMPPANIA
ENGINEER COMPANY

(INFANTRY COMPANY)

HEADQUARTERS

Pioneerikomppania HQ — 251

You must field one platoon from each box shaded black and may field one platoon from each box shaded grey.

COMBAT PLATOONS

INFANTRY
Pioneeri Platoon — 251

INFANTRY
Pioneeri Platoon — 251

INFANTRY
Pioneeri Platoon — 251

ALLIED PLATOONS

Heer Platoons in your force are Allies and follow the Allies rules in the rulebook.

DIVISIONAL SUPPORT PLATOONS

ARMOUR
Panssari Platoon — 239
Sturmi Platoon — 241
German StuG Platoon — 59

ANTI-TANK
Heavy Anti-tank Platoon — 252

INFANTRY
Jalkaväki Platoon — 247
Jalkaväki Scout Platoon — 252
German Grenadier Platoon — 27

ARTILLERY
Artillery Battery — 253

ARTILLERY
Artillery Battery — 253
Heavy Artillery Battery — 253

ANTI-AIRCRAFT
Anti-aircraft Platoon — 254

AIR SUPPORT
Air Support — 253
German Air Support — 172

MOTIVATION AND SKILL

The Pioneeri *are superbly trained and experienced, although some of the troops grow weary from five years of war. Most troops are rated as* **Confident Veteran.**

Elite *Pioneerikomppania are still as tough as ever. Units such as 4th Division were noteworthy examples. These veterans are soldiers of great experience and skill and fight with steely determination. Platoons of an Elite unit are rated* **Fearless Veteran.**

RELUCTANT	CONSCRIPT		RELUCTANT	CONSCRIPT
CONFIDENT	TRAINED		CONFIDENT	TRAINED
FEARLESS	**VETERAN**		**FEARLESS**	**VETERAN**
REGULAR			**ELITE**	

Any Divisional Support Platoons you take in your force must have the same unit symbol or no symbol.

HEADQUARTERS

PIONEERIKOMPPANIA HQ

HEADQUARTERS

Company HQ	40 points	50 points

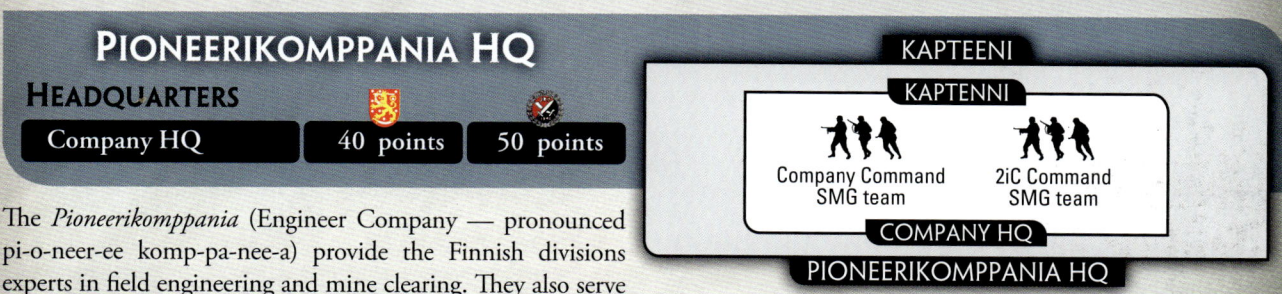

KAPTEENI
KAPTENNI

Company Command SMG team   2iC Command SMG team

COMPANY HQ

PIONEERIKOMPPANIA HQ

The *Pioneerikomppania* (Engineer Company — pronounced pi-o-neer-ee komp-pa-nee-a) provide the Finnish divisions experts in field engineering and mine clearing. They also serve as excellent assault troops for taking our enemy fortifications.

COMBAT PLATOONS

PIONEERI PLATOON

PLATOON

HQ Section with:

4 Pioneeri Squads	200 points	240 points
3 Pioneeri Squads	155 points	185 points
2 Pioneeri Squads	110 points	130 points

OPTIONS

- Replace the Command Pioneer Rifle team with a Command Pioneer SMG team for +5 points or with a Command Panzerfaust SMG team for +10 points.
- Replace all Pioneer Rifle teams with Pioneer SMG teams for +10 points per Pioneeri Squad.
- Add Pioneer Supply truck for +25 points.

You may replace up to two Pioneer Rifle or SMG teams in the platoon with a Flame-thrower team each at the start of the game before deployment.

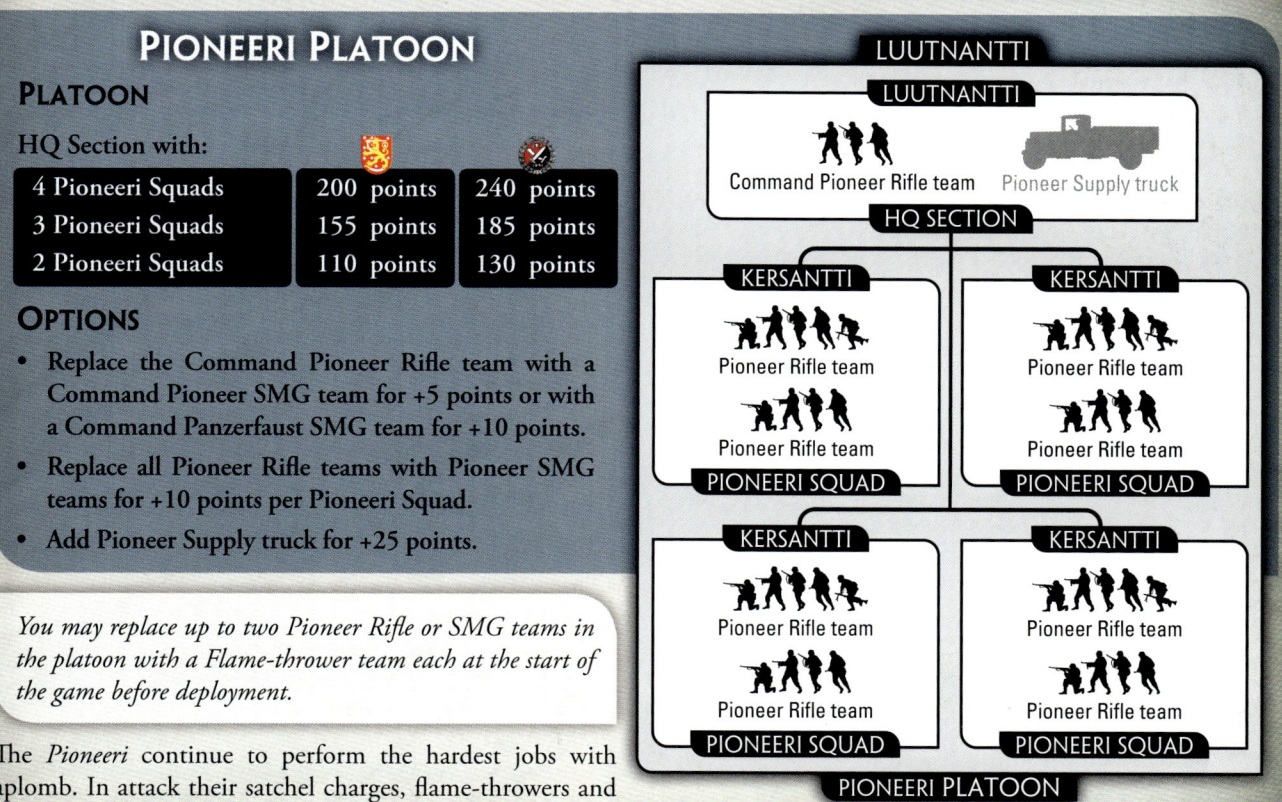

LUUTNANTTI
LUUTNANTTI

Command Pioneer Rifle team   Pioneer Supply truck

HQ SECTION

KERSANTTI   KERSANTTI

Pioneer Rifle team

Pioneer Rifle team

PIONEERI SQUAD   PIONEERI SQUAD

KERSANTTI   KERSANTTI

Pioneer Rifle team

Pioneer Rifle team

PIONEERI SQUAD   PIONEERI SQUAD

PIONEERI PLATOON

The *Pioneeri* continue to perform the hardest jobs with aplomb. In attack their satchel charges, flame-throwers and pioneer skills are invaluable. In defence they can lay the minefields that have so effectively slowed the Soviet advance.

FINNISH DIVISIONAL SUPPORT

MOTIVATION AND SKILL

All Divisional Support platoons are well trained and with officers often drawn from the regular army specialists. They have suffered least from the years of static warfare. Regular units, marked (), are rated **Confident Veteran**. Elite Units, marked (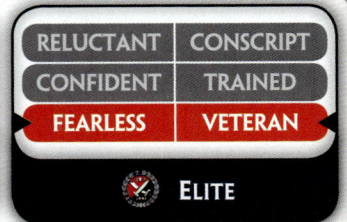), are rated **Fearless Veteran**.

RELUCTANT	CONSCRIPT
CONFIDENT	TRAINED
FEARLESS	**VETERAN**

REGULAR

RELUCTANT	CONSCRIPT
CONFIDENT	TRAINED
FEARLESS	**VETERAN**

ELITE

Any platoons you take in your force must have the same unit symbol or no symbol.

HEAVY ANTI-TANK PLATOON

PLATOON

HQ Section with:

2 50 PstK/38 guns	60 points	70 points
2 75 PstK/97-38 guns	65 points	75 points
2 75 PstK/40 guns	105 points	120 points
1 75 PstK/40 gun	55 points	60 points

Add:

1 Close-defence Rifle team	+20 points	+25 points
2 Close-defence Rifle teams	+40 points	+50 points

OPTION

• Replace Command SMG team with Command Panzerfaust SMG team for +10 points.

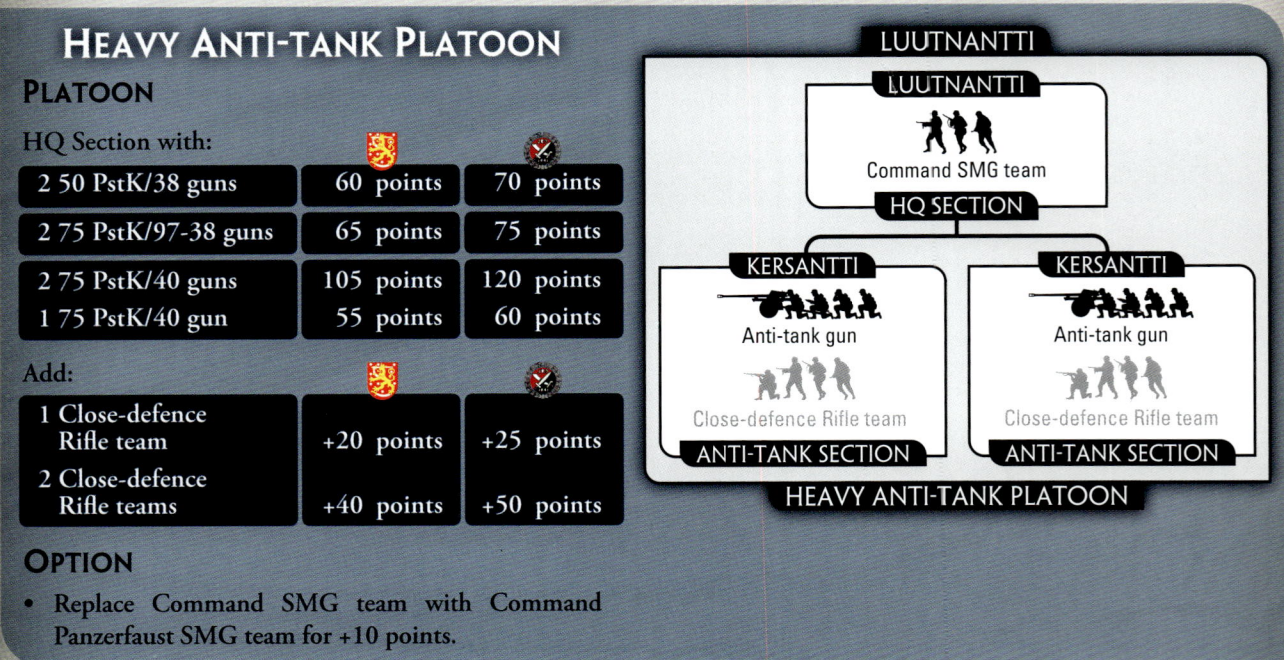

JALKAVÄKI SCOUT PLATOON

PLATOON

HQ Section with:

4 Scout Squads	235 points
3 Scout Squads	185 points
2 Scout Squads	135 points

OPTIONS

• Replace the Command Rifle team with a Command SMG team for +5 points or Command Close-defence SMG team for +10 points or with a Command Panzerfaust SMG team for +15 points.

• Replace all Rifle teams with SMG teams for +15 points per Scout Squad.

A *Jalkaväki Scout Platoon* is rated as **Fearless Veteran** regardless of what sort of Komppania it is fighting in.

FEARLESS	VETERAN

A Jalkaväki Scout Platoon is a Reconnaissance Platoon.

ARTILLERY BATTERY

PLATOON

HQ Section with:

		🦁	⚔
4 76 RK/27 guns		120 points	135 points
2 76 RK/27 guns		70 points	75 points
4 76 K/02 guns		135 points	150 points
2 76 K/02 guns		75 points	85 points
4 105 H/33 howitzers		205 points	230 points
2 105 H/33 howitzers		110 points	125 points
4 122 H/38 howitzers		205 points	230 points
2 122 H/38 howitzers		110 points	125 points

OPTIONS

- Add horse-drawn wagon and limbers for +5 points for the battery.
- Replace horse-drawn wagon with a 3-ton truck and limbers with tractors or 3-ton trucks at no cost.

Under General Nenonen's leadership and technical guidance, the artillery goes from strength to strength. It is now well trained, well equipped, and capable of responding quickly and accurately to fire requests. New fire control calculation cards automate range corrections and enable rapid coordination of many batteries.

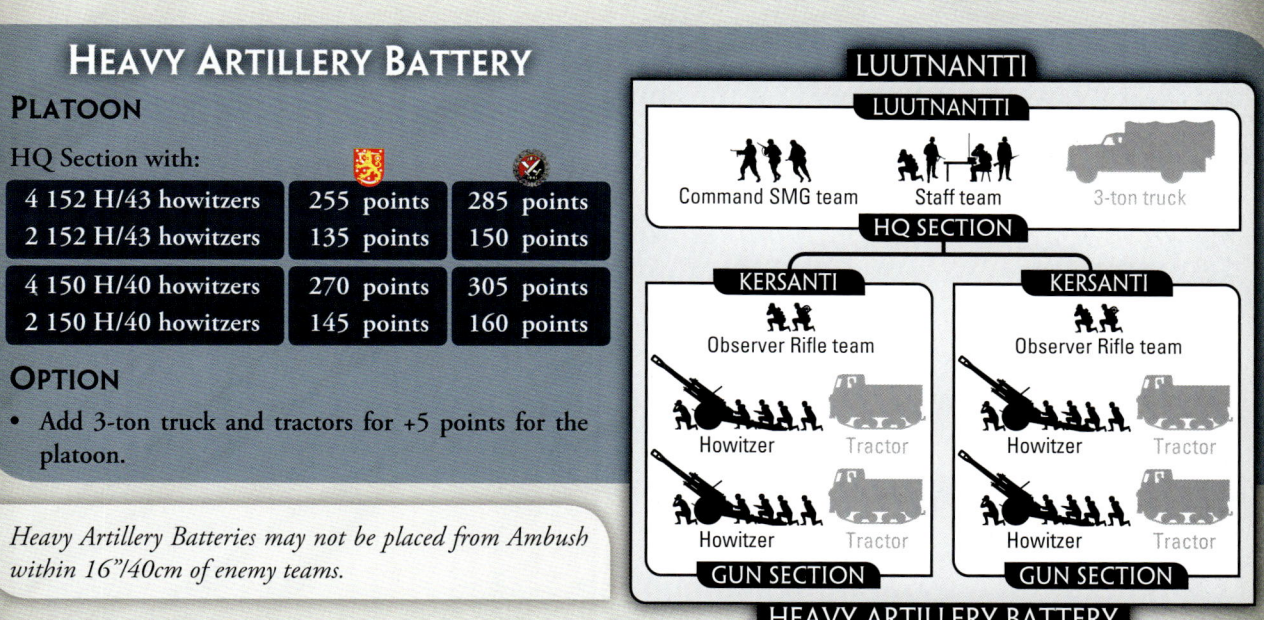

Artillery Batteries equipped with 105 H/33 or 122 H/38 howitzers may not be placed from Ambush within 16"/40cm of enemy teams.

HEAVY ARTILLERY BATTERY

PLATOON

HQ Section with:

		🦁	⚔
4 152 H/43 howitzers		255 points	285 points
2 152 H/43 howitzers		135 points	150 points
4 150 H/40 howitzers		270 points	305 points
2 150 H/40 howitzers		145 points	160 points

OPTION

- Add 3-ton truck and tractors for +5 points for the platoon.

Heavy Artillery Batteries may not be placed from Ambush within 16"/40cm of enemy teams.

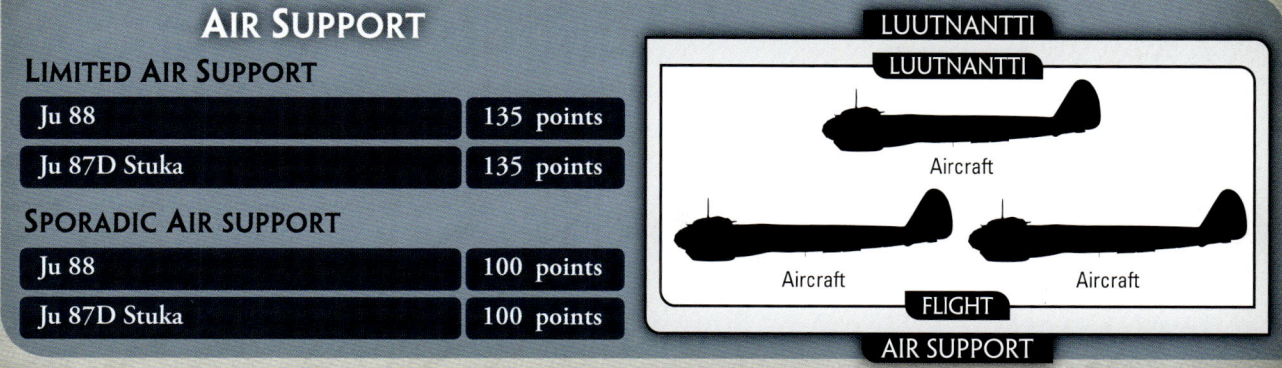

AIR SUPPORT

LIMITED AIR SUPPORT

Ju 88	135 points
Ju 87D Stuka	135 points

SPORADIC AIR SUPPORT

Ju 88	100 points
Ju 87D Stuka	100 points

ASSAULT GUN PLATOON

PLATOON

3 BT-42	-	210 points
2 BT-42	-	140 points

The BT-42 assault gun is now hopelessly obsolete, and has been placed in an independent company. Even so their crews have bravely set out to face Soviet heavy armour in the *Panssari* Division counterattack. They took heavy losses and were put into reserve.

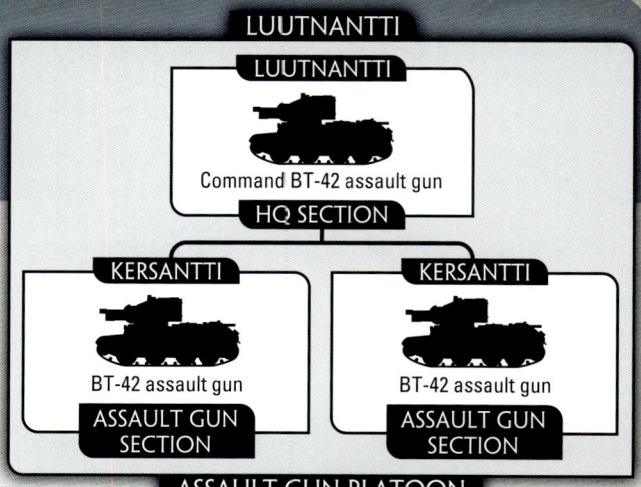

LUUTNANTTI

LUUTNANTTI

Command BT-42 assault gun

HQ SECTION

KERSANTTI

BT-42 assault gun

ASSAULT GUN SECTION

KERSANTTI

BT-42 assault gun

ASSAULT GUN SECTION

ASSAULT GUN PLATOON

SELF-PROPELLED ANTI-AIRCRAFT PLATOON

PLATOON

2 Landsverk Anti II	-	125 points

The Swedish made Landsverk Anti II self-propelled anti-aircraft guns are very serviceable weapons. Their 40mm Bofors cannon and high rate of fire make them very useful.

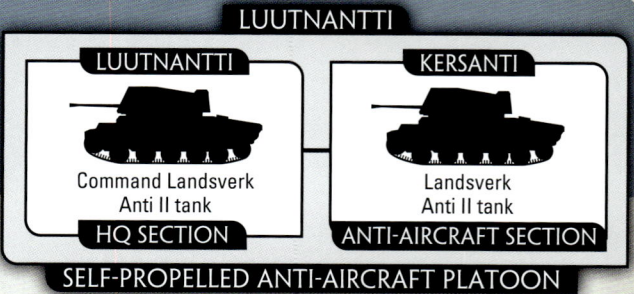

LUUTNANTTI

LUUTNANTTI

Command Landsverk Anti II tank

HQ SECTION

KERSANTI

Landsverk Anti II tank

ANTI-AIRCRAFT SECTION

SELF-PROPELLED ANTI-AIRCRAFT PLATOON

ANTI-AIRCRAFT PLATOON

PLATOON

HQ Section with:

2 Anti-aircraft Sections	65 points	70 points

OPTIONS

- Replace 40 LtK/38 guns with 20 LtK/38 guns for -15 points for the platoon.
- Add 3-ton trucks for +5 points for the platoon.

Finland no longer suffers from a lack of anti-aircraft guns. On the front line the durable and effective Bofors 40mm anti-aircraft gun continues to be the primary form of air defence, with German 20mm guns also still in use.

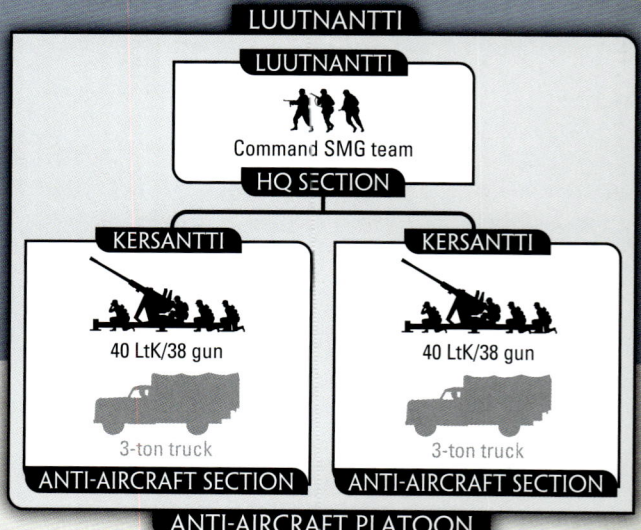

LUUTNANTTI

LUUTNANTTI

Command SMG team

HQ SECTION

KERSANTTI

40 LtK/38 gun

3-ton truck

ANTI-AIRCRAFT SECTION

KERSANTTI

40 LtK/38 gun

3-ton truck

ANTI-AIRCRAFT SECTION

ANTI-AIRCRAFT PLATOON

ARMOURED CAR PLATOON

PLATOON

3 BA-10	-	135 points
2 BA-10	-	90 points

An Armoured Car Platoon is a Reconnaissance Platoon.

Some of the many Soviet armoured cars captured in the Winter War are now used for reconnaissance in the *Panssari* Division. The BA-10 armoured car is the preferred model, with the lighter Soviet vehicles being discarded.

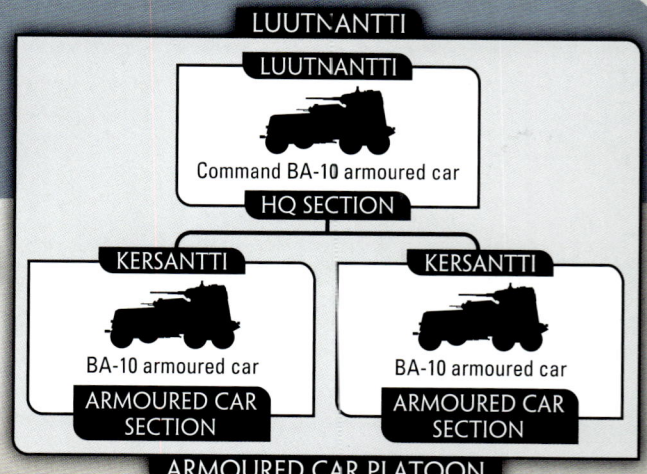

LUUTNANTTI

LUUTNANTTI

Command BA-10 armoured car

HQ SECTION

KERSANTTI

BA-10 armoured car

ARMOURED CAR SECTION

KERSANTTI

BA-10 armoured car

ARMOURED CAR SECTION

ARMOURED CAR PLATOON

FINNISH SPECIAL RULES

SELF SUFFICIENT

The Finnish Army is composed of farmers and hunters, used to an outdoor lifestyle and making their own decisions. Their training emphasised this self-sufficiency.

Finnish Platoons use the German Mission Tactics special rule (see page 183).

HUNTERS

Finland is a country of forests and lakes and the natives know their country well. They are able to quickly move through this difficult terrain with ease. Skiing is the national sport in Finland, as well as a necessity for travelling cross-country. Finnish platoons are ski-equipped, increasing their mobility during winter.

Finnish Infantry and Man-packed Gun teams may move At the Double through Woods and deep snow.

INTERCEPTED COMMUNICATIONS

During their fighting against the Soviets in 1944 the Finns would often intercept Soviet radio communications. This gave them prior knowledge of the Red Army's attacks and where their troops were massing. The Finns would use this information to bring down very accurate artillery bombardment on the enemy positions.

Because of the excellent quality of the Finnish radio interception, Artillery Bombardments fired by an artillery platoon with a Staff team may re-roll the third failed attempt to range in. The artillery platoon does not get a re-roll on their first and second attempts.

BREAKTHROUGH GUN

Some weapons are just so powerful that there is no chance of surviving a hit from them. These heavy guns are often mounted in tanks and self-propelled guns designed to break through enemy defensive lines.

Infantry teams, Gun teams and Unarmoured vehicles automatically fail their saves when hit by a Breakthrough Gun or Bunker Buster. This does not apply to Artillery Bombardments.

AUTOMATED FIRE CONTROL

General Vilho P Nenonen trained and reorganised the Finnish artillery. Now they have several innovations to improve its performance. Finnish artillery organisation is very flexible and allows any Forward Observer to call down fire from any battery in range. Major Unto Petäjä's new correction converter meant that artillery could respond to orders from any Observer very quickly, with a high degree of accuracy.

A Finnish Observer team from an artillery battery with a Staff team may act as the Spotting team for any artillery platoon, whether or not they have a Staff team.

When a Finnish artillery battery that has a Staff team fires an Artillery Bombardment, any other artillery batteries with Staff teams may join it before rolling to Range In and fire as a single combined Artillery Battery.

BITTER ENEMIES

The Finns are a stubborn people used to adversity. The survival of their country is at stake. All Finnish troops are prepared to fight at close quarters if needed to see off the invading Red Army.

When Finnish platoons fighting against any Soviet force take a Motivation Test to Counterattack in assaults, you may re-roll the die and apply the re-rolled result to Finnish platoons instead of the original result.

FINNISH ARSENAL

TANK TEAMS

Name _Weapon_	Mobility _Range_	Armour Front _ROF_	Side _Anti-tank_	Top _Firepower_	Equipment and Notes
TANKS					
T-26	Slow Tank	1	1	1	Co-ax MG, Limited vision, Unreliable.
45mm obr 1934 gun	_24"/60cm_	2	7	4+	
T-28	Slow Tank	4	3	2	Two Deck-turret MG, Turret-front MG, Turret-rear MG, Limited vision, Unreliable.
76mm L-10 gun	_24"/60cm_	2	7	3+	
T-34/76	Standard Tank	6	5	1	Co-ax MG, Fast tank, Hull MG, Limited vision, Wide-track.
76mm F-34 gun	_32"/80cm_	2	9	3+	
T-34/85	Standard Tank	7	5	1	Co-ax MG, Hull MG, Limited vision.
85mm D-5T gun	_32"/80cm_	2	12	3+	
KV-1e	Slow Tank	9	8	2	Co-ax MG, Hull MG, Turret-rear MG, Limited vision, Unreliable.
76mm F-34 gun	_32"/80cm_	2	9	3+	
ASSAULT GUNS					
Stu 40 G Sturmi	Standard Tank	7	3	1	Hull MG, Protected ammo, Improvised Armour.
7.5cm StuK40 gun	_32"/80cm_	2	11	3+	_Hull mounted._
BT-42	Standard Tank	1	1	1	Limited vision.
114 Psv.H/18 howitzer	_16"/40cm_	1	7	2+	_Breakthrough gun._
Firing bombardments	_48"/120cm_	-	4	4+	
ISU-152	Slow Tank	9	7	2	
152mm ML-20S gun	_32"/80cm_	1	13	1+	_Bunker buster, Hull mounted._
SELF-PROPELLED ANTI-AIRCRAFT					
Landsverk Anti II	Standard Tank	2	1	0	
40 ItK/38 gun	_24"/60cm_	4	6	4+	_Anti-aircraft._
ARMOURED CARS					
BA-10	Wheeled	1	0	0	Co-ax MG, Hull MG, Limited vision.
45mm obr 1934 gun	_24"/60cm_	2	7	4+	
VEHICLE MACHINE-GUNS					
Vehicle MG	_16"/40cm_	3	2	6	_ROF 1 if other weapons fire._

AIRCRAFT

Aircraft	Weapon	To Hit	Anti-tank	Firepower	Notes
Ju 87D Stuka	Bombs	4+	5	1+	
Ju 88	Bombs	4+	5	1+	

FORTIFICATIONS

Weapon	Mobility	Range	ROF	Anti-tank	Firepower	Notes
40mm Bofors Nest	Immobile	24"/60cm	4	6	4+	Anti-aircraft.
HMG Pillbox	Immobile	24"/60cm	6	2	6	
HMG Nest	Immobile	24"/60cm	6	2	6	

GUN TEAMS

Weapon	Mobility	Range	ROF	Anti-tank	Firepower	Notes
Maxim HMG	Man-packed	24"/60cm	6	2	6	
Lahti anti-tank rifle	Man-packed	16"/40cm	3	5	5+	
Tampella M/35 81mm mortar	Man-packed	24"/60cm	2	2	3+	Smoke, Minimum range 8"/20cm.
Firing bombardments		40"/100cm	-	2	6	Smoke bombardment.
Tampella M/40 120mm mortar	Light	56"/140cm	-	3	3+	Smoke bombardment.
37 PstK/37 gun	Light	24"/60cm	3	6	4+	Gun shield.
Firing Stielgranate		8"/20cm	1	12	5+	
45 PstK/37 gun	Light	24"/60cm	3	7	4+	Gun shield.
50 PstK/38 gun	Medium	24"/60cm	3	9	4+	Gun shield.
75 PstK/97-38 gun	Medium	24"/60cm	2	10	3+	Gun shield.
75 PstK/40 gun	Medium	32"/80cm	2	12	3+	Gun shield.
20 LtK/38 gun	Light	16"/40cm	4	5	5+	Anti-aircraft, Turntable.
40 LtK/38 gun	Immobile	24"/60cm	4	6	4+	Anti-aircraft, Turntable.
76 RK/27 gun	Light	16"/40cm	2	5	3+	Gun shield.
Firing bombardments		64"/160cm	-	3	6	
76 K/02 gun	Heavy	24"/60cm	2	8	3+	Gun shield, Smoke.
Firing bombardments		60"/160cm	-	3	6	Smoke bombardment.
105 H/33 howitzer	Immobile	24"/60cm	1	10	2+	Gun shield, Breakthrough gun, Smoke.
Firing bombardments		72"/180cm	-	4	4+	Smoke bombardment.
122 H/38 howitzer	Immobile	24"/60cm	1	7	2+	Gun shield, Breakthrough gun.
Firing bombardments		80"/200cm	-	4	3+	
152 H/43 howitzer	Immobile	24"/60cm	1	10	1+	Bunker buster, Gun shield.
Firing bombardments		80"/200cm	-	5	2+	
150 H/40 howitzer	Immobile	24"/60cm	1	13	1+	Bunker buster, Smoke.
Firing bombardments		80"/200cm	-	5	2+	Smoke bombardment.

INFANTRY TEAMS

Team	Range	ROF	Anti-tank	Firepower	Notes
Rifle team	16"/40cm	1	2	6	
Rifle/MG team	16"/40cm	2	2	6	
SMG team	4"/10cm	3	1	6	Full ROF when moving.
Light Mortar team	16"40cm	1	1	4+	Can fire over friendly teams.
Panzerschreck team	8"/20cm	2	11	5+	Tank Assault 5.
Flame-thrower team	4"/10cm	2	-	6	Flame-thrower.
Staff team	cannot shoot				Moves as a Heavy Gun team.

ADDITIONAL TRAINING AND EQUIPMENT

Panzerfaust	4"/10cm	1	12	5+	Tank Assault 6, Cannot shoot in the Shooting Step if moved in the Movement Step.

Close-defence and Pioneer teams are rated as Tank Assault 4.

TRANSPORT TEAMS

Vehicle	Mobility	Armour			Equipment and Notes
		Front	Side	Top	
TRUCKS					
3-ton truck	Wheeled	-	-	-	
Horse-drawn wagon	Horse-drawn	-	-	-	
TRACTORS					
Stalinets tractor	Slow Tank	-	-	-	
Horse-drawn limber	Horse-drawn	-	-	-	
RECOVERY VEHICLES					
Famo	Half-tracked	-	-	-	Recovery vehicle.
ISU Recovery	Slow Tank	9	7	2	Recovery vehicle.

Finnish Jääkäri light infantry supported by BT-42 assault guns.

Finnish Sturmi assault guns counterattack.

German StuG G assault guns.

The Finns still used a number of old Soviet T-28 tanks they captured in 1939.

PAINTING GUIDES

The following painting guide contains uniform, equipment and vehicle colours for the Germans, Hungarians and Finns on the Eastern Front in 1944 to 1945. Also provided are guides and tips on painting uniform camouflage and details.

All colour names and codes given are for the Vallejo range of Flames Of War paints, or Flames Of War spray cans, available from the online store and Flames Of War stockists. The online store, and more comprehensive painting and modelling guides can be found at:

www.FlamesOfWar.com

GERMAN HEER

GRENADIER

Green Grey (886)
Gas mask strap

Black (950)
Webbing, pouches

German Fieldgrey (830)
Tunic

German Cam. Dark Green (979)
Gas mask canister

Flat Brown (984)
Canteen

German Cam. Dark Green (979)
Helmet

Middlestone (882)
Alternative helmet

Flat Flesh (955)
Exposed skin

Middlestone (882)
Grenade head.

Beige Brown (875)
Rifle wood, grenade handle.

Gunmetal Grey (863)
Gun metal

German Camo Beige (821)
Bread bag, rifle sling

Green Grey (886)
Alternative Bread bag

Gunmetal Grey (863)
Mess tin, canteen top, tool head

Beige Brown (875)
Entrenching tool handle

German Camo Medium Brown (826)
Boots

German Fieldgrey (830)
Trousers

German Camo Beige (821)
Anklets

PANZERGRENADIER

	Splinter camouflage	
German Camo Medium Beige (821) Base colour	**Luftwaffe Camo Green (823)** Camo colour	**German Camo Medium Brown (826)** Camo colour

German Camo Dark Green (979)
Uncovered Helmet, Mess tin

Flat Flesh (955)
Exposed flesh

Black (950)
Belts and straps

German Camo Medium Beige (821)
Ammo pouches, Bread bag

Gunmetal (863)
Gun metal, cup

Flat Brown (984)
Water bottle

German Fieldgrey (830)
Trousers and tunic

Green Grey (886)
Gaiters

Black (950)
Boots

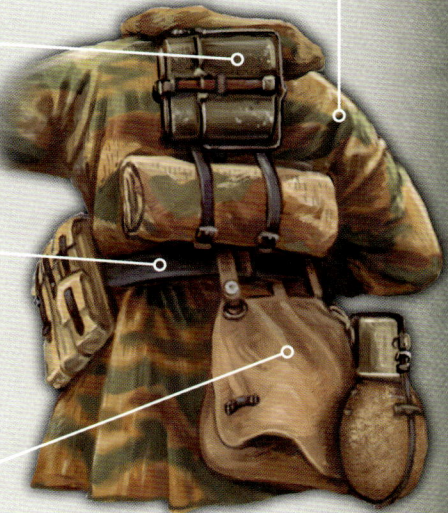

The soldiers of the *Feldherrnhalle,* and *SS* divisions wore a cuff title on their left sleeve. The example above has been painted with a line of Orange Ochre (824) with White (951) borders and a squiggle for the *Feldherrnhalle* title. For *SS* cuff titles, use Black (950) with White (951) borders and titles.

SPLINTER PATTERN CAMOUFLAGE

1. Start with a base of **German Camo Beige (821)**.

2. Paint irregular lines and patches of **German Camo Medium Brown (826)**. Try to create angular, zig-zag shapes; you should aim to cover approximately one-third of the base colour, at most.

3. Add small patches of **Luftwaffe Camo Green (823)** or **Flat Green (968)** between the brown areas.

This basic method ignores the vertical lines – Splinters – that give this camo pattern its name. Some people like to paint fine lines, but a more scale-appropriate method is to use a base colour which suggests the Splinters – **Green Grey (886)** works well – using small patches of **German Camo Beige (821)** to indicate the areas without Splinters.

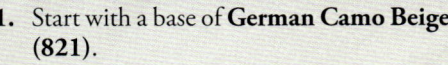

🔍**110%**

VEHICLE CREWS

German Field Grey (830)
StuG Field Cap
Black (950)
Panzer Field Cap

Flat Flesh (955)
Exposed flesh

German Field Grey (830)
StuG Trousers & tunic
Black (950)
Panzer Trousers & tunic

Black (950)
Webbing, boots

German Heer (Army) vehicle crew wore a standard uniform design specifically for their role. It is often referred to as the Panzer wrap, but variations were worn by assault gun, tank-hunter and self-propelled artillery crews. Assault gun, tank-hunter and self-propelled artillery crews uniform was *feldgrau* (German Field Grey 830) like that of the infantry. However Panzer (tank) crews wore a black version (Black 950).

SS vehicle crews also wore the Panzer wrap, but in addition to the *feldgrau* and black versions some were also issued SS Pea Dot camouflage items.

WAFFENFARBE

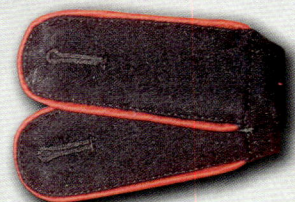 

Waffenfarbe is a German term for the coloured piping that identifies the branch of service that a soldier belongs to. It had a different colour depending on the profession that the soldier performed.

- **Grenadier:** White (951)
- **Artillery:** Flat Red (957)
- **Panzergrenadier:** Luftwaffe Camo Green (823)
- **Reconnaissance:** Light Brown (929)
- **Panzer and Panzerjäger:** 50/50 mix of White (951) and Flat Red (957)
- **Gebirgsjager:** German Camo Bright Green (833)
- **Pionier:** Black (950)

As part of the artillery StuG Crew and Begliet tank escorts attached to StuG batteries wore red *Waffenfarbe*.

GERMAN VEHICLES

The tanks, assault guns and vehicles of the German forces on the Eastern Front were painted in *Dunkelgelb* (Dark-yellow) with roughly painted *Rotbraun* (red brown) and *Olivgrün* (olive green) camouflage.

Most had national identification markings of black crosses on the superstructure sides along with the Company/Platoon/Tank vehicle identification number on the superstructure for self-propelled guns or turrets for tanks.

In winter tanks would be white washed roughly, often with a broom or such (White 951).

Middlestone (882) or German Armour Late (SP04)
Dunkelgelb: Vehicles, guns, equipment

German Camo Medium Brown (826)
Rotbraun: Camo colour

Reflective Green (890)
Olivgrün: Camo colour

Beige Brown (875)
Tool handles

Gunmetal Grey (863)
Tracks, machine gun, tool heads, exposed metal

WAFFEN-SS

SS-PANZERGRENADIER

German Fieldgrey (830)
Tunic

Gunmetal Grey (863)
Metal

Flat Flesh (955)
Skin

Black (950)
Leather

German Camo Beige (821)
SMG ammo pouches

Gunmetal Grey (863)
Metal

Flat Brown (984)
Canteen

German Camo Beige (821)
Anklets, bread bag

German Camo Medium Brown (826)
Boots

SS CAMOUFLAGE PATTERNS

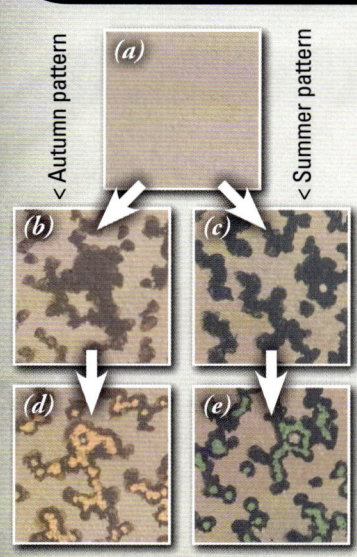

< Autumn pattern

< Summer pattern

OAK LEAF PATTERN

Start with **German Cam. Pale Brown (825)**, high-lighted with **USA Tan Earth (874)** *(a)*.

Apply mottled patches of **Chocolate Brown (872)** for the Autumn pattern *(b)*, or **German Cam. Dark Green (979)** for the Summer pattern *(c)*.

Carefully paint small dots inside the dark patches. Use **Light Brown (929)** for Autumn *(d)* or **German Cam. Bright Green (833)** for Summer *(e)*.

PEA DOT PATTERN

1. Start with a base of **US Field Drab (873)**.

2. Using **German Camo Dark Green (979)**, paint small dots and a small number of irregularly shaped patches.

3. Paint dots and patches of **German Camo Beige (821)**.

4. Add dots of **US Field Drab (873)** over the Dark Green and Beige patches.

5. Add dots of **German Camo Bright Green (833)**.

This four-colour scheme simplifies the five-colour Pea Dot pattern slightly. If you haven't been driven mad already, you can add dots of **Yellow Green (881)** for an even more historical finish.

HUNGARIAN

HONVÉD

Brown Violet (887)
Helmet

Flat Flesh (955)
Exposed flesh

Chocolate Brown (872)
Belts, straps and
rifle ammo pouches

Khaki (988)
Bread bag, SMG pouches

Gunmetal (863)
Gun metal

Beige Brown (875)
Rifle wood, tool handles

English Uniform (921)
Trousers, Tunics and Caps

Chocolate Brown (872)
Boots

Black (950)
Bayonet scabbard

Each branch of the Hungarian
Honvéd (Army) had unique uniform
collar tab colours.

- ● **Infantry:** Medium Olive (850)
- ● **Mobile troops (Armoured Divisions):** Dark Blue (930)
- ● **Artillery (including AA and assault guns):** Flat Red (957)
- ● **Engineers:** 50/50 mix of Dark Blue (930) and Medium Olive (850)

HUNGARIAN VEHICLES

German supplied vehicles were often supplied in *Dunkelgelb* (Middlestone 882), often with German three colour camouflage (see page 262). Many Hungarian vehicles were simply painted green (Reflective Green 890). The Hungarians also adopted three colour camouflage for their own manufactured vehicles. These were over a base of green (Reflective Green 890) with brown (German Camo Medium Brown 826) and ochre (Green Ochre 914) camouflage patches.

German Vehicles

Hungarian Vehicles

**Hungarian Vehicles
(Three colour camouflage)**

FINNISH

JALKAVÄKI

Brown Violet (887)
Helmet

Flat Flesh (955)
Exposed flesh

Gunmetal (863)
Gun metal

Chocolate Brown (872)
Belts, straps and
rifle ammo pouches

Medium Sea Grey (870)
Summer Tunics

German Camo Beige (821)
Bread bag, water bottle strap

Beige Brown (875)
Rifle wood, tool handles

Grey Green (866)
Trousers, Winter Tunics
and Caps

Black (950)
Boots

Each branch of the Finnish Army had unique
uniform collar tab colours.

- **Infantry:** Medium Olive (850)
- **Armoured Troops:** Black (950)
- **Artillery (including AA):** Flat Red (957)
- **Engineers:** Cavalry Brown (982)

FINNISH VEHICLES

Finnish captured Soviet vehicles and guns, were usually dark green (Russian Green 894). Some Finnish vehicles such BT-42 and Sturmi assault guns were painted in a unique hard edge three colour Finnish scheme. This was a base of moss green (Reflective Green 890) with sand brown (USA Tan Earth 874) and grey (Stone Grey 884) camouflage patches.

Soviet Vehicles

Finnish Vehicles

ATLANTIC
OCEAN

NORWAY

SWEDEN

BALTIC
SEA

KEY

German territories

Allied territories

Neutral countries

By December 1944, the Allied advance had
developed with pace. Romania now fought as an
ally of the Soviets, pushing into Hungary and
Slovakia. The Red Army was at the doorstep of
Budapest. In Poland, the Germans fought with
determination and clung to the Vistula River,
while in the north a fighting withdrawal saw them
defending Courland and East Prussia.

In the west the Allies had pushed into the
Netherlands and Germany itself following the
D-Day landings in France.

DENMARK

NORTH
SEA

Vistula River

Warsaw

UNITED
KINGDOM

THE
NETHERLANDS

Berlin

GERMANY

BELGIUM

Rhine River

LUXEMBOURG

SLOVAKIA

Paris

N

FRANCE

Danube River

Budapest

SWITZERLAND

0   Miles   100   200   300   400

0   KM   200   400   600

ITALY

YUGOSLAVIA